Advances in Industrial Control

Other titles published in this Series:

Digital Controller Implementation and Fragility
Robert S.H. Istepanian and
James F. Whidborne (Eds.)

Optimisation of Industrial Processes at Supervisory Level
Doris Sáez, Aldo Cipriano and
Andrzej W. Ordys

Robust Control of Diesel Ship Propulsion
Nikolaos Xiros

Hydraulic Servo-systems
Mohieddine Jelali and Andreas Kroll

Strategies for Feedback Linearisation
Freddy Garces, Victor M. Becerra,
Chandrasekhar Kambhampati and
Kevin Warwick

Robust Autonomous Guidance
Alberto Isidori, Lorenzo Marconi and
Andrea Serrani

Dynamic Modelling of Gas Turbines
Gennady G. Kulikov and Haydn A.
Thompson (Eds.)

Control of Fuel Cell Power Systems
Jay T. Pukrushpan, Anna G. Stefanopoulou
and Huei Peng

Fuzzy Logic, Identification and Predictive Control
Jairo Espinosa, Joos Vandewalle and
Vincent Wertz

Optimal Real-time Control of Sewer Networks
Magdalene Marinaki and Markos
Papageorgiou

Process Modelling for Control
Benoît Codrons

Computational Intelligence in Time Series Forecasting
Ajoy K. Palit and Dobrivoje Popovic

Modelling and Control of mini-Flying Machines
Pedro Castillo, Rogelio Lozano and
Alejandro Dzul

Rudder and Fin Ship Roll Stabilization
Tristan Perez

Hard Disk Drive Servo Systems (2nd Ed.)
Ben M. Chen, Tong H. Lee, Kemao Peng
and Venkatakrishnan Venkataramanan

Measurement, Control, and Communication Using IEEE 1588
John Eidson

Piezoelectric Transducers for Vibration Control and Damping
S.O. Reza Moheimani and Andrew J.
Fleming

Manufacturing Systems Control Design
Stjepan Bogdan, Frank L. Lewis, Zdenko
Kovačić and José Mireles Jr.

Windup in Control
Peter Hippe

Nonlinear H_2/H_∞ Constrained Feedback Control
Murad Abu-Khalaf, Jie Huang and
Frank L. Lewis

Practical Grey-box Process Identification
Torsten Bohlin

Modern Supervisory and Optimal Control
Sandor Markon, Hajime Kita, Hiroshi Kise
and Thomas Bartz-Beielstein

Wind Turbine Control Systems
Fernando D. Bianchi, Hernán De Battista
and Ricardo J. Mantz

Advanced Fuzzy Logic Technologies in Industrial Applications
Ying Bai, Hanqi Zhuang and Dali Wang
(Eds.)

Soft Sensors for Monitoring and Control of Industrial Processes
Luigi Fortuna, Salvatore Graziani,
Alessandro Rizzo and Maria Gabriella
Xibilia

Advanced Control of Industrial Processes
Piotr Tatjewski
Publication due October 2006

Adaptive Voltage Control in Power Systems
Giuseppe Fusco and Mario Russo
Publication due October 2006

Antonio Visioli

Practical PID Control

With 241 Figures

Springer

Antonio Visioli, PhD
Dipartimento di Elettronica per l'Automazione
Università degli Studi di Brescia
I-25123 Brescia
Italy

antonio.visioli@ing.unibs.it
www.ing.unibs.it/~visioli

British Library Cataloguing in Publication Data
Visioli, Antonio
 Practical PID control. - (Advances in industrial control)
 1. PID controllers
 I. Title
 629.8
ISBN-13: 9781846285851
ISBN-10: 1846285852

Library of Congress Control Number: 2006932289

Advances in Industrial Control series ISSN 1430-9491
ISBN-10: 1-84628-585-2   e-ISBN 1-84628-586-0   Printed on acid-free paper
ISBN-13: 978-1-84628-585-1

© Springer-Verlag London Limited 2006

Apart from any fair dealing for the purposes of research or private study, or criticism or review, as permitted under the Copyright, Designs and Patents Act 1988, this publication may only be reproduced, stored or transmitted, in any form or by any means, with the prior permission in writing of the publishers, or in the case of reprographic reproduction in accordance with the terms of licences issued by the Copyright Licensing Agency. Enquiries concerning reproduction outside those terms should be sent to the publishers.

The use of registered names, trademarks, etc. in this publication does not imply, even in the absence of a specific statement, that such names are exempt from the relevant laws and regulations and therefore free for general use.

The publisher makes no representation, express or implied, with regard to the accuracy of the information contained in this book and cannot accept any legal responsibility or liability for any errors or omissions that may be made.

9 8 7 6 5 4 3 2 1

Springer Science+Business Media
springer.com

Advances in Industrial Control

Series Editors

Professor Michael J. Grimble, Professor of Industrial Systems and Director
Professor Michael A. Johnson, Professor (Emeritus) of Control Systems and Deputy Director

Industrial Control Centre
Department of Electronic and Electrical Engineering
University of Strathclyde
Graham Hills Building
50 George Street
Glasgow G1 1QE
United Kingdom

Series Advisory Board

Professor E.F. Camacho
Escuela Superior de Ingenieros
Universidad de Sevilla
Camino de los Descobrimientos s/n
41092 Sevilla
Spain

Professor S. Engell
Lehrstuhl für Anlagensteuerungstechnik
Fachbereich Chemietechnik
Universität Dortmund
44221 Dortmund
Germany

Professor G. Goodwin
Department of Electrical and Computer Engineering
The University of Newcastle
Callaghan
NSW 2308
Australia

Professor T.J. Harris
Department of Chemical Engineering
Queen's University
Kingston, Ontario
K7L 3N6
Canada

Professor T.H. Lee
Department of Electrical Engineering
National University of Singapore
4 Engineering Drive 3
Singapore 117576

Professor Emeritus O.P. Malik
Department of Electrical and Computer Engineering
University of Calgary
2500, University Drive, NW
Calgary
Alberta
T2N 1N4
Canada

Professor K.-F. Man
Electronic Engineering Department
City University of Hong Kong
Tat Chee Avenue
Kowloon
Hong Kong

Professor G. Olsson
Department of Industrial Electrical Engineering and Automation
Lund Institute of Technology
Box 118
S-221 00 Lund
Sweden

Professor A. Ray
Pennsylvania State University
Department of Mechanical Engineering
0329 Reber Building
University Park
PA 16802
USA

Professor D.E. Seborg
Chemical Engineering
3335 Engineering II
University of California Santa Barbara
Santa Barbara
CA 93106
USA

Doctor K.K. Tan
Department of Electrical Engineering
National University of Singapore
4 Engineering Drive 3
Singapore 117576

Professor Ikuo Yamamoto
Kyushu University Graduate School
Marine Technology Research and Development Program
MARITEC, Headquarters, JAMSTEC
2-15 Natsushima Yokosuka
Kanagawa 237-0061
Japan

To Angela and Gianco

Series Editor's Foreword

The series *Advances in Industrial Control* aims to report and encourage technology transfer in control engineering. The rapid development of control technology has an impact on all areas of the control discipline. New theory, new controllers, actuators, sensors, new industrial processes, computer methods, new applications, new philosophies..., new challenges. Much of this development work resides in industrial reports, feasibility study papers and the reports of advanced collaborative projects. The series offers an opportunity for researchers to present an extended exposition of such new work in all aspects of industrial control for wider and rapid dissemination.

In February, 2006, *IEEE Control Systems Magazine* celebrated its first 25 years of publication and the special issue was devoted to the topic of PID control. It was fascinating to read of PID control developments in many of the departments of the magazine; these included several specialist PID control articles, a review of PID patents, software and industrial hardware, a new design software package for PID control and reviews of four substantial new books on different aspects of the PID control paradigm. The evidence from this special issue was that PID control continues to play a significant and important role in industrial control engineering. When seeking reasons for this industrial popularity, many cite the simplicity of the control law, the straight forwardness of its tuning procedures and so on but, perhaps a more fundamental point is that so many industrial control loops are easy to control and PID control is all that is needed. Then, the simplicity of the PID control law and the availability of pro-forma tuning procedures have real benefit particularly as these have been captured by automated tuning procedures in widely available software packages.

However, the converse of the above argument is also true and much of the science of PID control engineering has emerged from trying to understand and identify the exceptions, where PID control is not adequate for the complexities of the process, and the remedies that can be followed. One example of this type of new development is that of performance assessment and monitoring. This emerged from trying to find simple ways of determining whether the many PID control loops in an industrial plant (and often there are hundreds) had controller tunings that were fit for purpose. Questions like these on the practical aspects of PID control continue to motivate new developments for use in industrial practice.

The *Advances in Industrial Control* series of monographs has always sought to be abreast of developments in theory and applications that have an impact on the field of industrial control. During the late 1990s, there was a veritable clutch of titles in the series on PID control. C.C. Yu's monograph *Autotuning of PID Controllers: Relay Feedback Approach* was published in 1999 (and has since been republished as a second edition (ISBN: 1-84628-036-2) in 2006). The same year saw K.K. Tan and his colleagues develop, summarise and extend many new and existing concepts in a volume entitled *Advances in PID Control* (ISBN: 1-85233-614-5). This presented new methods for a fundamental understanding of the properties of PID controller tuning parameters. On a related subject, the series published the 1999 monograph *Performance Assessment of Control Loops* (ISBN: 1-85233-639-0) by B. Huang and S.L. Shah. This work grew from the seminal work of Professor Thomas Harris who sought ways of determining just how good an installed PID controller was. As if to capture this extensive ongoing research activity, PID control had its own conference event under the auspices of IFAC, for in 2000, a Workshop on Digital Control, *PID 2000* was held at Terrassa, Spain.

As the special issue of *IEEE Control Systems Magazine* shows, the industrial and academic interest in PID control continues and to continue the development of PID control from the millennium, *Advances in Industrial Control* welcomes *Practical PID Control* by Antonio Visioli of the University of Brescia, Italy. It is a very useful and pertinent addition because it focuses on the broader practical aspects of PID control other than those of how to select or tune the controller coefficients.

The new volume opens with an introductory chapter on the basics of PID controllers that establishes the notation, terminology, and structure of the controllers to be used in the text. Then Dr. Visioli presents chapters on derivative filter design, anti-windup strategies, the selection of set-point weightings, the use of feed-forward control, the implications of model identification and reduction for PID control, performance assessment procedures and, finally, the oft-neglected ratio control systems. In what is obviously a comprehensive set of contributions to PID control, Dr. Visioli also has a chapter on Plug & Control facilities that are often available in industrial SCADA and DCS software suites. Throughout the text, developments are illustrated with simulations and experimental results from two hardware process rigs, namely a level control system (the double tank apparatus from KentRidge Instruments) and a temperature control rig based on a laboratory-scale oven.

For those interested in the development of PID control, this monograph presents new perspectives to inspire new theoretical developments and experimental tests. The industrial engineer can use the book to investigate wider practical PID control problems and the research engineer will be able to initiate close study of many problems that often prevent PID control systems form reaching their full performance potential.

M.J. Grimble and M.A. Johnson
Glasgow, Scotland, U.K.

Preface

Although the new and effective theories and design methodologies being continually developed in the automatic control field, Proportional–Integral–Derivative (PID) controllers are still by far the most widely adopted controllers in industry owing to the advantageous cost/benefit ratio they are able to provide. In fact, although they are relatively simple to use, they are able to provide a satisfactory performance in many process control tasks. Indeed, their long history and the know-how that has been devised over the years has consolidated their usage as a standard feedback controller. However, the availability of high-performance microprocessors and software tools and the increasing demand of higher product quality at reduced costs still stimulates researchers to devise new methodologies for the improvement of performance and/or for an easier use of them. This is proven by the large number of publications on this topic (especially in recent years) and by the increasing number of products available on the market.

Actually, much of the effort of researchers has been concentrated on the development of new tuning rules for the selection of the values of the PID parameters. Although this is obviously a crucial issue, it is well-known that a key role in the achievement of high performance in practical conditions is also played by those functionalities that have to (or can) be added to the basic PID control law. Thus, in contrast to other books on PID control, this book focuses on some of these additional functionalities and on other practical problems that a typical practitioner has to face when implementing a PID controller (for scalar linear systems). Recent advances as well as more standard methodologies are presented in this context. To summarise, the book tries to answer the following questions:

- How can an effective filter on the PID action be implemented?
- How can an effective anti-windup strategy be implemented?
- How can the set-point weighting strategy be modified to improve performance?

- How can the identification (and model reduction) procedure be selected for the tuning of the parameters?
- How can an effective feedforward strategy be implemented?
- How can the achieved performance be assessed?
- How can PID-based control structures (ratio control and cascade control) be implemented effectively?

The aim of the following chapters is therefore to provide a comprehensive (although surely not exhaustive) review of approaches in the context outlined above and also aims at stimulating new ideas in the field.

The content of the book is organised as follows.

Chapter 1 provides an introduction to PID controllers, with the aim of making the book self-contained, of presenting the notation and of describing the practical issues that will be analysed in the following chapters. In particular, the three actions are described, the different controller structures are presented and the tuning issue is discussed.

In Chapter 2 the design of the low-pass filter that is necessary to make the controller transfer function proper is discussed. It is pointed out that this is indeed an important issue for the control performance and should be treated to all intents as a tuning parameter. Methodologies proposed recently in the literature in this context are described.

Chapter 3 presents and compares the different techniques that can be implemented to counteract the integrator windup effect due to the presence of a saturating actuator.

Chapter 4 addresses the use of the set-point weighting functionality. In particular, the standard technique of weighting the set-point for the proportional action (*i.e.*, of filtering the set-point of the closed-loop system) in order to reduce the overshoot is first reviewed. Then, the use of a variable set-point weight is also analysed in detail and it is shown that this might significantly improve the set-point following performances.

Chapter 5 further focuses on the use of a feedforward action to improve set-point following performance. In particular, a new design for a (causal) feedforward action is presented and it is compared to the standard approach. Further, two methodologies for the design of a noncausal feedforward action, based on input-output inversion, are explained. The design of feedforward action for disturbance rejection purposes is also briefly considered.

In Chapter 6 the recently developed Plug&Control strategy is described. It is shown that it represents a useful tool for the fast tuning of the controller at the start-up of the process.

Identification and model reduction techniques are analysed in Chapter 7. Different methods based on the open-loop step response or on the relay-feedback approach for the estimation of the parameters of first-order-plus-dead-time (FOPDT) or second-order-plus-dead-time (SOPDT) transfer functions are reviewed and compared in order to analyse and discuss their suitability of use in the context of PID control. Further, the use of model reduction techniques to be applied for the design of PID control of high-order processes is discussed. Chapter 8 presents methodologies for the assessment of the (stochastic and deterministic) performance obtained by a PID controller in the general framework of process monitoring.

Finally, Chapter 9 addresses control structures based on PID controllers. In particular, standard approaches together with recently proposed methodologies are presented for cascade control and ratio control.

A large number of simulation and experimental results are shown to analyse better each technique presented. Experimental results are obtained by means of two laboratory scale setups (described in the appendix), where a level control task and a temperature control task are implemented. Although true industrial plant data are not adopted, it is believed that these results are indeed significant for the evaluation of a methodology in a practical context. The book is therefore intended to be useful as a comprehensive review for academic researchers as well as for industrial practitioners who are looking for new methodologies to improve control systems performance while retaining their basic know-how and the ease of use and the low cost of the controller.

Readers are assumed to know the fundamentals of linear control systems, which are typically acquired in a basic course in automatic control at the university level. In particular, the description of a system through its transfer function is adopted over the whole book.

This book is a result of almost ten years of research in the field of PID control. I would like to thank Giovanna Finzi of the University of Brescia for having encouraged me in pursuing this research topic and for having always supported me with her friendship. It has been a privilege to work with Aurelio Piazzi of the University of Parma, I am indeed indebted with him for having shared his knowledge and experience with me. I am also grateful to Massimiliano Veronesi of Yokogawa Italia, Fausto Gorla of Paneutec and Michele Caselli of ER Sistemi for the useful discussions we had together. A particular thank is due to Claudio Scali of University of Pisa for having read the manuscript of the book and for his valuable comments. A special thank is due also to Leslie Mustoe of Loughborough University for the careful correction of the manuscript. Many experimental results have been obtained with the help of many students of the Faculty of Engineering of the University of Brescia. Their contribution is acknowledged. Many thanks also to Oliver Jackson of the publishing staff at Springer London, for his help during the preparation of the manuscript.

Finally, I would like to express my deep gratitude to my beloved wife Silvia and my dearest daughters Alessandra and Laura for their love, patience and support.

Dipartimento di Elettronica per l'Automazione *Antonio Visioli*
University of Brescia

Contents

1 **Basics of PID Control** 1
  1.1 Introduction 1
  1.2 Feedback Control 2
  1.3 On–Off Control 3
  1.4 The Three Actions of PID Control 3
      1.4.1 Proportional Action 3
      1.4.2 Integral Action 5
      1.4.3 Derivative Action 6
  1.5 Structures of PID Controllers 7
  1.6 Modifications of the Basic PID Control Law 8
      1.6.1 Problems with Derivative Action 8
      1.6.2 Set-point Weighting 11
      1.6.3 General ISA–PID Control Law 13
  1.7 Digital Implementation 13
  1.8 Choice of the Controller Type 15
  1.9 The Tuning Issue 16
  1.10 Automatic Tuning 18
  1.11 Conclusions and References 18

2 **Derivative Filter Design** 19
  2.1 Introduction 19
  2.2 The Significance of the Filter in PID Design 19
  2.3 Ideal *vs.* Series Form 22
  2.4 Simulation Results 27
  2.5 Four-parameters Tuning 33
  2.6 Conclusions 33

3 **Anti-windup Strategies** 35
  3.1 Introduction 35
  3.2 Integrator Windup 35
  3.3 Anti-windup Techniques 37

		3.3.1 Avoiding Saturation 37
		3.3.2 Conditional Integration 38
		3.3.3 Back-calculation 38
		3.3.4 Combined Approaches 41
		3.3.5 Automatic Reset Implementation 42
	3.4 Simulation Results 44
	3.5 Experimental Results 50
		3.5.1 Level Control 50
		3.5.2 Temperature Control 55
	3.6 Conclusions 60

4 Set-point Weighting 61
  4.1 Introduction 61
  4.2 Constant Set-point Weight Design 61
  4.3 Variable Set-point Weighting 63
    4.3.1 Methodology 63
    4.3.2 Simulation Results 67
  4.4 Fuzzy Set-point Weighting 72
    4.4.1 Methodology 72
    4.4.2 Tuning Procedure 73
    4.4.3 Simulation Results 75
    4.4.4 Experimental Results 82
  4.5 Discussion 87
  4.6 Conclusions 91

5 Use of a Feedforward Action 93
  5.1 Introduction 93
  5.2 Linear Causal Feedforward Action 93
  5.3 Nonlinear Causal Feedforward Action 96
    5.3.1 Simulation Results 98
    5.3.2 Experimental Results 103
  5.4 Noncausal Feedforward Action: Continuous-time Case 109
    5.4.1 Generalities 109
    5.4.2 Methodology 110
    5.4.3 Simulation Results 116
    5.4.4 Experimental Results 126
  5.5 Noncausal Feedforward Action: Discrete-time Case 130
    5.5.1 Methodology 130
    5.5.2 Simulation Results 133
    5.5.3 Experimental Results 136
  5.6 Feedforward Action for Disturbance Rejection 140
  5.7 Conclusions 143

6 Plug&Control ... 145
- 6.1 Introduction ... 145
- 6.2 Self-tuning Temperature Control ... 145
- 6.3 Time-optimal Plug&Control ... 149
  - 6.3.1 Methodology ... 149
  - 6.3.2 Algorithm ... 149
  - 6.3.3 Practical Considerations ... 152
  - 6.3.4 Simulation Results ... 153
  - 6.3.5 Experimental Results ... 157
  - 6.3.6 Discussion ... 163
- 6.4 Conclusions ... 163

7 Identification and Model Reduction Techniques ... 165
- 7.1 Introduction ... 165
- 7.2 FOPDT Systems ... 165
  - 7.2.1 Open-loop Identification Techniques ... 166
  - 7.2.2 Closed-loop Identification Techniques ... 173
- 7.3 SOPDT Systems ... 180
  - 7.3.1 Open-loop Identification Techniques ... 180
  - 7.3.2 Closed-loop Identification Techniques ... 191
- 7.4 Discussion ... 192
- 7.5 PID Control of High-order Systems ... 193
  - 7.5.1 Internal Model Control Design ... 194
  - 7.5.2 Process Model Reduction ... 195
  - 7.5.3 Controller Reduction ... 198
  - 7.5.4 Simulation Results ... 199
  - 7.5.5 Discussion ... 205
- 7.6 Conclusions and References ... 207

8 Performance Assessment ... 209
- 8.1 Introduction ... 209
- 8.2 Generalities ... 210
- 8.3 Stochastic Performance Assessment ... 210
  - 8.3.1 Minimum Variance Control ... 210
  - 8.3.2 Assessment of Performance ... 212
  - 8.3.3 Assessment of PID Control Performance ... 216
- 8.4 Deterministic Performance Assessment ... 222
  - 8.4.1 Useful Functionalities ... 223
  - 8.4.2 Optimal Performance for Single-loop Systems ... 232
  - 8.4.3 PID Tuning Assessment ... 234
- 8.5 Conclusions and References ... 250

9 Control Structures 251
9.1 Introduction 251
9.2 Cascade Control 251
9.2.1 Generalities 251
9.2.2 Relay Feedback Sequential Auto-tuning 253
9.2.3 Relay Feedback Simultaneous Auto-tuning 253
9.2.4 Simultaneous Identification Based on Step Response ... 257
9.2.5 Simultaneous Tuning of the Controllers 257
9.2.6 Tuning of the General Cascade Control Structure 260
9.2.7 Use of a Smith Predictor in the Outer Loop 263
9.2.8 Two Degree-of-freedom Control Structure 265
9.3 Ratio Control 267
9.3.1 Generalities 267
9.3.2 The Blend Station 268
9.3.3 Dynamic Blend Station 280
9.4 Conclusions 291

A Experimental Setups 295
A.1 Level Control Apparatus 295
A.2 Temperature Control Apparatus 296

References 299

Index 309

1
Basics of PID Control

1.1 Introduction

A Proportional–Integral–Derivative (PID) controller is a three-term controller that has a long history in the automatic control field, starting from the beginning of the last century (Bennett, 2000). Owing to its intuitiveness and its relative simplicity, in addition to satisfactory performance which it is able to provide with a wide range of processes, it has become in practice the standard controller in industrial settings. It has been evolving along with the progress of the technology and nowadays it is very often implemented in digital form rather than with pneumatic or electrical components. It can be found in virtually all kinds of control equipments, either as a stand-alone (single-station) controller or as a functional block in Programmable Logic Controllers (PLCs) and Distributed Control Systems (DCSs). Actually, the new potentialities offered by the development of the digital technology and of the software packages has led to a significant growth of the research in the PID control field: new effective tools have been devised for the improvement of the analysis and design methods of the basic algorithm as well as for the improvement of the additional functionalities that are implemented with the basic algorithm in order to increase its performance and its ease of use.
The success of the PID controllers is also enhanced by the fact that they often represent the fundamental component for more sophisticated control schemes that can be implemented when the basic control law is not sufficient to obtain the required performance or a more complicated control task is of concern.
In this chapter, the fundamental concepts of PID control are introduced with the aim of presenting the rationale of the control law and of describing the framework of the methodologies presented in the subsequent chapters. In particular, the meaning of the three actions is explained and the tuning issue is briefly discussed. The different forms for the implementation of a PID control law are also addressed.

1.2 Feedback Control

The aim of a control system is to obtain a desired response for a given system. This can be done with an open-loop control system, where the controller determines the input signal to the process on the basis of the reference signal only, or with a closed-loop control system, where the controller determines the input signal to the process by using also the measurement of the output (*i.e.*, the feedback signal).

Feedback control is actually essential to keep the process variable close to the desired value in spite of disturbances and variations of the process dynamics, and the development of feedback control methodologies has had a tremendous impact in many different fields of the engineering. Besides, nowadays the availability of control system components at a lower cost has favoured the increase of the applications of the feedback principle (for example in consumer electronics products).

The typical feedback control system is represented in Figure 1.1. Obviously, the overall control system performance depends on the proper choice of each component. From the purposes of controller design, the actuator and sensor dynamics are often neglected (although the saturation limits of the actuator have to be taken into account) and the block diagram of Figure 1.2 is considered, where P is the process, C is the controller, F is a feedforward filter, r is the reference signal, $e = r - y$ is the control error, u is the manipulated (control) variable, y is the process (controlled) variable, d is a load disturbance signal and n is a measurement noise signal.

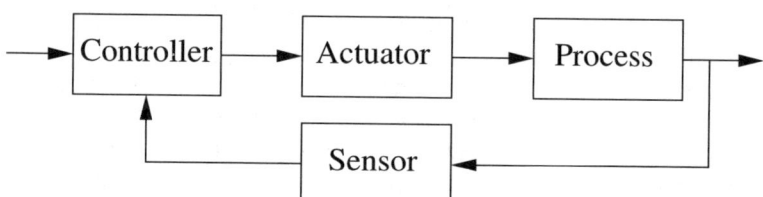

Fig. 1.1. Typical components of a feedback control loop

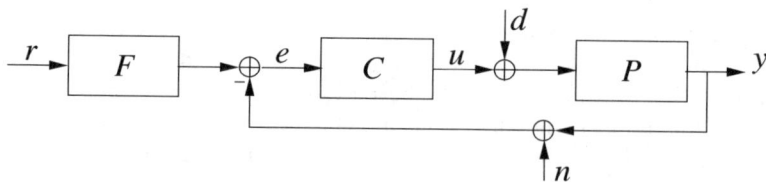

Fig. 1.2. Schematic block diagram of a feedback control loop

1.3 On–Off Control

One of the most adopted (and one of the simplest) controllers is undoubtedly the On–Off controller, where the control variable can assume just two values, u_{max} and u_{min}, depending on the control error sign. Formally, the control law is defined as follows:

$$u = \begin{cases} u_{max} & \text{if } e > 0 \\ u_{min} & \text{if } e < 0 \end{cases}, \qquad (1.1)$$

i.e., the control variable is set to its maximum value when the control error is positive and to its minimum value when the control error is negative. Generally, $u_{min} = 0$ (Off) is selected and the controller is usually implemented by means of a relay.

The main disadvantage of the On–Off controller is that a persistent oscillation of the process variable (around the set-point value) occurs. Consider for example the process described by the first-order-plus-dead-time (FOPDT) transfer function

$$P(s) = \frac{1}{10s+1} e^{-2s}$$

controlled by an On–Off controller with $u_{max} = 2$ and $u_{min} = 0$. The result of applying a unit step to the set-point signal is shown in Figure 1.3, where both the process variable and the control variable have been plotted.

Actually, in practical cases, the On–Off controller characteristic is modified by inserting a dead zone (this results in a three-state controller) or hysteresis in order to cope with measurement noise and to limit the wear and tear of the actuating device. The typical controller functions are shown in Figure 1.4.

Because of its remarkable simplicity (there are no parameters to adjust), the On–Off controller is indeed suitable for adoption when no tight performance is required, since it is very cost-effective in these cases. For this reason it is generally available in commercial industrial controllers.

1.4 The Three Actions of PID Control

Applying a PID control law consists of applying properly the sum of three types of control actions: a proportional action, an integral action and a derivative one. These actions are described singularly hereafter.

1.4.1 Proportional Action

The proportional control action is proportional to the current control error, according to the expression

$$u(t) = K_p e(t) = K_p(r(t) - y(t)), \qquad (1.2)$$

where K_p is the proportional gain. Its meaning is straightforward, since it implements the typical operation of increasing the control variable when the

4   1 Basics of PID Control

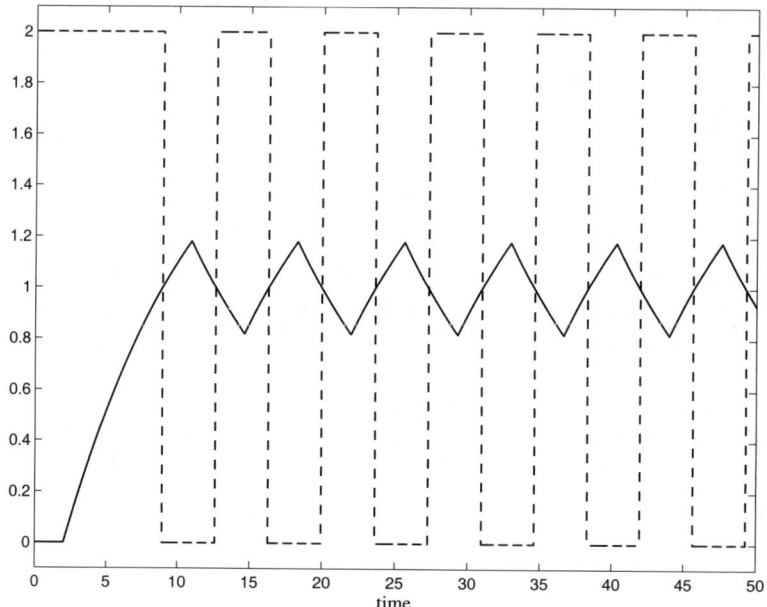

Fig. 1.3. Example of an On–Off control application. Solid line: process variable; dashed line: control variable.

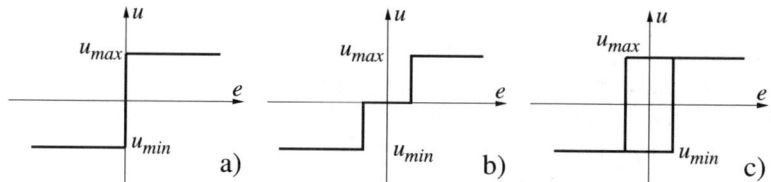

Fig. 1.4. Typical On–Off controller characteristics. a) ideal; b) modified with a dead zone; c) modified with hysteresis.

control error is large (with appropriate sign). The transfer function of a proportional controller can be derived trivially as

$$C(s) = K_p. \tag{1.3}$$

With respect to the On–Off controller, a proportional controller has the advantage of providing a small control variable when the control error is small and therefore to avoid excessive control efforts. The main drawback of using a pure proportional controller is that it produces a steady-state error. It is worth noting that this occurs even if the process presents an integrating dynamics (*i.e.*, its transfer function has a pole at the origin of the complex plane), in case a constant load disturbance occurs. This motivates the addition of a bias (or reset) term u_b, namely,

$$u(t) = K_p e(t) + u_b. \tag{1.4}$$

The value of u_b can be fixed at a constant level (usually at $(u_{max} + u_{min})/2$) or can be adjusted manually until the steady-state error is reduced to zero. It is worth noting that in commercial products the proportional gain is often replaced by the proportional band PB, that is the range of error that causes a full range change of the control variable, *i.e.*,

$$PB = \frac{100}{K_p}. \tag{1.5}$$

1.4.2 Integral Action

The integral action is proportional to the integral of the control error, *i.e.*, it is

$$u(t) = K_i \int_0^t e(\tau)d\tau, \tag{1.6}$$

where K_i is the integral gain. It appears that the integral action is related to the past values of the control error. The corresponding transfer function is:

$$C(s) = \frac{K_i}{s}. \tag{1.7}$$

The presence of a pole at the origin of the complex plane allows the reduction to zero of the steady-state error when a step reference signal is applied or a step load disturbance occurs. In other words, the integral action is able to set automatically the correct value of u_b in (1.4) so that the steady-state error is zero. This fact is better explained in Figure 1.5, where the resulting transfer function is

$$C(s) = K_p \left(1 + \frac{1}{T_i s}\right), \tag{1.8}$$

i.e., a PI controller results. For this reason the integral action is also often called *automatic reset*.

Thus, the use of a proportional action in conjunction to an integral action, *i.e.*, of a PI controller, solves the main problems of the oscillatory response associated to an On–Off controller and of the steady-state error associated to a pure proportional controller.

It has to be stressed that when integral action is present, the so-called *integrator windup* phenomenon might occur in the presence of saturation of the control variable. This aspect will be thoroughly analysed in Chapter 3.

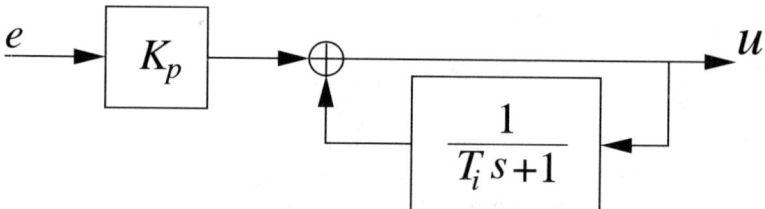

Fig. 1.5. PI controller in automatic reset configuration

1.4.3 Derivative Action

While the proportional action is based on the current value of the control error and the integral action is based on the past values of the control error, the derivative action is based on the predicted future values of the control error. An ideal derivative control law can be expressed as:

$$u(t) = K_d \frac{de(t)}{dt}, \tag{1.9}$$

where K_d is the derivative gain. The corresponding controller transfer function is

$$C(s) = K_d s. \tag{1.10}$$

In order to understand better the meaning of the derivative action, it is worth considering the first two terms of the Taylor series expansion of the control error at time T_d ahead:

$$e(t + T_d) \simeq e(t) + T_d \frac{de(t)}{dt}. \tag{1.11}$$

If a control law proportional to this expression is considered, *i.e.*,

$$u(t) = K_p \left(e(t) + T_d \frac{de(t)}{dt} \right), \tag{1.12}$$

this naturally results in a PD controller. The control variable at time t is therefore based on the predicted value of the control error at time $t + T_d$. For this reason the derivative action is also called *anticipatory control*, or *rate action*, or *pre-act*.

It appears that the derivative action has a great potentiality in improving the control performance as it can anticipate an incorrect trend of the control error and counteract for it. However, it has also some critical issues that makes it not very frequently adopted in practical cases. They will be discussed in the following sections.

1.5 Structures of PID Controllers

The combination of the proportional, integral, and derivative actions can be done in different ways. In the so-called *ideal* or *non-interacting* form, the PID controller is described by the following transfer function:

$$C_i(s) = K_p \left(1 + \frac{1}{T_i s} + T_d s\right), \tag{1.13}$$

where K_p is the proportional gain, T_i is the integral time constant, and T_d is the derivative time constant. An alternative representation is the *series* or *interacting form*:

$$C_s(s) = K_p' \left(1 + \frac{1}{T_i' s}\right)(T_d' s + 1) = K_p' \left(\frac{T_i' s + 1}{T_i' s}\right)(T_d' s + 1), \tag{1.14}$$

where the fact that a modification of the value of the derivative time constant T_d' affects also the integral action justifies the nomenclature adopted.
It has to be noted that a PID controller in series form can be always represented in ideal form by applying the following formulae:

$$K_p = K_p' \frac{T_i' + T_d'}{T_i'}$$

$$T_i = T_i' + T_d' \tag{1.15}$$

$$T_d = \frac{T_i' T_d'}{T_i' + T_d'}$$

Conversely, it is not always possible to convert a PID controller in series form into a PID controller in ideal form. This can be done only if

$$T_i \geq 4 T_d \tag{1.16}$$

through the following formulae:

$$K_p' = \frac{K_p}{2}\left(1 + \sqrt{1 - 4\frac{T_d}{T_i}}\right)$$

$$T_i' = \frac{T_i}{2}\left(1 + \sqrt{1 - 4\frac{T_d}{T_i}}\right) \tag{1.17}$$

$$T_d' = \frac{T_i}{2}\left(1 - \sqrt{1 - 4\frac{T_d}{T_i}}\right)$$

1 Basics of PID Control

It is worth noting that a PID controller has two zeros, a pole at the origin and a gain (the fact that the transfer function is not proper will be discussed in Section 1.6). When $T_i = 4T_d$ the resulting zeros of $C_i(s)$ are coincident, while when $T_i < 4T_d$ they are complex conjugates. Thus, the ideal form is more general than the series form since it allows the implementation of complex conjugate zeros.

The reason for preferring the series form to the ideal form is that the series form was the first to be implemented in the last century with pneumatic technology. Then, many manufacturers chose to retain the know-how and to avoid changing the form of the PID controller. Further, it is sometimes claimed that a PID controller in series form is more easy to tune.

Another way to implement a PID controller is in *parallel form* [1], *i.e.*,

$$C_p(s) = K_p + \frac{K_i}{s} + K_d s. \tag{1.18}$$

In this case the three actions are completely separated. Actually, the parallel form is the most general of the different forms, as it allows to exactly switch off the integral action by fixing $K_i = 0$ (in the other cases the value of the integral time constant should tend to infinity). The conversion between the parameters of the parallel PID controller and those of the ideal one can be done trivially by means of the following formulae:

$$K_i = \frac{K_p}{T_i} \tag{1.19}$$

$$K_d = K_p T_d$$

1.6 Modifications of the Basic PID Control Law

The expressions (1.13), (1.14) and (1.18) of a PID controller given in the previous section are actually not adopted in practical cases because of a few problems that can be solved with suitable modifications of the basic control law. These are analysed in this section.

1.6.1 Problems with Derivative Action

From Expressions (1.13), (1.14) and (1.18) it appears that the controller transfer function is not proper and therefore it can not be implemented in practice.

---

[1] Actually, the term *parallel* PID controller is often adopted also for expression (1.13) (see for example (Tan et al., 1999; Seborg et al., 2004)). However, here it is preferred to use the nomenclature of (Åström and Hägglund, 1995; Ang et al., 2005) for the sake of clarity and in order to distinguish better the three considered forms.

1.6 Modifications of the Basic PID Control Law

This problem is evidently caused by the derivative action. Indeed, the high-frequency gain of the pure derivative action is responsible for the amplification of the measurement noise in the manipulated variable. Consider for example a sinusoidal signal

$$n(t) = A\sin(\omega t)$$

which represents measurement noise in the control scheme of Figure 1.2. If the derivative action only is considered, the control variable term due to this measurement noise is

$$u(t) = AK_d\omega\cos(\omega t).$$

It can be easily seen that the amplification effect is more evident when the frequency of the noise is high. In practical cases, a (very) noisy control variable signal might cause a damage of the actuator. The problems outlined above can be solved by filtering the derivative action with (at least) a first-order low-pass filter. The filter time constant should be selected in order to filter suitably the noise and to avoid to influence significantly the dominant dynamics of the PID controller.

In this context, the PID control laws (1.13), (1.14) and (1.18) are usually modified as follows. The ideal form becomes:

$$C_{i1a}(s) = K_p\left(1 + \frac{1}{T_i s} + \frac{T_d s}{\frac{T_d}{N}s + 1}\right), \quad (1.20)$$

or, alternatively (Gerry and Shinskey, 2005),

$$C_{i1b}(s) = K_p\left(1 + \frac{1}{T_i s} + \frac{T_d s}{1 + \frac{T_d}{N}s + 0.5\left(\frac{T_d}{N}s\right)^2}\right). \quad (1.21)$$

The series form becomes:

$$C_s(s) = K'_p\left(1 + \frac{1}{T'_i s}\right)\left(\frac{T'_d s + 1}{\frac{T'_d}{N'}s + 1}\right) = K'_p\left(\frac{T'_i s + 1}{T'_i s}\right)\left(\frac{T'_d s + 1}{\frac{T'_d}{N'}s + 1}\right), \quad (1.22)$$

where N generally assumes a value between 1 and 33, although in the majority of the practical cases its setting falls between 8 and 16 (Ang et al., 2005). The expression of the parallel form can be straightforwardly derived as well. It is worth noting that an alternative expression for the ideal form is to filter the overall control variable, i.e., to use the following controller:

$$C_{i2a}(s) = K_p \left(1 + \frac{1}{T_i s} + T_d s\right) \frac{1}{T_f s + 1}, \qquad (1.23)$$

or, alternatively (Åström and Hägglund, 2004),

$$C_{i2b}(s) = K_p \left(1 + \frac{1}{T_i s} + T_d s\right) \frac{1}{(T_f s + 1)^2}. \qquad (1.24)$$

The block diagrams of the most adopted controllers are shown in Figures 1.6–1.8. Note that if the PI part of a series controller is in the automatic reset configuration, then the corresponding series PID controller is reported in Figure 1.9.

While these modifications are those that can be usually found in the literature (see for example (Luyben, 2001a)), it has to be stressed that the filter to be adopted is a critical issue and therefore this design aspect will be thoroughly analysed in Chapter 2.

Another issue related to the derivative action that has to be considered is the so-called *derivative kick*. In fact, when an abrupt (stepwise) change of the set-point signal occurs, the derivative action is very large and this results in a spike in the control variable signal, which is undesirable. A simple solution to avoid this problem is to apply the derivative term to the process output only instead of the control error. In this case the ideal (not filtered) derivative action becomes:

$$u(t) = -K_d \frac{dy(t)}{dt}. \qquad (1.25)$$

It is worth noting that when the set-point signal is constant, applying the derivative term to the control error or to the process variable is equivalent. Thus, the load disturbance rejection performance is the same in the two cases.

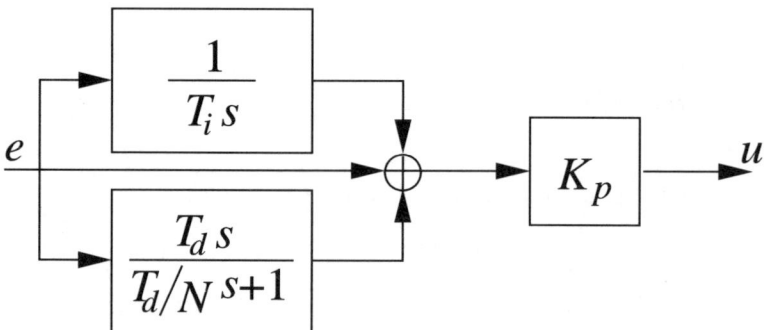

Fig. 1.6. Block diagram of a PID controller in ideal form

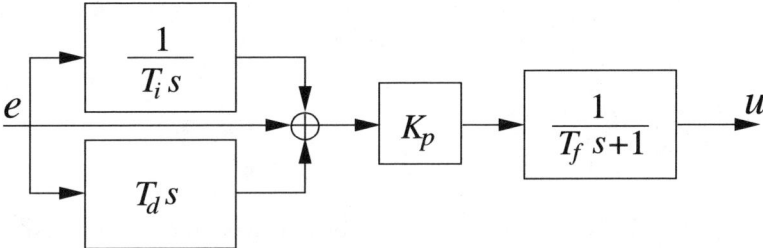

Fig. 1.7. Alternative block diagram of a PID controller in ideal form

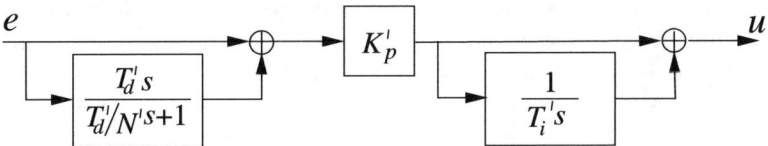

Fig. 1.8. Block diagram of a PID controller in series form

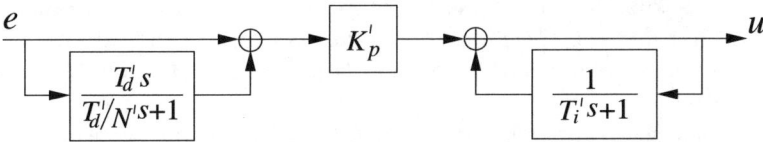

Fig. 1.9. Block diagram of a PID controller in series form with the PI part in automatic reset configuration

1.6.2 Set-point Weighting

A typical problem with the design of a feedback controller is to achieve at the same time a high performance both in the set-point following and in the load disturbance rejection performance. Roughly speaking, a fast load disturbance rejection is achieved with a high-gain controller, which gives an oscillatory set-point step response on the other side. This problem can be approached by designing a two-degree-of-freedom control architecture, namely, a combined feedforward/feedback control law.

In the context of PID control this can be achieved by weighting the set-point signal for the proportional action, that is, to define the proportional action as follows:

$$u(t) = K_p(\beta r(t) - y(t)), \qquad (1.26)$$

where the value of β is between 0 and 1.

In this way, the control scheme represented in Figure 1.10 is actually implemented, where

$$C(s) = K_p \left(1 + \frac{1}{T_i s} + T_d s\right) \tag{1.27}$$

and

$$C_{sp}(s) = K_p \left(\beta + \frac{1}{T_i s} + T_d s\right) \tag{1.28}$$

(the filter of the derivative action has not been considered for the sake of simplicity). It appears that the load disturbance rejection task is decoupled from the set-point following one and obviously it does not depend on the weight β. Thus, the PID parameters can be selected to achieve a high load disturbance rejection performance and then the set-point following performance can be recovered by suitably selecting the value of the parameter β. An equivalent control scheme is shown in Figure 1.11, where

$$F(s) = \frac{1 + \beta T_i s + T_i T_d s^2}{1 + T_i s + T_i T_d s^2}. \tag{1.29}$$

Here it is more apparent that the function of the set-point weight is to smooth the (step) set-point signal in order to damp the response to a set-point change. Note also that if $\beta = 0$ the *proportional kick* is avoided. Indeed, many industrial controllers implement this solution (Åström and Hägglund, 1995, page 110).

The use of the set-point weighting and of other feedforward control strategies for the improvement of performances will be analysed thoroughly in Chapters 4 and 5.

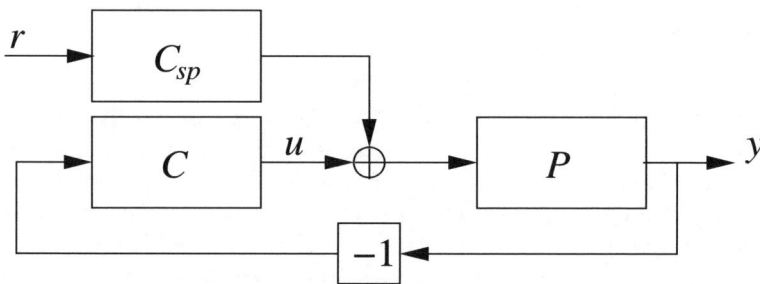

Fig. 1.10. Two-degree-of-freedom PID control scheme

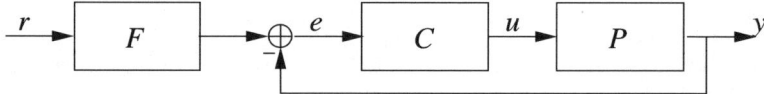

Fig. 1.11. Equivalent two-degree-of-freedom PID control scheme

1.6.3 General ISA–PID Control Law

If all the modifications of the basic control law previously addressed are considered, the following general PID control law can be derived:

$$u(t) = K_p \left(\beta r(t) - y(t) + \frac{1}{T_i} \int_0^t e(\tau) d\tau + T_d \left(\frac{d(\gamma r(t) - y_f(t))}{dt} \right) \right) \quad (1.30)$$

$$\frac{T_d}{N} \frac{dy_f(t)}{dt} = y(t) - y_f(t)$$

where, in general, it is $0 \leq \beta \leq 1$ and $0 \leq \gamma \leq 1$, although the value of γ is usually either 0 (the derivative action is entirely applied to the process output) or 1 (the derivative action is entirely applied to the control error), as explained in Section 1.6.1.

The previous one is usually called a PID controller in ISA form or, alternatively, a *beta-gamma* controller. Often, if $\beta = 1$ and $\gamma = 0$ the controller is indicated as PI–D, while if $\beta = 0$ and $\gamma = 0$ it is indicated as I–PD. The block diagram corresponding to an ISA–PID controller is the same as in Figure 1.11, where in this case

$$C(s) = C_{i1a}(s) = K_p \left(1 + \frac{1}{T_i s} + \frac{T_d s}{\frac{T_d}{N} s + 1} \right) \quad (1.31)$$

and

$$F(s) = \frac{1 + \left(\beta T_i + \frac{T_d}{N} \right) s + T_i T_d \left(\gamma + \frac{\beta}{N} \right) s^2}{1 + \left(T_i + \frac{T_d}{N} \right) s + T_i T_d \left(1 + \frac{1}{N} \right) s^2}. \quad (1.32)$$

1.7 Digital Implementation

If a digital implementation of the PID controller is adopted, then the previously considered control laws have to be discretised. This can be done with any of the available discretisation method (Åström and Wittenmark, 1997). For the sake of clarity and for future reference (see Chapter 8), an example is shown hereafter. Consider the continuous time expression of a PID controller in ideal form:

$$u(t) = K_p \left(e(t) + \frac{1}{T_i} \int_0^t e(\tau) d\tau + T_d \frac{de(t)}{dt} \right), \quad (1.33)$$

and define a sampling time Δt. The integral term in (1.33) can be approximated by using backward finite differences as

$$\int_0^{t_k} e(\tau)d\tau = \sum_{i=1}^{k} e(t_i)\Delta t, \qquad (1.34)$$

where $e(t_i)$ is the error of the continuous time system at the ith sampling instant. By applying the backward finite differences also to the derivative term it results:

$$\frac{de(t_k)}{dt} = \frac{e(t_k) - e(t_{k-1})}{\Delta t}. \qquad (1.35)$$

Then, the discrete time control law becomes:

$$u(t_k) = K_p \left(e(t_k) + \frac{\Delta t}{T_i} \sum_{i=1}^{k} e(t_i) + \frac{T_d}{\Delta t}(e(t_k) - e(t_{k-1})) \right). \qquad (1.36)$$

In this way, the value of the control variable is determined directly. Alternatively, the control variable at time instant t_k can be calculated based on its value at the previous time instant $u(t_{k-1})$. By subtracting the expression of $u(t_{k-1})$ from that of $u(t_k)$, we obtain:

$$u(t_k) = u(t_{k-1}) +$$

$$K_p \left[\left(1 + \frac{\Delta t}{T_i} + \frac{T_d}{\Delta t}\right) e(t_k) + \left(-1 - \frac{2T_d}{\Delta t}\right) e(t_{k-1}) + \frac{T_d}{\Delta t} e(t_{k-2}) \right]. \qquad (1.37)$$

For an obvious reason, the control algorithm (1.37) is called *incremental algorithm* or *velocity algorithm*, while that expressed in (1.36) is called *positional algorithm*.

Expression (1.37) can be rewritten more compactly as:

$$u(t_k) - u(t_{k-1}) = K_1 e(t_k) + K_2 e(t_{k-1}) + K_3 e(t_{k-2}), \qquad (1.38)$$

where

$$K_1 = K_p \left(1 + \frac{\Delta t}{T_i} + \frac{T_d}{\Delta t}\right),$$

$$K_2 = -K_p \left(1 + \frac{2T_d}{\Delta t}\right), \qquad (1.39)$$

$$K_3 = K_p \frac{T_d}{\Delta t}.$$

By defining q^{-1} as the backward shift operator, i.e.,

$$q^{-1} u(t_k) = u(t_{k-1}), \qquad (1.40)$$

the discretised PID controller in velocity form can be expressed as

$$C(q^{-1}) = \frac{K_1 + K_2 q^{-1} + K_3 q^{-2}}{1 - q^{-1}}, \qquad (1.41)$$

where K_1, K_2 and K_3 can be viewed as the tuning parameters.

1.8 Choice of the Controller Type

For a given control task, it is obviously not necessary to adopt all the three actions. Thus, the choice of the controller type is an integral part of the overall controller design, taking into account that the final aim is to obtain the best cost/benefit ratio and therefore the simplest controller capable to obtain a satisfactory performance should be preferred.

In this context it is worth analysing briefly some guidelines on how the controller type (P, PI, PD, PID) has to be selected. As already mentioned, a P controller has the disadvantage, in general, of giving a non zero steady-state error. However, in control tasks where this is not of concern, such as for example in surge tank level control or in inner (secondary) loops of cascade control architectures, where the zero steady-state error is ensured by the integral action adopted in the outer (primary) controller (see Chapter 9), a P controller can be the best choice, as it is simple to design (indeed, if the process has a low-order dynamics the proportional gain can be set to a high value in order to provide a fast response and a low steady-state error). Further, if an integral component is present in the system to be controlled (such as in mechanical servosystems or in surge vessels where the manipulated variable is the difference between inflow and outflow) and no load disturbances are likely to occur, then there is no need of an integral action in the controller to provide a zero steady-state control error. In this case the control performance can be usually improved by adding a derivative action, *i.e.*, by adopting a PD controller. In fact, the derivative action provides a phase lead that allows to increase the bandwidth of the system and therefore to speed up the response to a set-point change.

If the zero steady-state error is an essential control requirement, then the simplest choice is to use a PI controller. Actually, a PI controller is capable to provide an acceptable performance for the vast majority of the process control tasks (especially if the dominant process dynamics is of first order) and it is indeed the most adopted controller in the industrial context. This is also due to the problems associated with the derivative actions, namely the need of properly filtering the measurement noise and the difficulty in selecting an appropriate value of the derivative time constant.

In any case, the use of the derivative action, that is, of a PID controller, provides very often the potentiality of significantly improve the performance. For example, if the process has a second-order dominant dynamics, the zero introduced in the controller by the derivative action can be adopted to cancel the fastest pole of the process transfer function (see, for example,

(Skogestad, 2003)). However, it is also often claimed that if the process has a significant (apparent) dead time, then the derivative action should be disconnected. Actually, the usefulness of the derivative action has been the subject of some investigation (Åström and Hägglund, 2000b). Recent contributions to the literature have shown that the performance improvement given by the use of the derivative action decreases as the ratio between the apparent dead time and the effective time constant increases but it can be very beneficial if this ratio is not too high (about two) (Åström and Hägglund, 2004; Kristiansson and Lennartson, 2006).

Finally, it is worth noting that for processes affected by a large dead time (with respect to the dominant time constant) the use of a dead-time compensator controller, such as a Smith predictor based scheme (Palmor, 1996) or the so-called PID-deadtime controller (where the time-delay compensation is added to the integral feedback loop of the PID controller in automatic reset configuration) (Shinskey, 1994), can be essential in obtaining a satisfactory control performance (Ingimundarson and Hägglund, 2002).

1.9 The Tuning Issue

The selection of the PID parameters, *i.e.*, the tuning of the PID controllers, is obviously the crucial issue in the overall controller design. This operation should be performed in accordance to the control specifications. Usually, as already mentioned, they are related either to the set-point following or to the load disturbance rejection task, but in some cases both of them are of primary importance. The control effort is also generally of main concern as it is related to the final cost of the product and to the wear and life-span of the actuator. It should be therefore kept at a minimum level. Further the robustness issue has to be taken into account.

A major advantage of the PID controller is that its parameters have a clear physical meaning. Indeed, increasing the proportional gain leads to an increasing of the bandwidth of the system and therefore a faster but more oscillatory response should be expected. Conversely, increasing the integral time constant (*i.e.*, decreasing the effect of the integral action) leads to a slower response but to a more stable system. Finally, increasing the derivative time constant gives a damping effect, although much care should be taken in avoiding to increase it too much as an opposite effect occurs in this case and an unstable system could eventually result.

The problem associated with tuning of the derivative action can be better understood with the following analysis (Ang *et al.*, 2005). Suppose that the process to be controlled is described by a general FOPDT transfer function

$$P(s) = \frac{K}{Ts+1}e^{-Ls}. \tag{1.42}$$

Suppose also that an ideal PD controller is adopted, *i.e.*,

$$C(s) = K_p \left(1 + T_d s\right).\tag{1.43}$$

The gain of the open-loop transfer function is determined as

$$|C(j\omega)P(j\omega)| = KK_p \sqrt{\frac{1 + T_d^2 \omega^2}{1 + T^2 \omega^2}} \geq KK_p \min\left(1, \frac{T_d}{T}\right),\tag{1.44}$$

where the inequality is justified by the fact that $\sqrt{(1+T_d^2\omega^2)/(1+T^2\omega^2)}$ is monotonic with ω. It can be easily determined that if $T_d \leq T$ and $KK_p \geq 1$ or if $T_d \geq T$ and $T_d \geq T/(KK_p)$, then the crossover frequency ω_c is at infinity, i.e., the magnitude of the open-loop transfer function is not less than 0 dB. As a consequence, since the phase decreases when the frequency increases because of the time delay, the closed-loop system will be unstable.

To illustrate this fact, consider an example where the process (1.42) with $K = 2$, $T = 1$ and $L = 0.2$ is controlled by a PID controller in series form (1.14) with $K_p = 1$ and $T_i = 1$. Then, if it is selected $T_d = 0.01$ the gain margin results to be 12.3 dB and the phase margin results to be 68.2 deg. Increasing the derivative time constant to $T_d = 0.05$ yields an increase of the gain margin and of the phase margin to 13.2 dB and 72.7 deg, respectively. Thus, in this case, increasing the derivative action implies that a more sluggish response and a more robust system is obtained. However, if the derivative time constant is raised to 0.5 the system stability is lost.

The aforementioned concepts allow the operator to manually tune the controller in a relatively easy way, although the trial-and-error operation can be very time consuming and the final result can be far from the optimum and heavily depends on the operator's skill.

In order to help the operator in tuning the controller correctly and with a small effort, starting with the well-known Ziegler–Nichols formulae (Ziegler and Nichols, 1942), a large number of tuning rules have been devised in the last sixty years (Åström and Hägglund, 1995; O'Dwyer, 2006). They try to address the possible different control requirements and they are generally based on a simple model of the plant. They have been derived empirically or analytically. The operator has therefore to obtain a suitable model of the plant and to select the most convenient tuning rule with respect to the given control requirements. It has to be noted that the obtained PID parameters (that is, the selected tuning rule) have to be appropriate for the adopted controller structure (ideal, series, etc.), otherwise they have to be converted (see Expressions (1.15), (1.17) and (1.19)).

Finally, it is worth highlighting that many software packages have been developed and are available on the market which assist practitioners in designing the overall controller, namely, to identify an accurate process model based on available data, to tune the controller according to the given requirements, to perform a *what-if* analysis and so on. A review of them can be found in (Ang et al., 2005).

1.10 Automatic Tuning

The functionality of automatically identifying the process model and tuning the controller based on that model is called automatic tuning (or, simply, auto-tuning). In particular, an identification experiment is performed after an explicit request of the operator and the values of the PID parameters are updated at the end of it (for this reason the overall procedure is also called *one-shot automatic tuning* or *tuning-on-demand*). The design of an automatic tuning procedure involves many critical issues, such as the choice of the identification procedure (usually based on an open-loop step response or on a relay feedback experiment (Yu, 1999)), of the *a priori* selected (parametric or non parametric) process model and of the tuning rule. An excellent presentation of this topic can be found in (Leva *et al.*, 2001).

The one-shot automatic tuning functionality is available in practically all the single-station controllers available on the market. Advanced (more expensive) control units might provide a *self-tuning* functionality, where the identification procedure is continuously performed during routine process operation in order to track possible changes of the system dynamics and the PID parameters values are modified adaptively. In this case all the issues related to adaptive control have to be taken into account (Åström and Wittenmark, 1995).

1.11 Conclusions and References

In this chapter the fundamental concepts of PID controllers have been introduced. The main practical problems connected with their use have been outlined and the most adopted controller structures have been presented. In the following chapters different aspects that have been considered will be further developed.

Basic concepts of PID controllers can be found in almost every book of process control (see for example (Shinskey, 1994; Ogunnaike and Ray, 1994; Luyben and Luyben, 1997; Marlin, 2000; Corripio, 2001; Bequette, 2003; Seborg *et al.*, 2004; Corriou, 2004; Ellis, 2004; Altmann, 2005)). For a detailed treatment, see (Åström and Hägglund, 1995) and (Åström and Hägglund, 2006) where all the methodological as well as technological aspects are covered. An excellent collection of tuning rules can be found in (O'Dwyer, 2006). Recent advances are presented in (Tan *et al.*, 1999).

2

Derivative Filter Design

2.1 Introduction

It is a matter of fact that the derivative action is seldom adopted in practical cases (actually, 80% of the employed PID controllers have the derivative part switched-off (Ang et al., 2005)), although it has been shown that it is possible to provide a significant improvement of the control performance (note that this improvement becomes less important as the ratio between the apparent time delay and the effective time constant increases (Kristiansson and Lennartson, 2006; Åström and Hägglund, 2004)). This is due to a number of reasons, one of them being certainly that it is the most difficult to tune, as explained in Section 1.9. Indeed, the stability regions for PID controllers are more complex than those for PI controllers and therefore the tuning of a PID controller is more difficult (Åström and Hägglund, 2000b). Also, the inherent amplification of the measurement noise represents a significant technological problem, because, if not properly filtered, it might cause a damage to the actuator.

In this chapter it is shown that part of the problem is due also to the structure of the PID controller (see (1.20)–(1.24)), in particular if a PID controller in ideal form with a fixed derivative filter parameter N is adopted.

2.2 The Significance of the Filter in PID Design

It is interesting to evaluate how the presence of a filter of the derivative action changes the location of the zeros in the PID controller. It is trivial to derive that if the PID controller is in series form (1.22) or in ideal form (1.23)–(1.24) with the filter applied to the control variable, then the addition of the filter does not alter the position of the zeros of the controller. Hence, the interesting case to analyse is that related to the PID controller in ideal form (1.20) (or (1.31)).

If the derivative filter is not applied, the zeros of the PID controller (1.13) are the solution of the equation

$$T_i T_d s^2 + T_i s + 1 = 0. \tag{2.1}$$

They can be easily derived as:

$$z_{1,2} = \frac{1}{2} \frac{-T_i \pm \sqrt{T_i^2 - 4T_i T_d}}{T_i T_d}. \tag{2.2}$$

If the derivative filter is applied, the zeros of the controller are the solution of the equation

$$T_i T_d \left(1 + \frac{1}{N}\right) s^2 + \left(T_i + \frac{T_d}{N}\right) s + 1 = 0. \tag{2.3}$$

It results:

$$\bar{z}_{1,2} = \frac{1}{2} \frac{-T_i N - T_d \pm \sqrt{(T_i N - T_d)^2 - 4T i T_d N^2}}{T_i T_d (1 + N)}. \tag{2.4}$$

A sensitivity analysis can be performed in order to evaluate the influence of the parameter N, *i.e.*, of the filter, on the location of zeros (Leva and Colombo, 2001). The relative perturbation of the ith zero can be calculated as:

$$e_{r,i} := \frac{|\bar{z}_i - z_i|}{|z_i|}. \tag{2.5}$$

To evaluate it quantitatively with an example, T_i is fixed to be 100 and the value of $e_{r,i}$ has been determined by varying T_d from 1 to 100, *i.e.*, by varying the ratio T_d/T_i from 0.01 to 1. Results related to the case $N = 5$ and $N = 20$ are shown in Figures 2.1 and 2.2. It can be seen that the relative error can be greater than 30% and a high value appears when $T_i = 4T_d$ (*i.e.*, when the two zeros are real and coincident), which is a very relevant case, as this relation is adopted in many tuning rules such as the Ziegler–Nichols one .

This analysis is coherent with the results presented in (Kristiansson and Lennartson, 2006), where the performance achieved by a PI(D) controller is evaluated by considering both its capability in the load disturbance rejection task and the corresponding control activity. It is shown that, in general, the proper use of the derivative action allows to significantly increase the load disturbance rejection performance with a modest increase of the control effort. However, if T_i is fixed to be $4T_d$ and N to be 10, then a (slight) increase of the load disturbance rejection performance can be made only at the expense of a much increased control effort (with respect to an optimal PI controller). All these results confirm that the presence of the derivative filter in a PID controller in ideal form cannot be neglected in general in the controller design phase (Leva and Colombo, 2001). Other practical issues concerning the presence of the derivative filter are addressed in the following sections.

2.2 The Significance of the Filter in PID Design

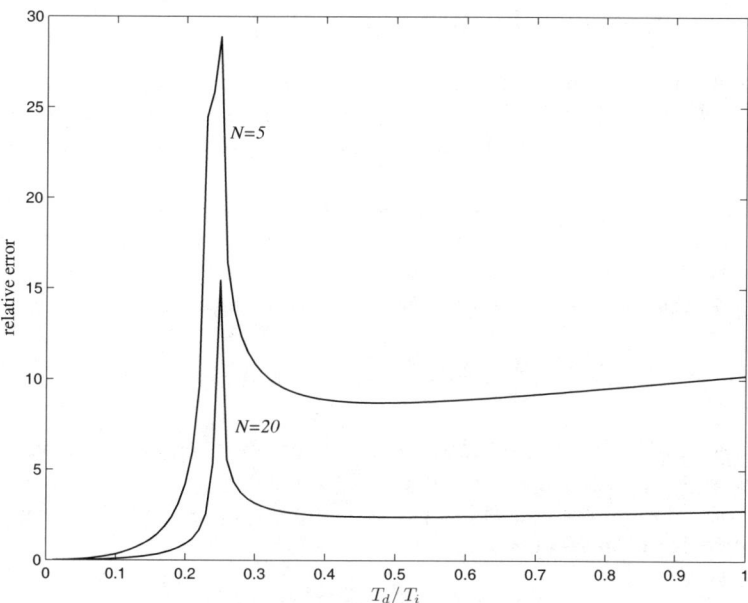

Fig. 2.1. Relative error of the controller zero z_1 due to the presence of the derivative filter in an ideal form PID controller

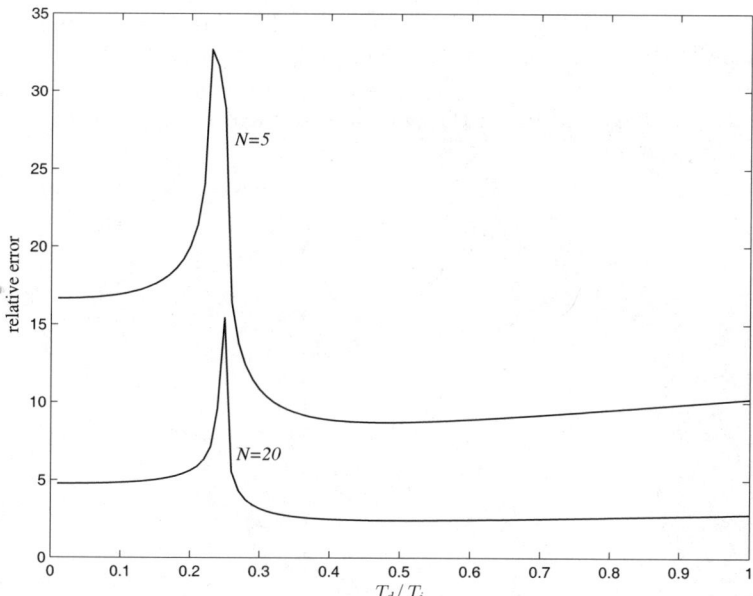

Fig. 2.2. Relative error of the controller zero z_2 due to the presence of the derivative filter in an ideal form PID controller

2.3 Ideal vs. Series Form

From another point of view with respect to the approach made in Section 2.2, the PID controllers in the ideal and or in the series form are compared, according to the analysis and the examples presented in (Isaksson and Graebe, 2002).

In particular, the role of the controller structure in the classical lead-lag design or in the pole-placement design is outlined by means of the following examples. Suppose that the control of a tank level with a first-order actuator has to be performed. The process is described by the following transfer function

$$P(s) = \frac{Y(s)}{U(s)} = \frac{K}{s(\tau s + 1)}, \quad K = 0.1, \quad \tau = 2, \tag{2.6}$$

where the input $u(t)$ is the valve position set-point and the output $y(t)$ is the tank level. A classical controller design leads to the following controller transfer function, in the context of the typical unitary-feedback control scheme (see Figure 1.2 with $F(s) = 1$):

$$C(s) = 1.06 \frac{(3s+1)(8s+1)}{3s(2s+1)}. \tag{2.7}$$

This assures a crossover frequency of 0.3 rad/s and a phase margin of slightly more than 45 deg. The Bode diagram of the open-loop transfer function $C(s)P(s)$ is shown in Figure 2.3. The designed controller corresponds to a PID controller in series form (1.22) where $K'_p = 1.06$, $T'_i = 3$, $T'_d = 8$, and $N' = 4$ or, equivalently, $K'_p = 2.83$, $T'_i = 8$, $T'_d = 3$, and $N' = 1.5$. These controllers can be converted in a PID controller in ideal form (1.20) by applying the following formulae:

$$
\begin{aligned}
T_i &= T'_i + \left(1 - \frac{1}{N'}\right) \\
K_p &= K'_p \frac{T_i}{T'_i} \\
T_d &= T'_d \left(\frac{T'_i}{T_i} - \frac{1}{N'}\right) \\
N &= \frac{T_d N'}{T'_d}
\end{aligned}
\tag{2.8}
$$

In both cases, it follows that $K_p = 3.18$, $T_i = 9$, $T_d = 0.67$ and $N = 1/3$. It can be seen that N and N' are not within the typical range of $5 \div 20$ and they do have a significant role in the overall controller design procedure. Indeed, setting $N = 1/3$ in the ideal PID controller means that the additional pole introduced by the derivative filter is still at a higher frequency than the two controller zeros (the two zeros are at $s = -0.121$ and $s = -1.026$, while the introduced pole is at $s = -4.48$). The fact that the derivative part provides a phase lead is actually evident in the series controller, since N' is grater than one.

Similar considerations apply if a pole-placement technique is adopted. Suppose that an ideal PID controller (1.20) is applied to the tank level process (2.6). The following characteristic equation results:

$$\tau \frac{T_d}{N} s^4 + \left(\tau + \frac{T_d}{N}\right) s^3 + \left(1 + K_p K T_d \left(1 + \frac{1}{N}\right)\right) s^2$$
$$+ K K_p \left(1 + \frac{T_d}{N T_i}\right) s + \frac{K K_p}{T_i} = 0 \qquad (2.9)$$

Assume now that the location of the desired closed-loop poles is such as there are two complex poles at a distance λ from the origin and with the same complex and real part (i.e., $s = (-1 \pm j)/(\lambda\sqrt{2})$) so that they have a corresponding damping factor of $\sqrt{2}/2$. Then, the two remaining poles are placed in the same position on the real axis at a distance of $-1/\lambda$ from the origin. In this way the desired characteristic equation is

$$\left(s^2 + \frac{\sqrt{2}}{\lambda}s + \frac{1}{\lambda^2}\right)\left(s + \frac{1}{\lambda}\right)^2 =$$
$$s^4 + \frac{2+\sqrt{2}}{\lambda}s^3 + \frac{2+2\sqrt{2}}{\lambda^2}s^2 + \frac{2+\sqrt{2}}{\lambda^3}s + \frac{1}{\lambda^4} = 0 \qquad (2.10)$$

Comparing the polynomial coefficients, the following PID parameters can be determined by fixing $\lambda = 3$:

$$K_p = 3.36, \quad T_i = 8.68, \quad T_d = 0.463, \quad N = 0.296. \qquad (2.11)$$

It turns out that the value of N is significantly outside the typical range also in this case, but this corresponds to a series controller with phase lead (i.e., with $N' > 1$). Actually, the parameters of the corresponding PID controller in series form are:

$$K'_p = 3.12, \quad T'_i = 8.08, \quad T'_d = 2.18, \quad N' = 1.39. \qquad (2.12)$$

and the resulting zeros of the controller are $s = -0.12$ and $s = -0.46$ while the poles are at $s = 0$ and $s = -0.64$. It is worth stressing that the choice of $\lambda = 3$ results in a control system that has, as in the previous case, a crossover frequency of about 0.3 rad/s and a phase margin of about 45 deg. The Bode diagram of the open-loop system $C(s)P(s)$ is presented in Figure 2.4. The similarity with the previous one is evident. In order to verify the improve-

24   2 Derivative Filter Design

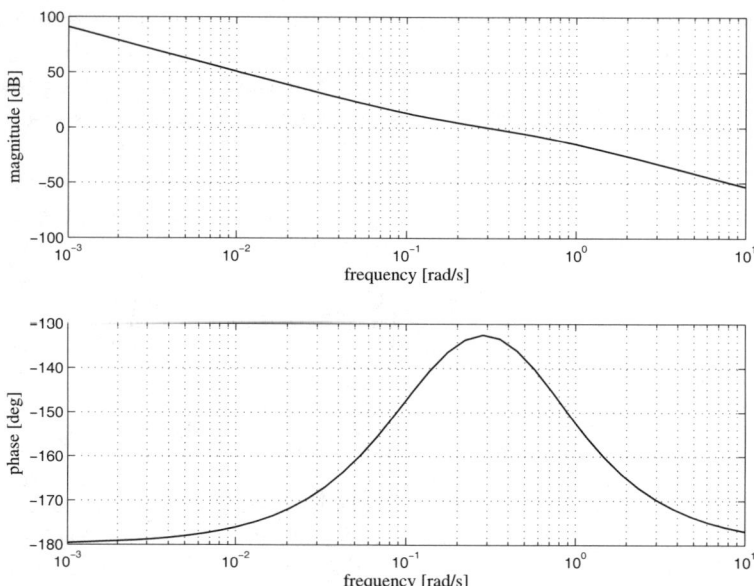

Fig. 2.3. Bode plot of the open-loop transfer function $C(s)P(s)$ resulting from the lead-lag design (Process (2.6))

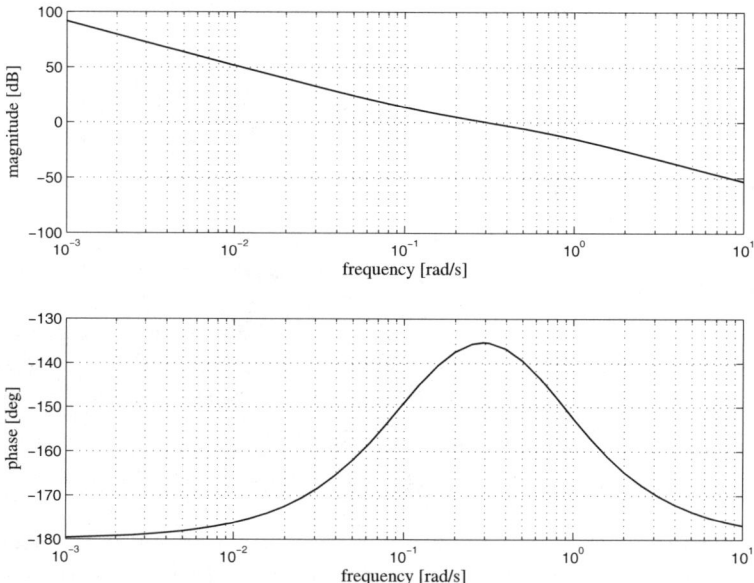

Fig. 2.4. Bode plot of the open-loop transfer function $C(s)P(s)$ resulting from the pole-placement design (Process (2.6))

ment in the performance given by the derivative action, the pole-placement approach is applied also with a PI controller (1.8). The characteristic equation is in this case:

$$\tau s^3 + s^2 + KK_p s + \frac{KK_p}{T_i} = 0. \tag{2.13}$$

It has to be noted that there are three poles to be placed but only two design parameters, while in the previous case there were four conditions for four parameters, because of the presence of the derivative filter parameter N (N'). Thus, a dominant pole design strategy is adopted, namely, only the location of the two dominant poles is selected, while the location of the third pole is checked at the end. In this context, the two dominant poles are chosen as in the previous case at $s = (-1 \pm j)/(\lambda\sqrt{2})$. Denoting as δ the third time constant, the desired characteristic equation is:

$$\left(s^2 + \frac{\sqrt{2}}{\lambda}s + \frac{1}{\lambda^2}\right)\left(s + \frac{1}{\delta}\right)$$
$$= s^3 + \left(\frac{\sqrt{2}}{\lambda} + \frac{1}{\delta}\right)s^2 + \left(\frac{1}{\lambda^2} + \frac{\sqrt{2}}{\lambda\delta}\right)s + \frac{1}{\lambda^2\delta} = 0. \tag{2.14}$$

By comparing the coefficients of Equations (2.13) and (2.14) it follows that:

$$\frac{\sqrt{2}}{\lambda} + \frac{1}{\delta} = \frac{1}{\tau} \tag{2.15}$$

$$\frac{KK_p}{\tau} = \frac{1}{\lambda^2} + \frac{\sqrt{2}}{\delta\lambda} \tag{2.16}$$

$$\frac{KK_p}{\tau T_i} = \frac{1}{\lambda^2\delta} \tag{2.17}$$

From Equation (2.15) it turns out that the smaller λ is the higher δ is and therefore the system cannot be made arbitrarily fast. Indeed, it is $\delta > 0$ (*i.e.*, the system is asimptotically stable) if $\lambda < \sqrt{2}\tau$ and therefore there is a clear limitation in the nominal performance. The value of $\lambda = 3.5$ (that implies $\delta = 10.4$) is eventually selected in order to achieve the best performance (Isaksson and Graebe, 2002). The resulting PI parameters are $K_p = 2.41$ and $T_i = 15.4$.

Set-point step responses and load disturbance responses obtained by the two designed PID controllers and the PI controller are shown in Figures 2.5 and 2.6. It appears that the two PID controllers give very similar responses and they outperform the PI controller in the load disturbance rejection task. Thus, the benefits of the derivative action appears in this case.

Summarising, from the examples presented, it can be deduced that, for a PID controller in series form, it can be sensible to choose a fixed derivative factor

26   2 Derivative Filter Design

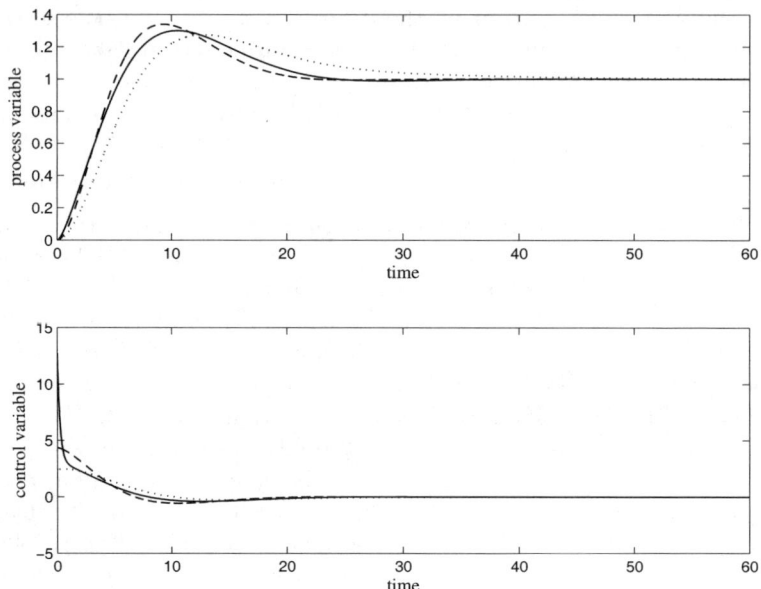

Fig. 2.5. Set-point step response for the designed controllers (Process (2.6)). Solid line: phase-lag PID; dashed line: pole-placement PID; dotted line: PI.

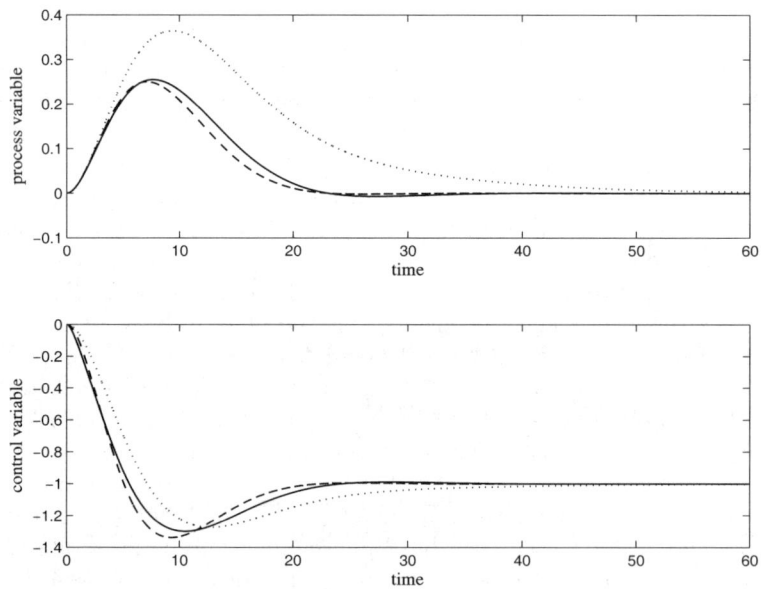

Fig. 2.6. Load disturbance step response for the designed controllers (Process (2.6)). Solid line: phase-lag PID; dashed line: pole-placement PID; dotted line: PI.

$N' > 1$, as a controller with a phase lead might result (note that the maximum phase lead depends only on N' and it is achieved when $N' = 10$). Conversely, for a PID controller in ideal form $C_{i1a}(s)$ (1.20), the necessary phase lead might be achieved with values of N also less than one and therefore fixing it to a constant value greater than one (in the range from 8 to 16 as is done in the vast majority of the industrial implementations (Ang et al., 2005)) can represent an unnecessary limitation of the performance.

It is worth stressing that if the alternative output-filtered form of the ideal controller $C_{i2a}(s)$ (1.23) (or $C_{i2b}(s)$ (1.24)) is adopted, the reasoning related to the series form has to be applied, since the filter is in series with the overall controller transfer function. Thus, if this structure is adopted, the choice of the value of the filter time constant T_f is more intuitive.

In any case, it appears from this analysis that the tuning of a PID controller should involve four parameters, since the derivative filter plays a major role in the overall control system performance.

2.4 Simulation Results

In order to understand better the previously described problems associated with the design of the derivative filter, some simulation results are given. Consider the process

$$P(s) = \frac{1}{s+1} e^{-0.2s}. \tag{2.18}$$

Then, consider a PID controller whose parameters are selected according to the Ziegler–Nichols rules based on the frequency response (note that the ultimate gain K_u is equal to 8.5 and the ultimate period is $P_u = 1.34$). Both the ideal form (1.20) and the series form (1.22) are evaluated. The controller parameters are reported in Table 2.1, where the conversion between the ideal and series structure has been performed by means of formulae (1.17), i.e., without taking into account the derivative filter. Note that $T_i = 4T_d$, that is, the two controller zeros are in the same position for the series controller and for the ideal one if the derivative filter is not considered. The derivative filter time

Table 2.1. Parameters for the ideal and series PID controller for the examples of Section 2.4

	Ziegler–Nichols	Kappa–Tau
K_p	5.00	5.74
T_i	0.672	0.66
T_d	0.168	0.15
K'_p	2.50	3.75
T'_i	0.336	0.43
T'_d	0.336	0.23

28   2 Derivative Filter Design

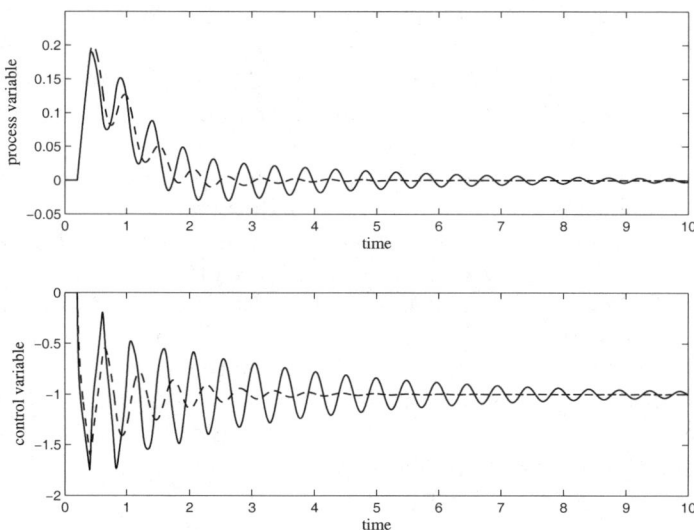

Fig. 2.7. Load disturbance step response for the PID controllers with Ziegler–Nichols parameters (Process (2.18)). Solid line: ideal form with derivative filter; dashed line: series form with derivative filter.

constant has been selected as $N = N' = 10$. The control system responses when a load disturbance unitary step is applied in both cases are plotted in Figure 2.7. The significantly different behaviour of the control system appears. This is due to the fact that the actual zeros of the ideal controller are in $s = -2.77 \pm j0.60$, while they should be the same as those of the series controller that are both in $s = -2.98$. Note that the phase margin of the resulting ideal controller is 44.2 deg (the crossover frequency is $\omega_c = 8.57$ rad/s), while that of the series one is 55.1 deg (the crossover frequency is $\omega_c = 6.19$ rad/s). The same reasoning is applied by considering the Kappa–Tau tuning rules proposed in (Åström and Hägglund, 1995). The parameters obtained are reported in Table 2.1, while the load disturbance unitary step responses are plotted in Figure 2.8. Also in this case the two responses are significantly different. The series controller assures a phase margin of 41.7 deg ($\omega_c = 7.83$ rad/s), while the ideal one, because of the presence of the derivative filter, provides a phase margin of just 15.9 deg ($\omega_c = 11.2$ rad/s). These results confirm the issues discussed in the previous sections that imply the fact that the design of the derivative filter should be considered carefully. The filtering of the measurement noise is also considered hereafter. Consider the same process (2.18) with the following controllers:

- a PI controller with $K_p = 4$ and $T_i = 1$;
- a derivative-filtered PID controller in ideal form (1.20) with $K_p = 4$, $T_i = 1$, $T_d = 0.1$ and $N = 10$;

- a derivative-filtered PID controller in ideal form (1.20), where the derivative filter is a second-order system, with again $K_p = 4$, $T_i = 1$, $T_d = 0.1$ and $N = 10$;
- an output-filtered PID controller in ideal form (1.23) with $K_p = 4$, $T_i = 1$, $T_d = 0.1$ and $T_f = 0.1$;
- an output-filtered PID controller in ideal form (1.24), where the filter is a second-order system with $K_p = 4$, $T_i = 1$, $T_d = 0.1$ and $T_f = 0.1$;
- a derivative-filtered PID controller in series form with $K'_p = 3.55$, $T'_i = 0.89$, $T'_d = 0.11$, $N' = 10$ (note that these parameters have been found by converting the parameters of the controllers in ideal form).

In all the cases a measurement white noise whose amplitude is in the range $[-5 \cdot 10^{-3}, 5 \cdot 10^{-3}]$ is applied to the control system. The resulting process variables and the control variables are plotted in Figures 2.9–2.14. It can be seen that the control variable is less noisy for the output-filtered PID structures. This is somewhat obvious, since the proportional action is also responsible for the amplification of the measurement noise and therefore the filter applied to the whole control variable is more effective than that applied to the derivative action only. If a second-order filter is adopted, the reduction of the noise effect is more evident. However, if the value of T_f in an output-filtered PID controller in ideal form is such that the additional poles are not at a much higher frequency with respect to the zeros (for a more effective filtering), then the presence of the second-order filter might influence the control performance.

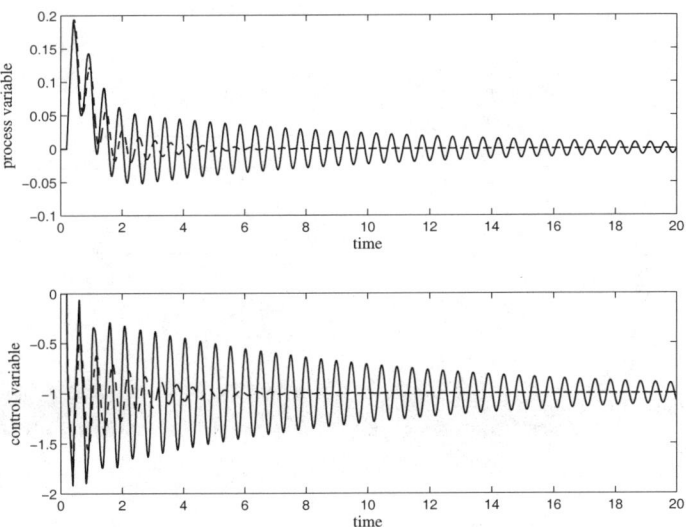

Fig. 2.8. Load disturbance step response for the PID controllers with Kappa–Tau parameters (Process (2.18)). Solid line: ideal form with derivative filter; dashed line: series form with derivative filter.

30   2 Derivative Filter Design

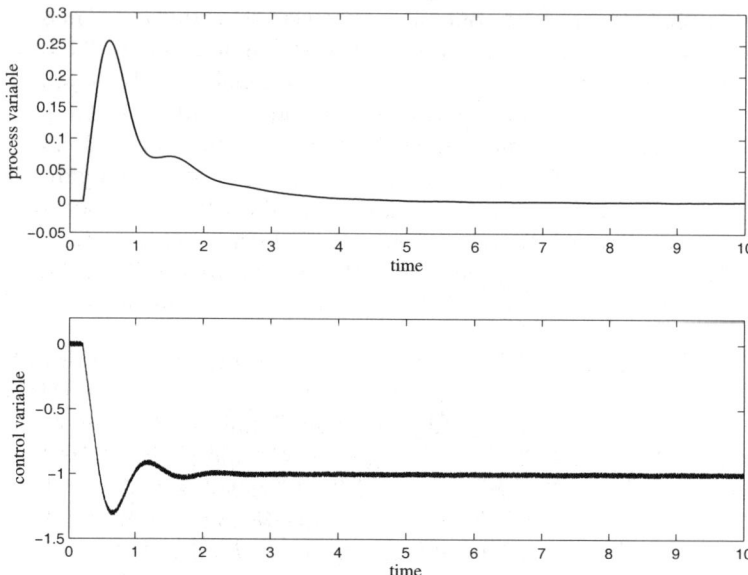

Fig. 2.9. Load disturbance step response (with noise measurement) for the PI controller

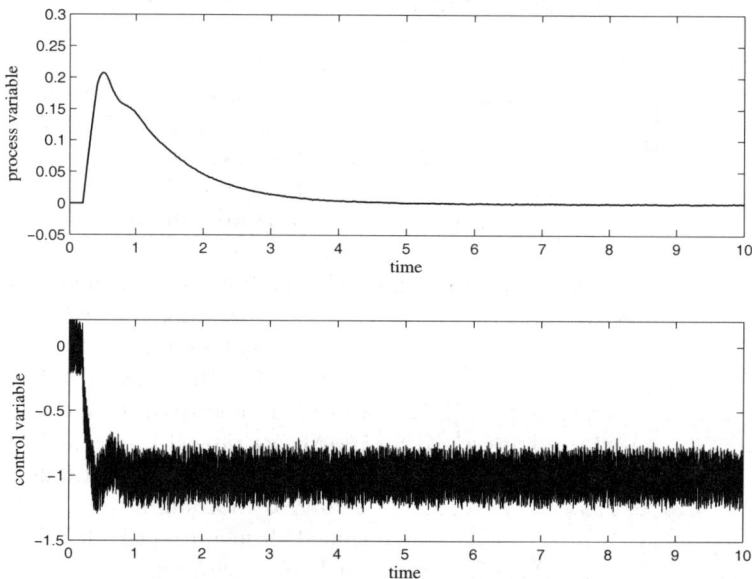

Fig. 2.10. Load disturbance step response (with noise measurement) for the ideal PID controller with a first-order derivative filter

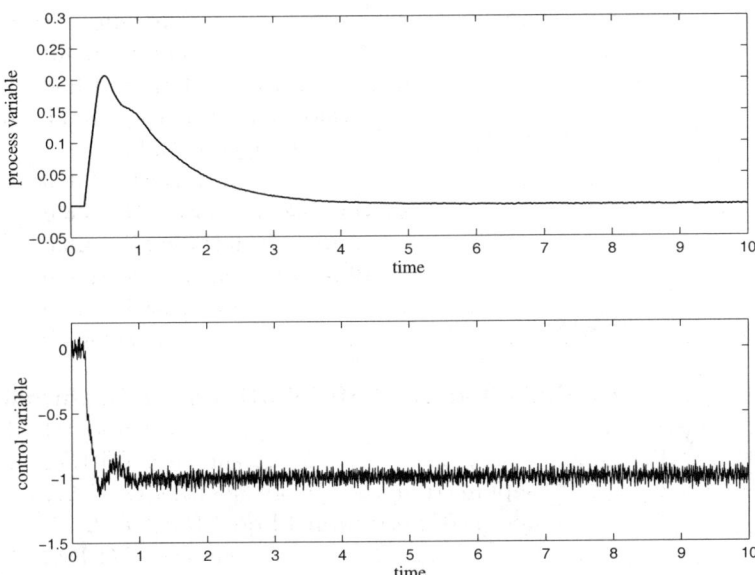

Fig. 2.11. Load disturbance step response (with noise measurement) for the ideal PID controller with a second-order derivative filter

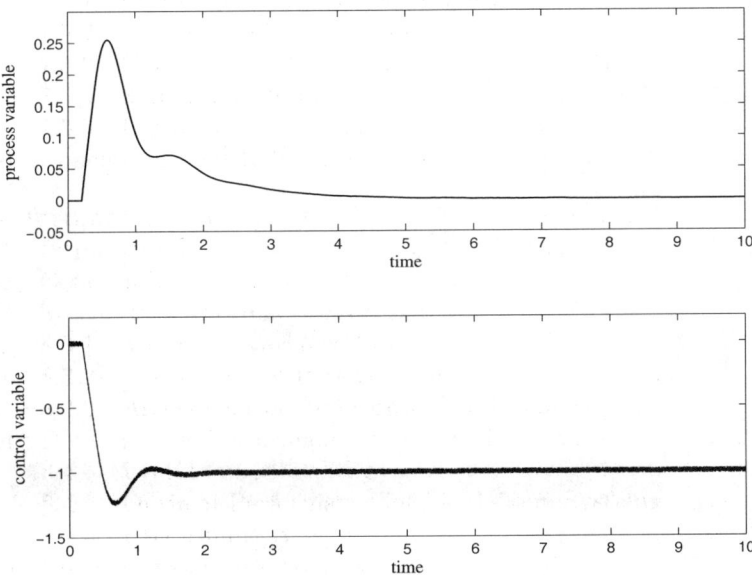

Fig. 2.12. Load disturbance step response (with noise measurement) for the ideal PID controller with a first-order output filter

32   2 Derivative Filter Design

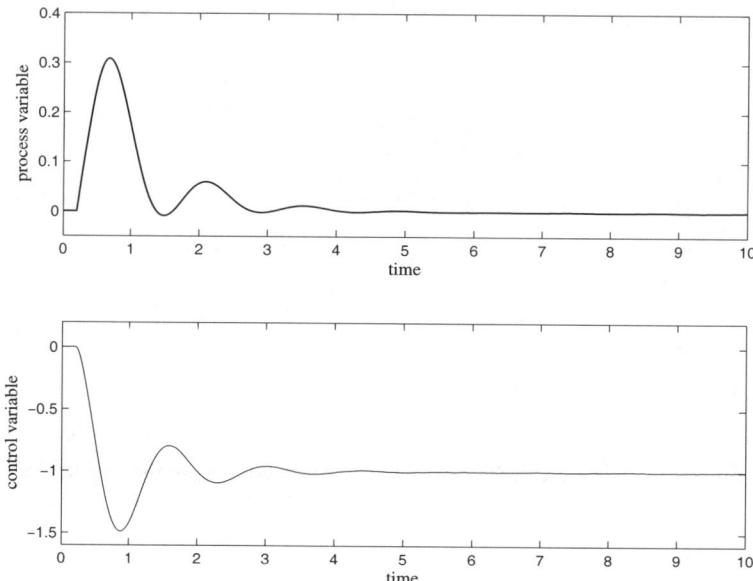

Fig. 2.13. Load disturbance step response (with noise measurement) for the ideal PID controller with a second-order output filter

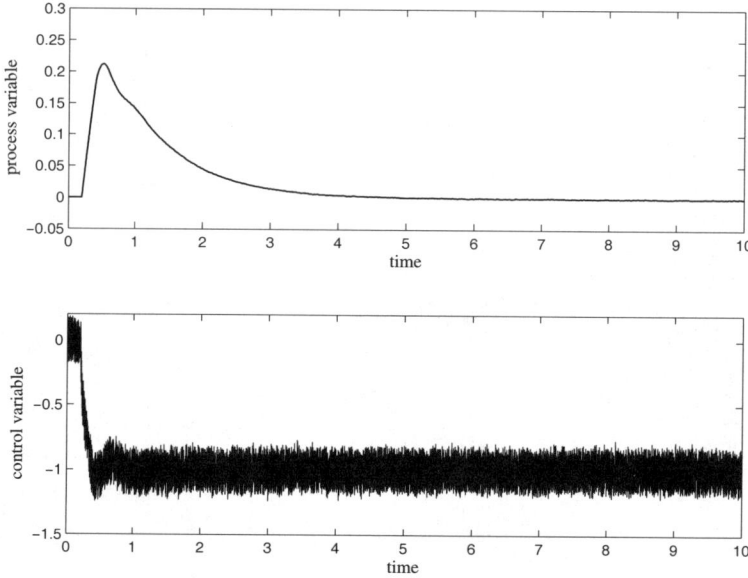

Fig. 2.14. Load disturbance step response (with noise measurement) for the series PID controller with a first-order derivative filter

2.5 Four-parameters Tuning

In the previous sections it has been underlined that problems associated with the derivative action that prevent a wide use of it are not just due to the noise. Indeed, tuning rules for a PID controller should involve four parameters, as also stressed in (Luyben, 2001a).
The most well-known design method that provides the values of all the four parameters of an ideal output-filtered PID controller is surely that based on the Internal Model Control (IMC) approach (Rivera *et al.*, 1986; Morari and Zafiriou, 1989). It can be remarked that a user-chosen parameter allows the handling of the trade-off between aggressiveness and robustness. The effectiveness of this tuning methodology has been shown in the literature; however, it has to be borne in mind that, being based on a pole-zero cancellation, it is not suitable for lag-dominant processes for which a very sluggish load disturbance response occurs (Shinskey, 1994; Shinskey, 1996). In this context an effective modification has been proposed in (Skogestad, 2003).
Recently, tuning rules that comprises also the derivative filter has been proposed in (Åström and Hägglund, 2004). They are based on the maximisation of the integral gain (so that the integrated error when a load disturbance occurs is minimised), subject to a robustness constraint. It is also stressed that the appropriate value of the ratio between the integral time constant and the derivative time constant should vary depending on the process dynamics (in particular, depending on the relative dead time of the process) and in most cases is less than four.
Similar conclusions are drawn in (Kristiansson and Lennartson, 2006). There, four-parameters tuning rules are proposed which take into account the trade-off between load disturbance rejection performance (in terms of integrated absolute error) and control effort, with a constraint on the generalised maximum sensitivity, which is a measure of the robustness of the control system. It is shown that the benefits of the derivative action can be severely limited if the ratio between the integral time constant and the derivative time constant is fixed to four and if the derivative filter factor is fixed in a PID controller in ideal form $C_{i1a}(s)$ (1.20) (or $C_{i1b}(s)$ (1.21)).
For this PID controller, it is suggested to set $T_i/T_d = 2.5$. Further, it is shown that considering the derivative filter time constant as a true tuning parameter allows a significant improvement of the overall performance.

2.6 Conclusions

In this chapter the design of the derivative filter has been discussed. Although the analysis provided and the examples presented are certainly not exhaustive, they are sufficient to show that the choice of the controller structure and of the derivative filter factor is indeed a critical issue and the PID controller should be considered as a four-parameters controller. In fact, the derivative action is a key factor in improving the control system performance and the reason for being rarely adopted in practice is not only the amplification of the measurement noise.

In particular, it has been shown that predefining the derivative filter factor in an ideal form controller $C_{i1a}(s)$ (1.20) (or $C_{i1b}(s)$ (1.21)) might severely limit the performance. If a series controller $C_s(s)$ (1.22) is adopted, then the filter does not influence the location of the controller zeros. However, in this case the two zeros have to be real and this factor might limit the performance as well. Thus, the most convenient choice appears to be the use of an output-filtered ideal form PID controller $C_{i2a}(s)$ (1.23) (or $C_{i2b}(s)$ (1.24)) since this is the most general expression and the drawbacks of the other two forms are avoided. Further, effective tuning rules for the selection of the four parameters K_p, T_i, T_d, and T_f are available in this case.

3

Anti-windup Strategies

3.1 Introduction

One of the most well-known possible source of degradation of performance is surely the so-called integrator windup phenomenon, which occurs when the controller output saturates (thus, this problem is of particular concern at the process start-up).
Strategies for limiting this effect are illustrated and compared in this chapter.

3.2 Integrator Windup

The integrator windup effect is explained in this section. When a set-point change is applied, the control variable might attain the actuator limit during the transient response. In this case the system operates as in the open-loop case, since the actuator is at its maximum (or minimum) limit, independently of the process output value. The control error decreases more slowly as in the ideal case (where there is no saturation limits) and therefore the integral term becomes large (it *winds up*). Thus, even when the value of the process variable attains that of the reference signal, the controller still saturates due to the integral term and this generally leads to large overshoots and settling times.
The situation is illustrated in the following example. Consider the control scheme depicted in Figure 3.1 which is similar to that of Figure 1.2 but in this case the controller output u differs in general from the process input u' because of the presence of an actuator saturation with a upper limit u_{max} and an lower limit u_{min}. In this context the process

$$P(s) = \frac{1}{10s+1}e^{-4s} \qquad (3.1)$$

is controlled by an ideal PID controller (the derivative filter is not adopted for simplicity) with $K_p = 3$, $T_i = 8$ and $T_d = 2$ (note that these are the param-

3 Anti-windup Strategies

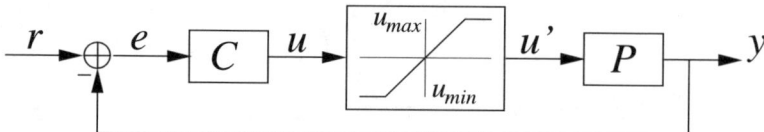

Fig. 3.1. General control scheme with saturation

eters obtained by employing the Ziegler–Nichols tuning rules). The actuator saturation limits are $u_{min} = 0$ and $u_{max} = 1.5$. The set-point unitary step response (starting from null initial conditions) is plotted in Figure 3.2. It can be seen that at time $t = 15$ the process output attains the set-point value but, despite this, the process input still remains (for quite a long time) at the maximum level because of the high value of the integral term. This causes a significant overshoot which is recovered after a long time, that is, when the integral term decrement is sufficient for the control variable to be lower than the saturation limit.

From this example it is clear that the nonlinear dynamics of the actuator can be detrimental for the performance and has therefore to be somehow taken into account in the design of the PID controller.

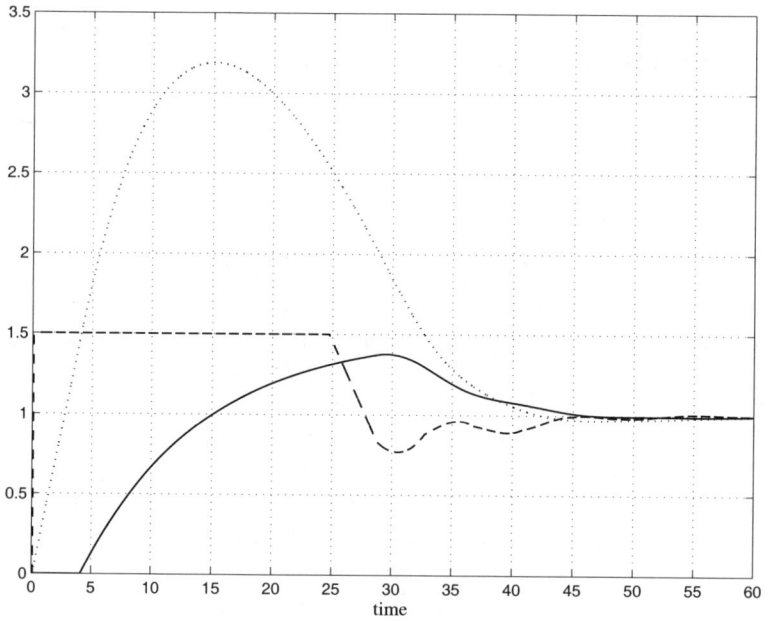

Fig. 3.2. Set-point step response illustrative of the integrator windup phenomenon. Solid line: process output; dashed line: process input; dotted line: integral term.

It has to be noted that the integrator windup occurs mainly when a step is applied to the reference set-point signal rather than to the manipulated variable (*i.e.*, in the presence of a load disturbance) (Vrancic, 1997). Furthermore, the most significant effects of the integrator windup take place when the process is of low order (and the dead time is small with respect to the time constant). For these reasons, in the following part of this chapter the analysis will be restricted to the set-point response of first-order-plus-dead-time systems.

3.3 Anti-windup Techniques

In order to cope with the presence of the actuator saturation, two design approaches can be followed in general. In the first one the nonlinearity is considered explicitly from the beginning of the design phase and the control law is derived in the context of the nonlinear control theory. Although this is a more rigorous approach, it might be too complicated to be applied in practical cases where the cost (and the fast commissioning) of the controller is of primary importance. In other words, the advantages provided by the use of a standard PID control law are no more exploited (note that this is not entirely true, as it will be shown in Chapter 5). In the second approach, on the other hand, the control law is designed disregarding the actuator nonlinearity, so that a PID controller can be adopted. Then, the detrimental effects due to the integrator windup are compensated by conveniently adopting an additional functionality designed for this purpose (Kothare *et al.*, 1994). These anti-windup techniques are analysed hereafter.

It is worth noting at this point that if the PID controller is implemented in incremental form (1.38) (see Section 1.7), the windup effect is naturally avoided as the integral action is 'outside' the PID control law (note that there is no accumulation of the error in Equation (1.38)).

3.3.1 Avoiding Saturation

The most intuitive way of avoiding the integrator windup is to avoid the saturation of the control variable. This can be done by limiting or smoothing the set-point changes and/or by detuning the controller (*i.e.*, by selecting a more sluggish controller). It is obvious that this requires a somewhat significant additional effort in the controller design and, most of all, it might imply (especially if the controller is detuned) an unacceptable decrement of the performance. Thus, this approach is not advisable in practical cases, although it has to be stressed that very effective methodologies have been recently developed for the determination of a convenient reference signal to be adopted for the achievement of a satisfactory transient response without exceeding the saturation limits (see Chapter 5).

3.3.2 Conditional Integration

A classical effective methodology is the so-called conditional integration. It consists of switching off the integration (in other words, the error to be integrated is set to zero) when a certain condition is verified. For this reason this method is also called *integrator clamping*. The following options can be implemented:

- the integral term is limited to a predefined value;
- the integration is stopped when the error is greater than a predefined threshold, namely, when the process variable value is far from the set-point value;
- the integration is stopped when the control variable saturates, *i.e.*, when $u \neq u'$;
- the integration is stopped when the control variable saturates and the control error and the control variable have the same sign (*i.e.*, when $u \cdot e > 0$).

The first two methods, which appear to be particularly suitable for the start-up phase of batch processes, have the disadvantage that they might result in a steady-state error. Actually, in the first case the limitation of the integral term must not prevent the attainment of the set-point value and in the second case it has to be avoided that the controller gets stuck at a (steady-state) value such that the control error is still greater than the threshold. It appears that both methods require an additional design parameter that has to be carefully selected in order to properly handle the trade-off between the need of avoiding the integrator windup and the need of assuring a zero steady-state error. This indeed is a major drawback as it somehow limits the ease of use of the PID controller, which is always of concern.

These problems are avoided in the third and fourth techniques. However, with respect to the third method, the fourth one has the great advantage that the integrator is not inhibited when it helps to push the control variable away from the saturation. For this reason it should be preferred to the others, as also stated in (Hansson *et al.*, 1994).

A technique that is slightly different from the previous ones is the so-called *preloading* (Shinskey, 1994; Shinskey, 1996). It consists of setting the integral term to a predefined value I_{max} or I_{min} during saturation. This value should be normally chosen smaller than the steady-state value of the integral term in order to avoid an excessive overshoot. Hence, there is an additional design parameter whose proper selection might require the knowledge of the gain of the process. Further, the presence of load disturbances can represent a severe drawback of the method.

3.3.3 Back-calculation

A valuable alternative approach to conditional integration is the so-called back-calculation, which consists of recomputing the integral term when the

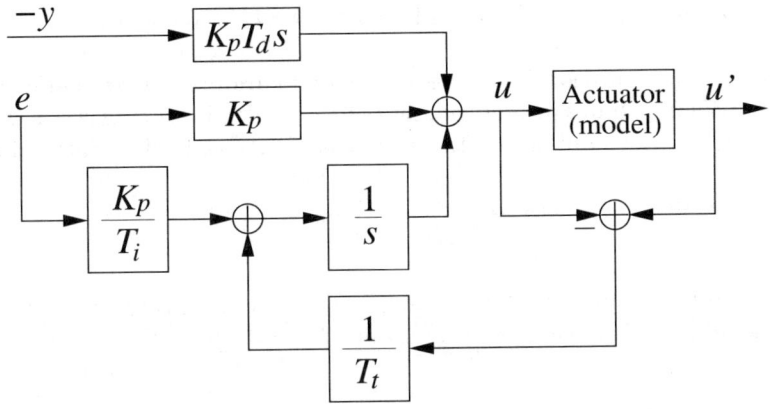

Fig. 3.3. Anti-windup scheme with back-calculation

controller saturates. In particular, the integral value is reduced or increased (when the controller output is greater than its upper limit u_{max} or when it is less than its lower limit u_{min}, respectively) by feeding back the difference of the saturated and unsaturated control signal, as shown in Figure 3.3, where T_t is called the tracking time constant (Åström and Hägglund, 1995). Formally, denoting by e_i the integrator input, it is:

$$e_i = \frac{K_p}{T_i} e + \frac{1}{T_t}(u' - u) \qquad (3.2)$$

It has to be noted that the back-calculation technique has an inherent observer property (Walgama and Sternby, 1990). In particular, it aims at estimating the correct state of the controller when it does not correspond to the input of the process (because of the saturation of the actuator).

The value of T_t clearly determines the rate at which the integral term is reset and therefore its choice determines the performance of the overall control scheme. In order to help the operator in this context, tuning rules for the tracking time constant have been proposed. In (Åström and Hägglund, 1995) it is suggested that

$$T_t = \sqrt{T_i T_d}. \qquad (3.3)$$

The main drawback of this formula is that it cannot be adopted in PI control where $T_d = 0$. Alternatively, in (Bohn and Atherton, 1995) it is suggested that

$$T_t = T_i. \qquad (3.4)$$

It has to be stressed at this point that the back-calculation technique includes the so-called conditioning technique (Hanus et al., 1987; Walgama et al., 1991), which represent a general anti-windup and bumpless transfer method. Indeed, both the commutation between manual and automatic mode and the integrator windup effect can be seen in the general context of dealing with a situation when the actual input of the process is different from the controller output. The conditioning technique assures that when the difference between the actual control variable and the desired one disappears, then the process output attains the steady-state with the same dynamics as the unconstrained closed-loop system. In this framework the suggestion is to set

$$T_t = K_p. \tag{3.5}$$

The correlation between anti-windup and bumpless transfer between manual and automatic mode is depicted in Figure 3.4. It appears that, as stated in (Åström and Hägglund, 1995), the controller can be interpreted as having a control mode, when it operates like an ordinary controller, and a tracking mode, when the integrator tracks a specified signal (note that the tracking mode is automatically disconnected when the signal to be tracked is the controller output).

Note that the saturation block in Figure 3.3 (and 3.4) might represent the true actuator or, alternatively, a model of it if a measure of the actuator output is not available. An alternative scheme, based on the use of a dead-zone, that exploits a model of the actuator and that implements the same back-calculation approach is shown in Figure 3.5 (Bohn and Atherton, 1995). It can be seen that if the control variable is in the range $[u_{min}, u_{max}]$, then the integrator output depends on the error signal only, but if the control variable is outside this range (*i.e.*, the actuator is saturating), then the integrator output is modified by a quantity that depends on the dead-zone gain (which has the same physical meaning of the tracking time constant).

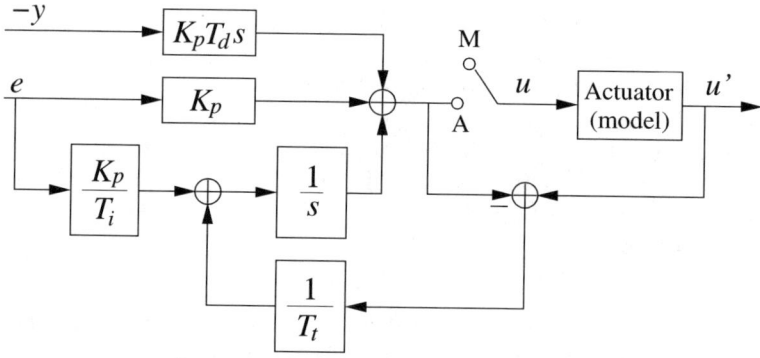

Fig. 3.4. Control scheme for anti-windup and bumpless transfer between manual (M) and automatic (A) mode

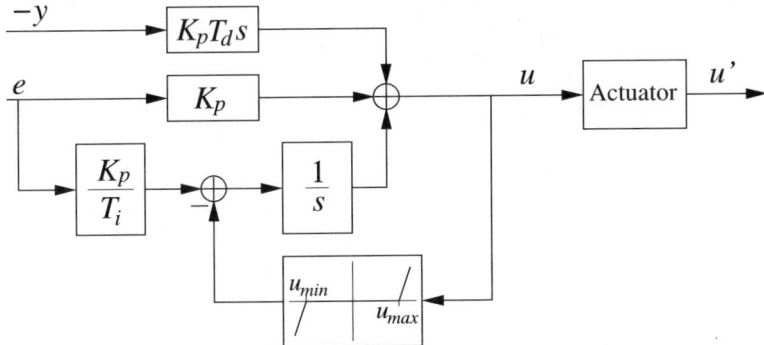

Fig. 3.5. Alternative anti-windup scheme with back-calculation

3.3.4 Combined Approaches

Methods that combine the conditional integration and the back-calculation approach have been also presented in the literature. In particular, in (Bohn and Atherton, 1995) it is proposed to apply an additional limit to the proportional-derivative part of the manipulated variable adopted to generate the anti-windup feedback signal. Thus, there is the significant drawback that an additional parameter has to be selected by the user.
Alternatively, in (Hodel and Hall, 2001) the summing junction that performs the feedback in the integral term is activated by a switch that is closed when a certain condition is met, namely, when

$$u \neq u' \quad \text{and} \quad e(u - \bar{u}) > 0 \tag{3.6}$$

where

$$\bar{u} := \frac{u_{max} + u_{min}}{2} \tag{3.7}$$

The control scheme is shown in Figure 3.6, where it appears that α denotes the design parameter. This technique has been denominated Variable-Structure PID (VSPID) anti-windup method and aims at keeping u as close as possible to u' during saturation so that the controller returns as fast as possible to linear operation. In this context, as a rule of thumb, it is suggested that α be set in such a way that the integrator feedback loop, during saturation, settles from two to five times faster than the closed-loop.
Finally, in (Visioli, 2003a), an alternative switching condition has been proposed, namely, the back-calculation is employed when the controller saturates, the control error has the same sign of the manipulated variable and the process output has left its previous set-point value. Assuming that the required control task is to perform an output transition from a set-point value y_0 to a set-point value y_1, the integrated input can be formally expressed as

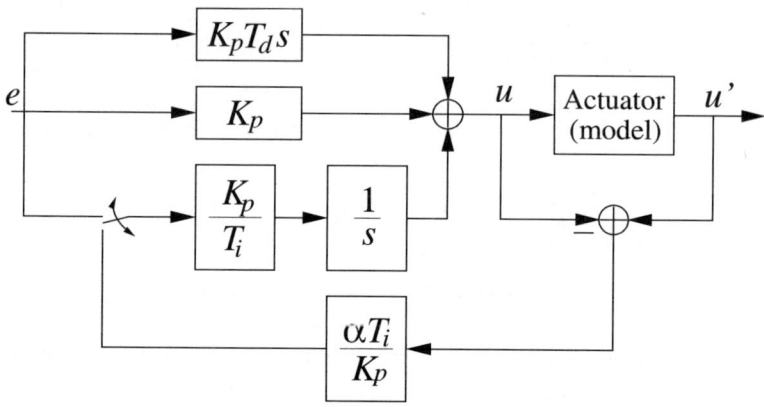

Fig. 3.6. Variable-structure PID anti-windup scheme

$$e_i = \begin{cases} \dfrac{K_p}{T_i}e + \dfrac{1}{T_t}(u' - u) & \text{if } u \neq u' \text{ and } u \cdot e > 0 \text{ and } \begin{cases} y > y_0 \text{ if } y_1 > y_0 \\ y < y_0 \text{ if } y_1 < y_0 \end{cases} \\ \dfrac{K_p}{T_i}e & \text{otherwise} \end{cases}$$

(3.8)

The rationale of this method is to avoid stopping the integration at the beginning of the transient response when the saturation is actually caused by the proportional action (this is particularly important when the ratio between the dead time and dominant time constant of the process is high) and at the same time to allow the decrease of the value of the tracking-time constant in order to have a smaller overshoot when the dead-time of the process is small. The suggested value for the tracking-time constant with this method is

$$T_t = 0.03 T_i. \tag{3.9}$$

It should be noted that in practical cases the condition $y > y_0$ (or, equivalently, $y < y_0$) has to cope with the measurement noise. A simple sensible solution is to define a noise band NB (Åström et al., 1993) (whose amplitude should be equal to the amplitude of the measurement noise) and to rewrite the condition as $y > y_0 + NB$.

3.3.5 Automatic Reset Implementation

When the PI control part is in the automatic reset configuration (see Figures 1.5 and 1.9), the anti-windup scheme can be implemented very easily by inserting the saturation function in the control scheme as depicted in Figures 3.7 and 3.8.

In the first case the model of the saturation limits the overall control variable,

3.3 Anti-windup Techniques   43

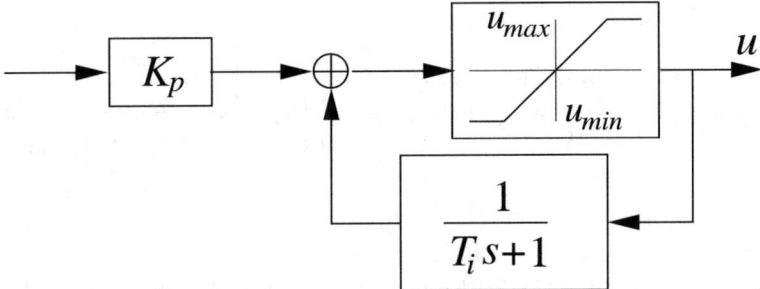

Fig. 3.7. Anti-windup scheme with PI controller in automatic reset configuration

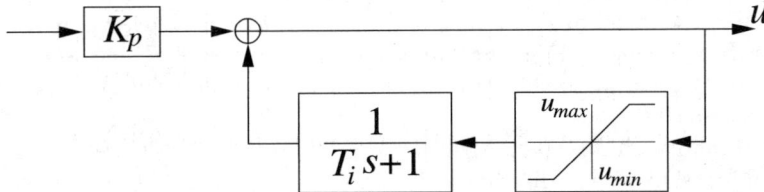

Fig. 3.8. Alternative anti-windup scheme with PI controller in automatic reset configuration

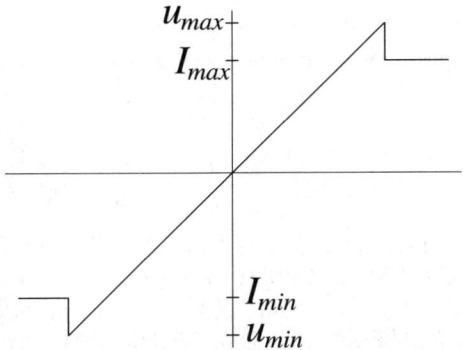

Fig. 3.9. Nonlinear function for implementing the preloading anti-windup technique with PI controller in automatic reset configuration

while in the second case it limits the integral action only. It is worth noting that the preloading technique can be easily implemented in the latter case by replacing the saturation function with the nonlinear function shown in Figure 3.9.

3.4 Simulation Results

In order to understand better the problems due to the integrator windup and to analyse the different antiwindup methods, the following results are presented. For the sake of clarity, the considered techniques are summarised and denoted as follows:

- CI: the conditional integration technique where the integral term is frozen when the control variable saturates and $u \cdot e > 0$;
- BC1: the back-calculation technique where $T_t = \sqrt{T_i T_d}$;
- BC2: the back-calculation technique where $T_t = T_i$;
- BC3: the back-calculation technique where $T_t = K_p$ (conditioning technique);
- CI-BC: the combined conditional integration and back-calculation approach proposed in (Visioli, 2003a) (see (3.8)), where $T_t = 0.03T_i$;
- VSPID: the variable-structure PID anti-windup method (see (3.6)–(3.7)) with $\alpha = 10$;
- S1: the anti-windup technique for the PID controller in automatic reset configuration where the saturation model is applied to the whole control variable (see Figure 3.7);
- S2: the anti-windup technique for the PID controller in automatic reset configuration where the saturation model is applied to the reset term (see Figure 3.8);
- PR: the preload technique (with $I_{max} = 0.8$) implemented by means of the nonlinear function of Figure 3.9.

The following two processes are then considered:

$$P(s) = \frac{1}{10s+1} e^{-Ls} \qquad L = 2, 8 \qquad (3.10)$$

In both cases, the Ziegler–Nichols tuning rules have been adopted (the resulting parameters have been properly converted for the series controllers). It results $K_p = 6$, $T_i = 4$ and $T_d = 1$ for $L = 2$, and $K_p = 1.5$, $T_i = 16$ and $T_d = 4$ for $L = 8$. The derivative action has been applied to the process output. Then, it has been fixed $u_{max} = 1.5$ and $u_{min} = 0$.

In order to understand better the following results, related to a unitary step set-point change, the cases where no saturation is present and where the saturation is present and no anti-windup strategy has been adopted are shown in Figures 3.10 and 3.11 respectively. Indeed, it can be seen that the windup effect is not present when $L = 8$. Thus, in this case it is essential that the anti-windup method would not decrement the performance (note that this confirms that the windup is more significant when the dead time is small with respect to the time constant).

Simulation results for the different anti-windup strategies for both $L = 2$ and $L = 8$ are plotted in Figures 3.12–3.20. In order to analyse better the performance, three indices, namely, the rise time (*i.e.*, the time required for the

3.4 Simulation Results

Table 3.1. Performance indices for the different anti-windup methods for $L = 2$

	Rise time	Overshoot [%]	Settling time
CI	8.48	2.1	12.20
BC1	8.48	4.8	12.12
BC2	8.48	12.2	19.24
BC3	8.48	17.9	21.39
CI-BC	8.48	2.1	12.20
VSPID	9.02	2.1	12.74
S1	8.47	4.7	12.12
S2	8.47	4.7	12.12
PR	8.48	2.1	12.21

Table 3.2. Performance indices for the different anti-windup methods for $L = 2$

	Rise time	Overshoot [%]	Settling time
CI	23.88	0	48.70
BC1	18.37	2.8	29.85
BC2	10.11	5.0	36.54
BC3	22.81	0.2	47.42
CI-BC	9.00	7.8	43.10
VSPID	23.89	0	48.71
S1	18.36	2.8	29.85
S2	18.36	2.8	29.85
PR	23.50	0	48.36

process output to change from 10% to 90 % of the final steady-state value), the maximum overshoot and the 5% settling time are reported in Tables 3.1 and 3.2.

From the results presented it can be deduced that the presence of the saturation and the choice of the anti-windup strategy influence significantly the results and that it is difficult to find a methodology that always provides the best results.

Actually, the amount of integral term that is present when the control variable is no more saturated determines the rise time and the overshoot of the process response. This is evident for example for the conditional integration technique CI that stops the integration for a somewhat long interval at the beginning of the transient response during the saturation and this might caused a high rise time (see Figure 3.12).

For the same reason, when the back-calculation technique is applied the results depend obviously on the value of the tracking time constant. The most suitable tuning seems to be $T_t = \sqrt{T_i T_d}$, but it has to be remembered that it cannot be applied in the PI control context. The combined approach provides satisfactory results for both $L = 2$ and $L = 8$ (see Figure 3.16), since it properly decreases the integral term when when a large overshoot is likely to occur.

The variable-structure PID methodology does not seem to provide a significant improvement with respect to the other methods. Note that the performance achieved cannot be improved significantly by selecting other values for the design parameter α.

The two series controller performs similarly and better than the preload technique. Indeed, the solution is simple and effective, although it has to be remembered that the series controller does not admit complex zeros and this might represent a serious limitation in the tuning phase.

In fact, in general, the achieved performance depends significantly on the tuning of the PID parameters in addition to the selected anti-windup strategy. However, the results presented show in any case the pros and cons of the different considered techniques and give a clear idea of how the problem can be approached.

It is worth noting that a comparison between different methodologies could be performed by considering the realisable reference concept, namely, by determining the (new) reference signal that should be applied to the control system in order to obtain a control variable u equal to the process input u' obtained by applying the step reference signal (Peng et al., 1996). However, this is a more theoretical approach that does not always provide a clear idea of the performance achieved.

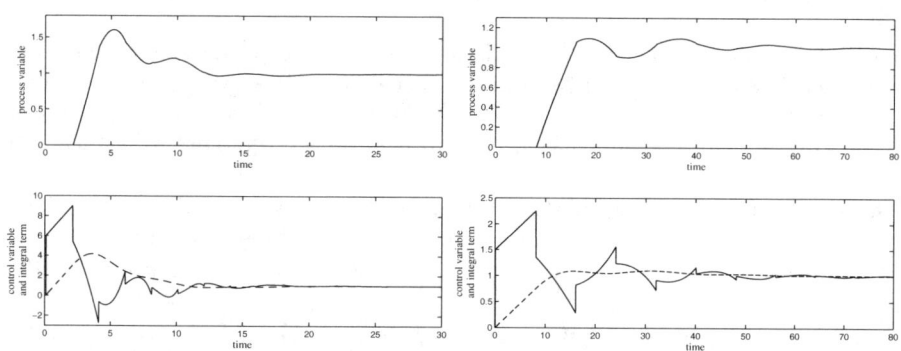

Fig. 3.10. Process variable, control variable (solid line) and integral term (dashed line) when no saturation is present for $L = 2$ (left) and $L = 8$ (right)

3.4 Simulation Results   47

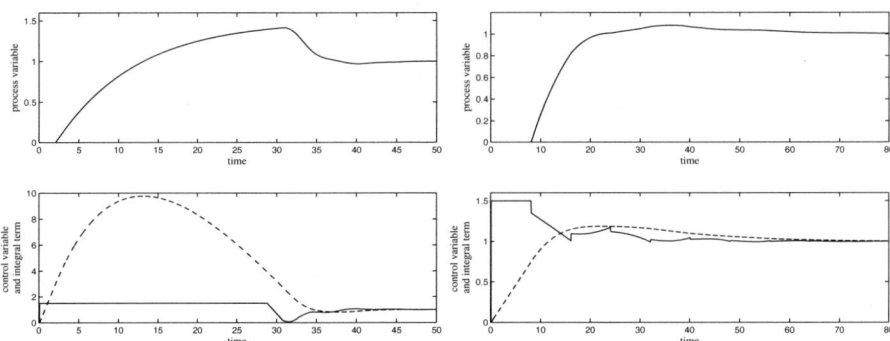

Fig. 3.11. Process variable, control variable (solid line) and integral term (dashed line) when no anti-windup is present for $L = 2$ (left) and $L = 8$ (right)

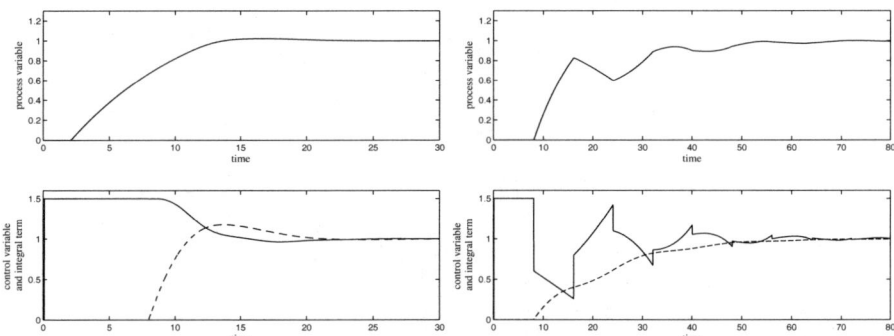

Fig. 3.12. Process variable, control variable (solid line) and integral term (dashed line) for the conditional integration method CI for $L = 2$ (left) and $L = 8$ (right)

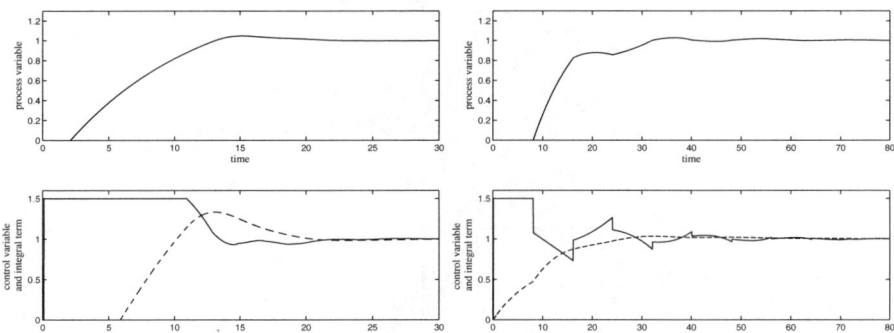

Fig. 3.13. Process variable, control variable (solid line) and integral term (dashed line) for BC1 with $T_t = 2$ for $L = 2$ (left) and with $T_t = 8$ for $L = 8$ (right)

48   3 Anti-windup Strategies

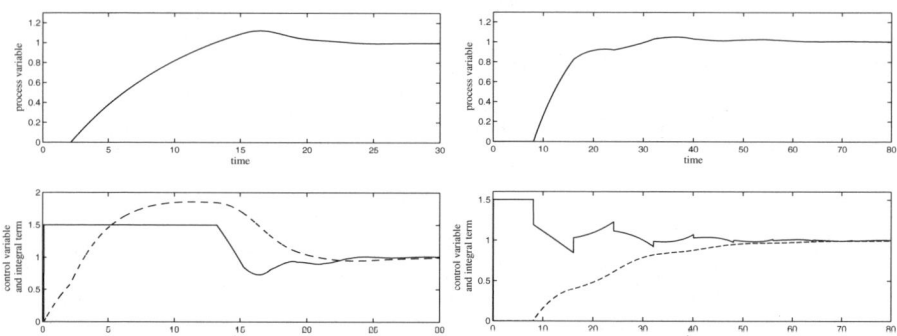

Fig. 3.14. Process variable, control variable (solid line) and integral term (dashed line) for BC2 with $T_t = 4$ for $L = 2$ (left) and with $T_t = 16$ for $L = 8$ (right)

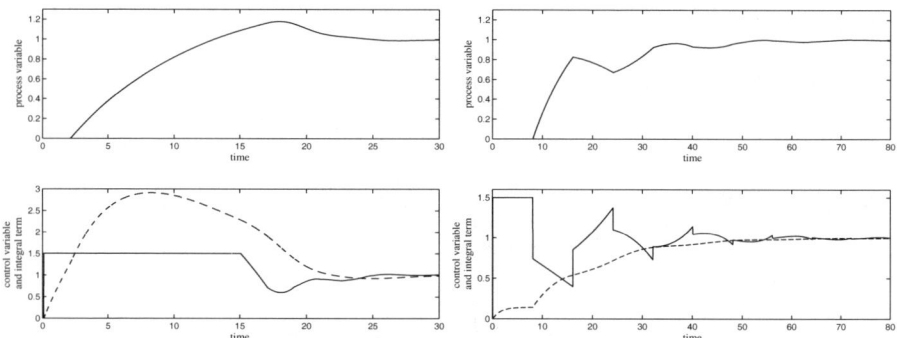

Fig. 3.15. Process variable, control variable (solid line) and integral term (dashed line) for BC3 with $T_t = 6$ for $L = 2$ (left) and with $T_t = 1.5$ for $L = 8$ (right)

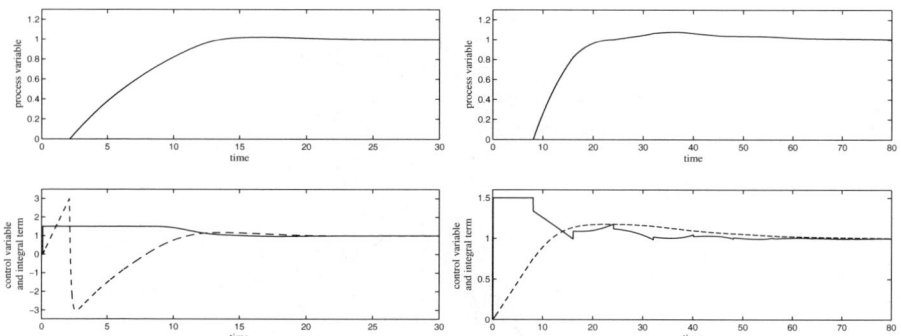

Fig. 3.16. Process variable, control variable (solid line) and integral term (dashed line) for CI-BC with $T_t = 0.12$ for $L = 2$ and with $T_t = 0.48$ for $L = 8$ (right)

3.4 Simulation Results   49

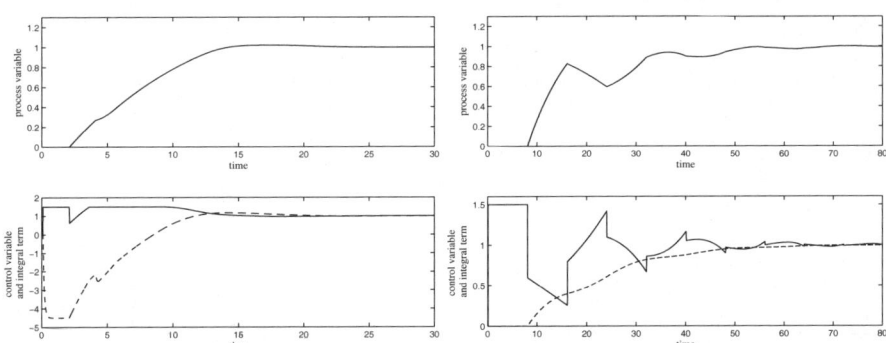

Fig. 3.17. Process variable, control variable (solid line) and integral term (dashed line) for the VSPID with $\alpha = 10$ for $L = 2$ (left) and for $L = 8$ (right)

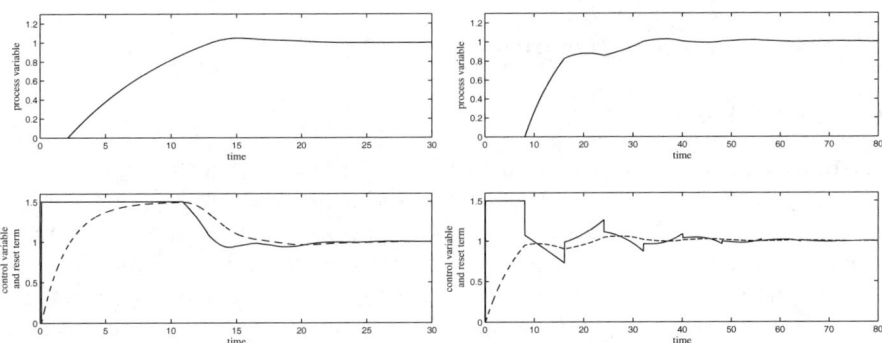

Fig. 3.18. Process variable, control variable (solid line) and reset term (dashed line) for the automatic reset configuration S1 for $L = 2$ (left) and for $L = 8$ (right)

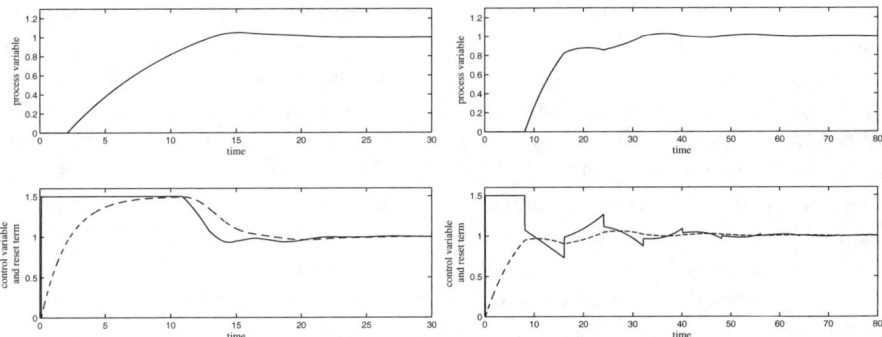

Fig. 3.19. Process variable, control variable (solid line) and reset term (dashed line) for the automatic reset configuration S2 for $L = 2$ (left) and for $L = 8$ (right)

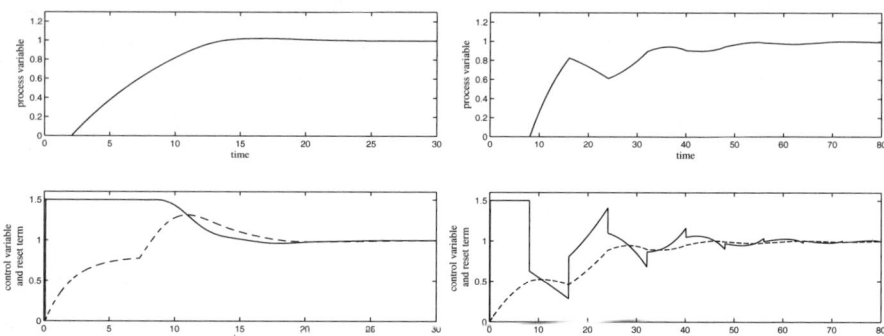

Fig. 3.20. Process variable, control variable (solid line) and reset term (dashed line) for the preload method PR for $L = 2$ (left) and for $L = 8$ (right)

3.5 Experimental Results

3.5.1 Level Control

The different anti-windup techniques have been tested with the experimental setup for the level control task described in Section A.1. A set-point change from 1.5 V to 3.5 V is considered and the saturation levels have been fixed to $u_{max} = 4.5$ V and $u_{min} = 0$ V. A PI controller with $K_p = 8$ and $T_i = 10$ has been used in all cases (note that, since $T_d = 0$, the back-calculation technique BC1 where $T_t = \sqrt{T_i T_d}$ has not been evaluated). In order to evaluate the integrator windup effect, the result obtained with no anti-windup strategy is shown in Figure 3.21. Then, the results obtained with the anti-windup techniques listed in Section 3.4 are shown in Figures 3.22–3.29. Note that a noise band of 0.02 V has been fixed for the combined approach CI-BC. Further, α has been fixed to 0.02 for the VSPID method (other values can be selected without changing significantly the process response) and $I_{max} = 2$ for the preload technique PR. By evaluating the results it turns out that all the proposed techniques are indeed effective in providing a good performance, by reducing significantly the overshoot caused by the integrator windup and, in general, without impairing the rise time. More specifically, the conditional integration method allows to achieve a small rise time without overshoot. A small rise time is achieved also by the anti-windup techniques for the PI controller in automatic reset configuration but in this case a slight overshoot occurs. As expected, the performance obtained by the back-calculation method largely depends on the value of the selected time constant. Indeed, the somewhat sluggish response obtained by the combined technique is due to the high value of T_t, since the dead time of the process is rather small and therefore the combined approach actually behaves as a standard back-calculation technique. Similar results are obtained also for the other methods, with a slight overshoot that occurs with the automatic reset configurations S1 and S2.

3.5 Experimental Results 51

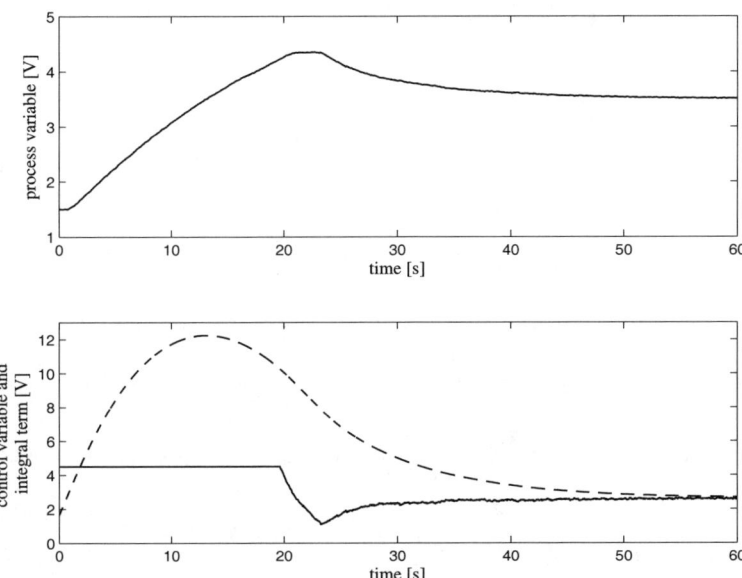

Fig. 3.21. Process variable, control variable (solid line) and integral term (dashed line) when no anti-windup is present for the level control task

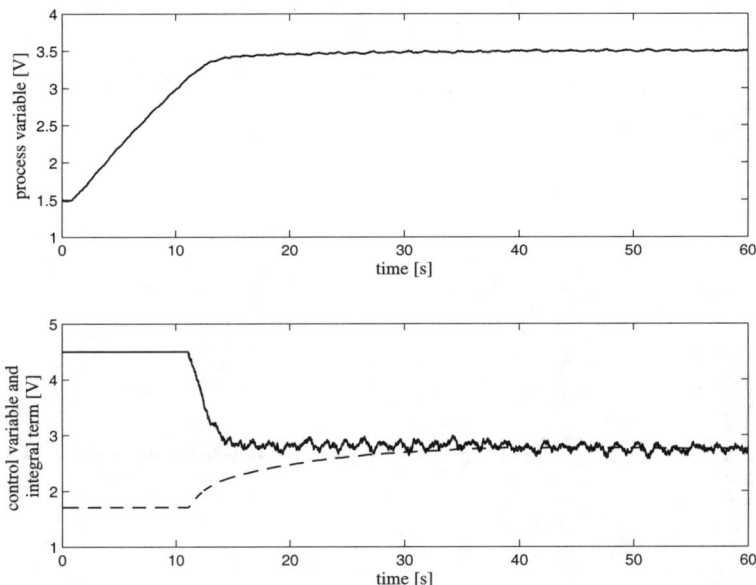

Fig. 3.22. Process variable, control variable (solid line) and integral term (dashed line) for CI for the level control task

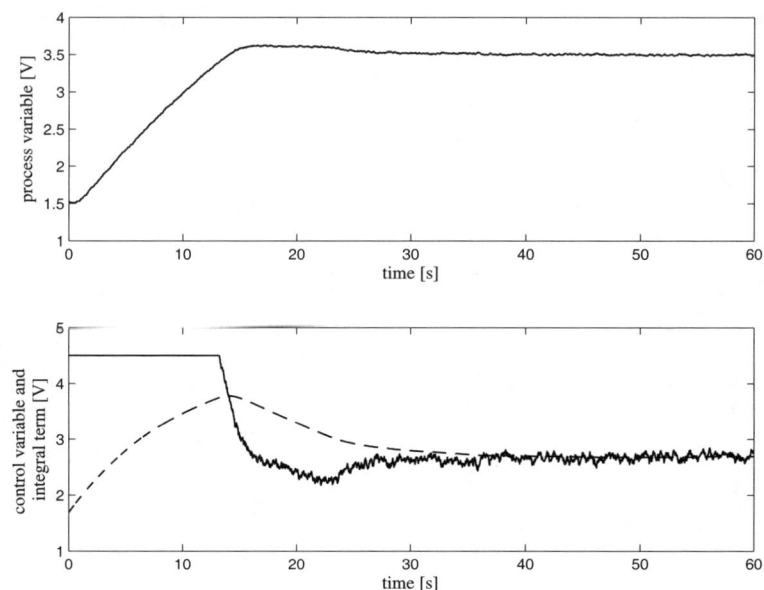

Fig. 3.23. Process variable, control variable (solid line) and integral term (dashed line) for BC2 with $T_t = T_i = 10$ for the level control task

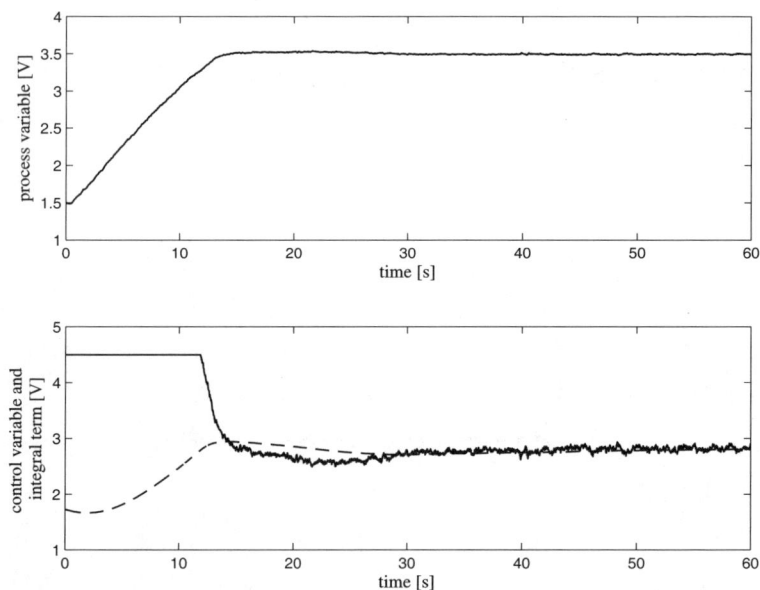

Fig. 3.24. Process variable, control variable (solid line) and integral term (dashed line) for BC3 with $T_t = K_p = 8$ for the level control task

3.5 Experimental Results   53

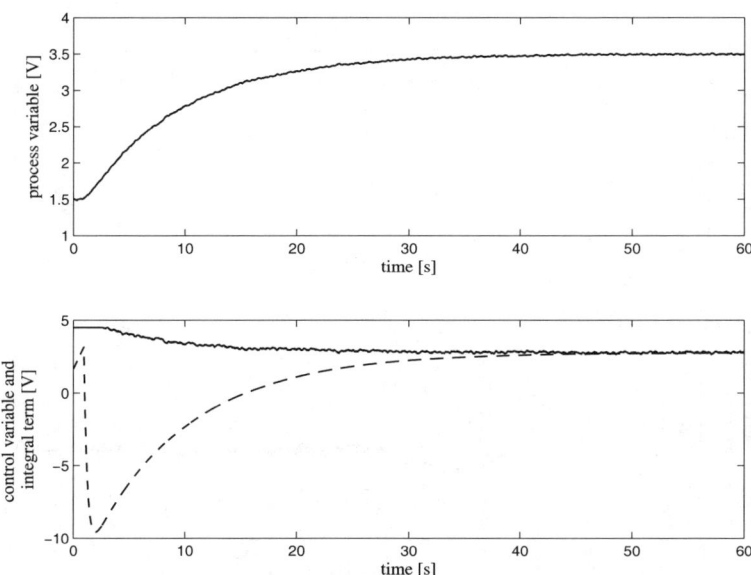

Fig. 3.25. Process variable, control variable (solid line) and integral term (dashed line) for CI-BC with $T_t = 0.03T_i = 0.3$ for the level control task

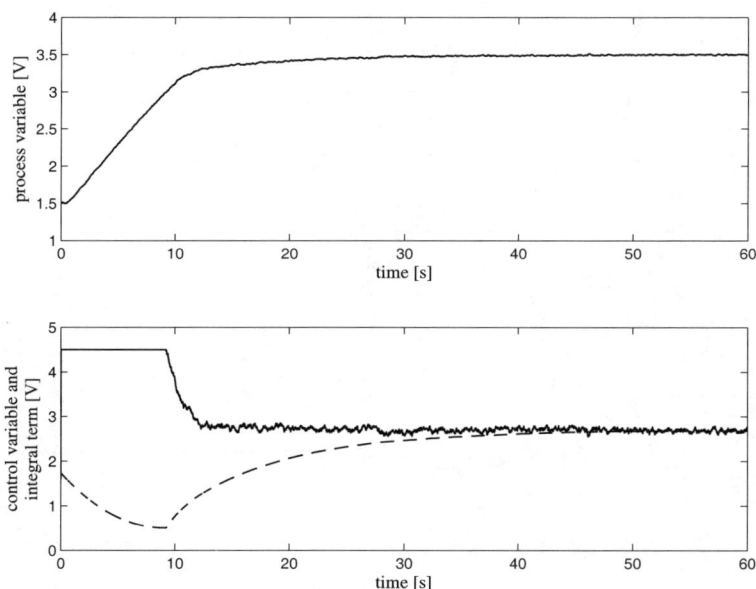

Fig. 3.26. Process variable, control variable (solid line) and integral term (dashed line) for VSPID with $\alpha = 10$ for the level control task

54   3 Anti-windup Strategies

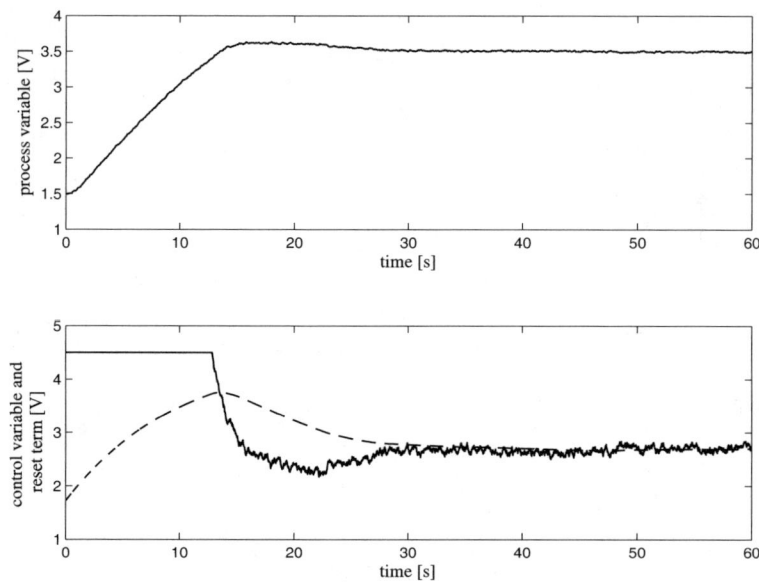

Fig. 3.27. Process variable, control variable (solid line) and reset term (dashed line) for the automatic reset configuration S1 for the level control task

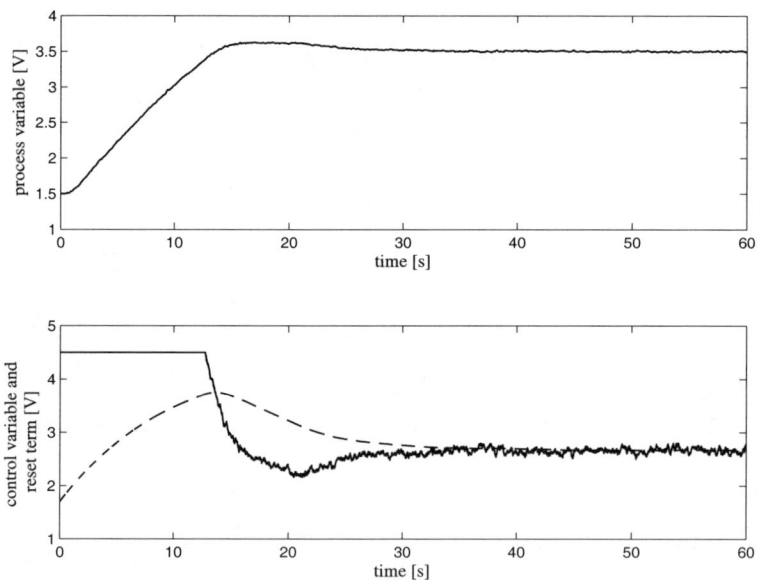

Fig. 3.28. Process variable, control variable (solid line) and reset term (dashed line) for the automatic reset configuration S2 for the level control task

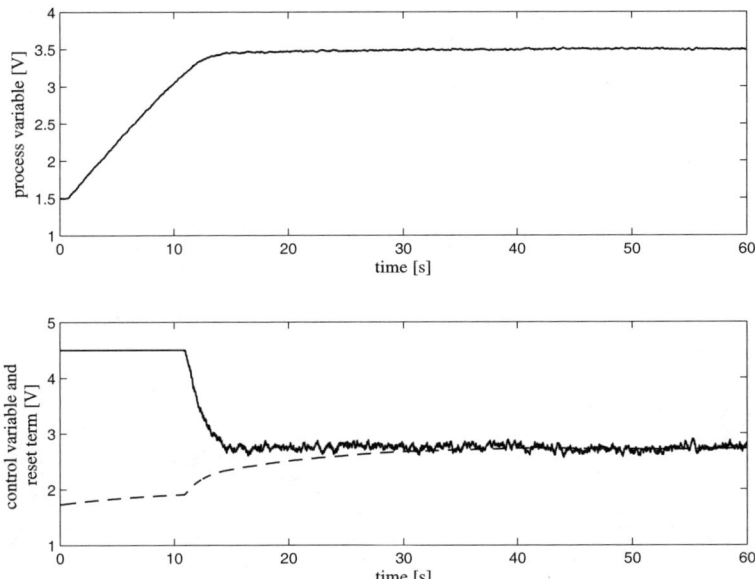

Fig. 3.29. Process variable, control variable (solid line) and reset term (dashed line) for the preload method PR for the level control task

3.5.2 Temperature Control

A temperature control task has been also considered by means of the laboratory setup described in Section A.2. In particular, a set-point value of 3 V has been imposed (starting from the room temperature, which corresponds to about 0.5 V) and the actuator limits have been fixed to $u_{max} = 4.5$ V and $u_{min} = 0$ V.

A PI controller has been tuned in order to achieve deliberately a very significant integrator windup so that the effectiveness of the anti-windup techniques can be better evaluated. Thus, the values $K_p = 1$ and $T_i = 100$ have been fixed. The corresponding process response (no anti-windup) is shown in Figure 3.30. It can be seen that the windup phenomenon occurs with respect to both the actuator limits. Results related to the application of the different considered anti-windup methods are shown in Figures 3.31–3.38. Note that, as with the level control task, the value $NB = 0.02$ V has been fixed for the combined approach CI-BC, $\alpha = 3 \cdot 10^{-3}$ for the VSPID method (performance does not change significantly by varying this value) and $I_{max} = 1.5$ for the preload technique PR. It appears that all the methods allow to avoid the integrator windup effect (it is worth stressing again that the detrimental effect of the integrator windup occurs when the control variable saturates even when the process variable attains the set-point value) and they basically provide the same performance.

56   3 Anti-windup Strategies

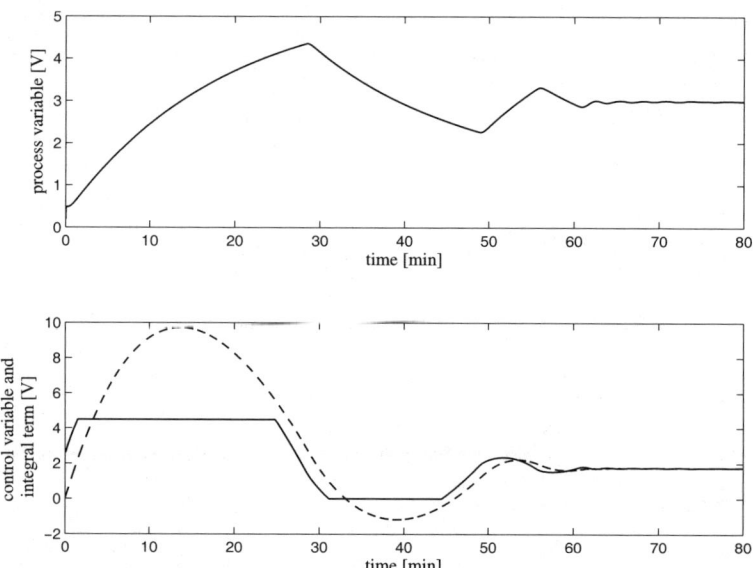

Fig. 3.30. Process variable, control variable (solid line) and integral term (dashed line) when no anti-windup is present for the temperature control task

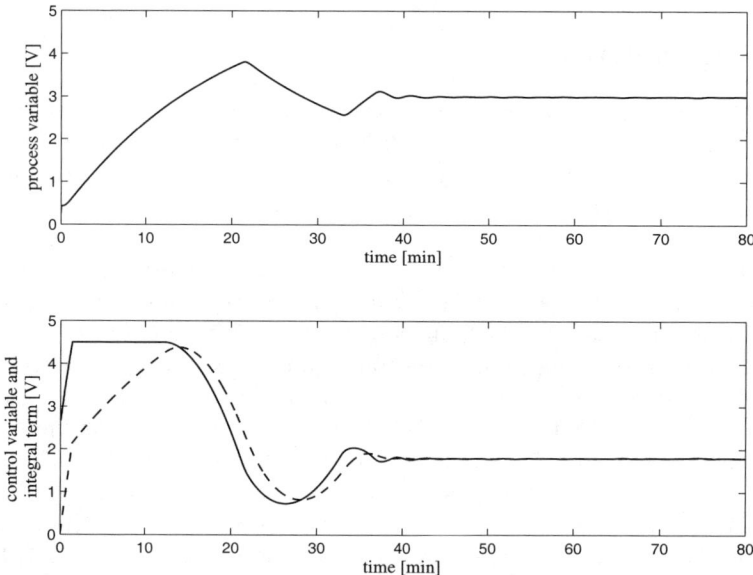

Fig. 3.31. Process variable, control variable (solid line) and integral term (dashed line) for CI for the level temperature task

3.5 Experimental Results 57

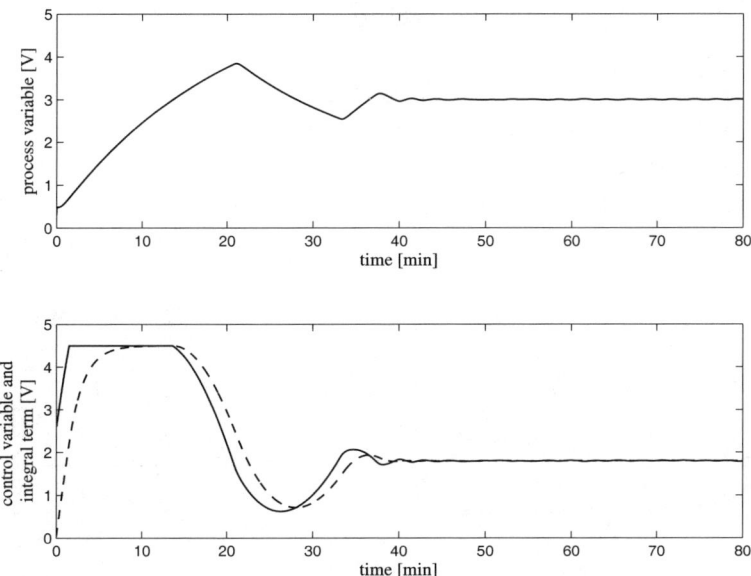

Fig. 3.32. Process variable, control variable (solid line) and integral term (dashed line) for BC2 with $T_t = T_i = 10$ for the temperature control task

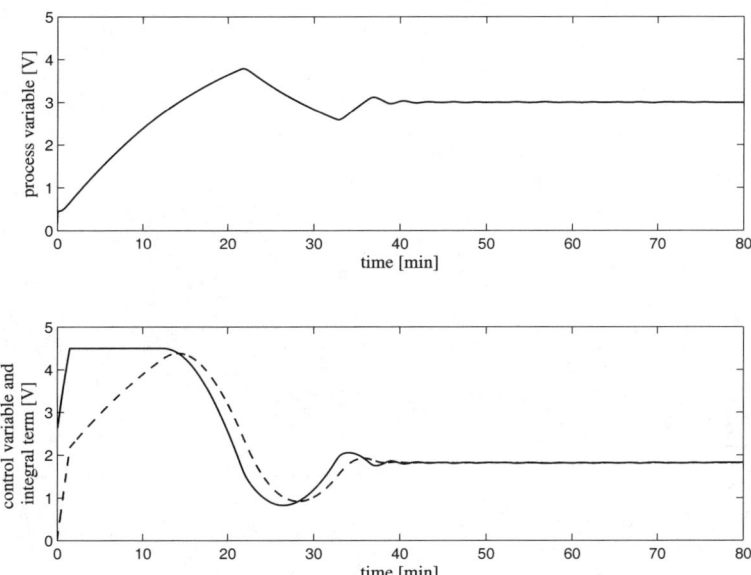

Fig. 3.33. Process variable, control variable (solid line) and integral term (dashed line) for BC3 with $T_t = K_p = 8$ for the temperature control task

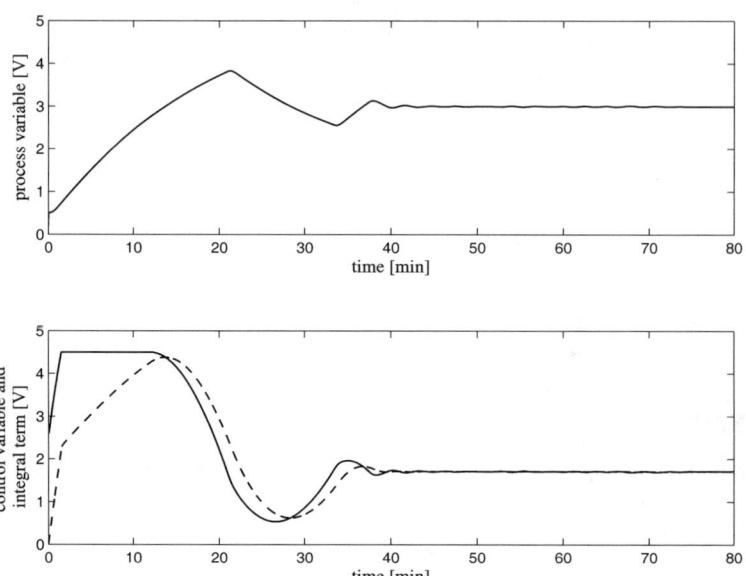

Fig. 3.34. Process variable, control variable (solid line) and integral term (dashed line) for CI-BC with $T_t = 0.03 T_i = 0.3$ for the temperature control task

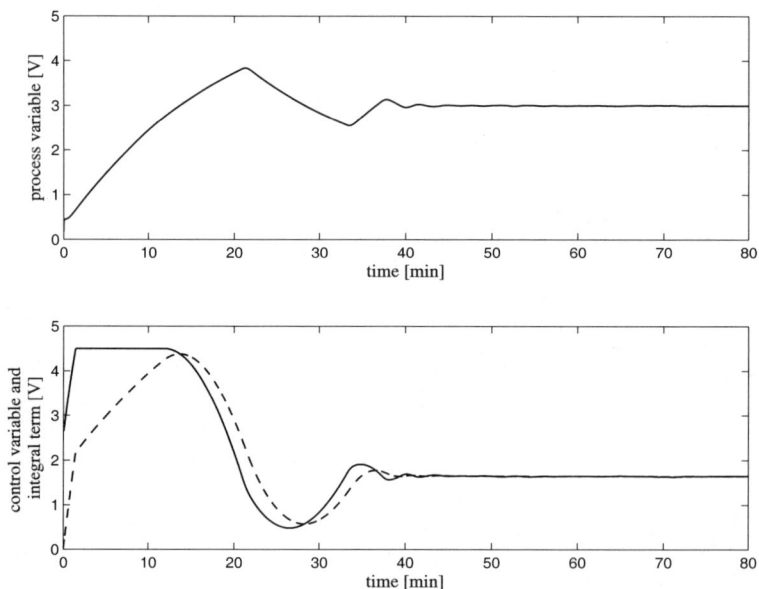

Fig. 3.35. Process variable, control variable (solid line) and integral term (dashed line) for VSPID with $\alpha = 3 \cdot 10^{-3}$ for the temperature control task

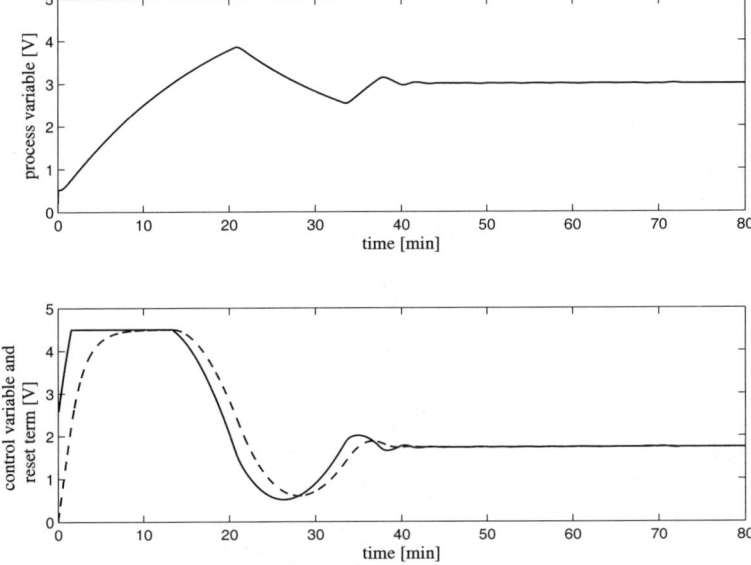

Fig. 3.36. Process variable, control variable (solid line) and reset term (dashed line) for the automatic reset configuration S1 for the temperature control task

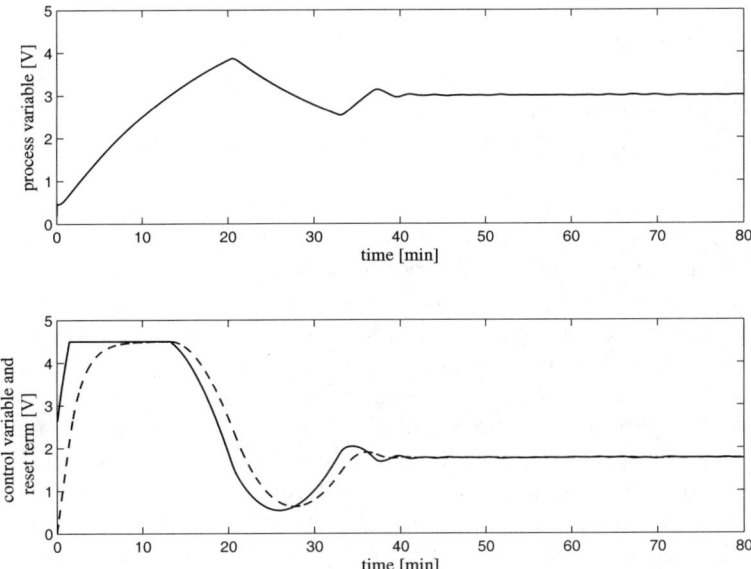

Fig. 3.37. Process variable, control variable (solid line) and reset term (dashed line) for the automatic reset configuration S2 for the temperature control task

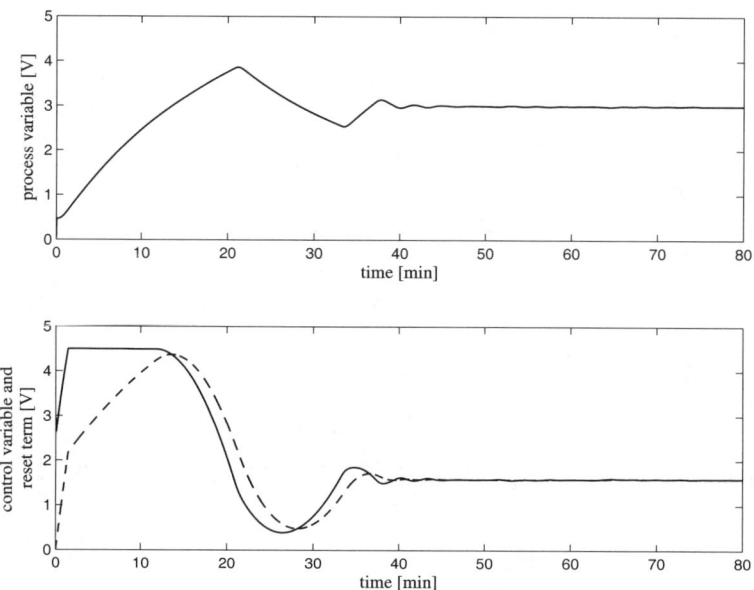

Fig. 3.38. Process variable, control variable (solid line) and integral term (dashed line) for the preload method PR for the level temperature task

3.6 Conclusions

In this chapter the integrator windup effect in the context of PID controllers has been analysed and different anti-windup techniques have been presented and compared.
Actually, all the considered methods are effective and each one has particular features that should be taken into account in a given application (indeed, there is not a technique that performs better than the others for all the kind of processes, PID parameters and actuator limits). From one point of view, the conditional integration approach has the advantage of being without an additional tuning parameter, but, from another point of view, the back-calculation methodology provides the capability to influence the transient response through the tuning of the tracking time constant (namely, a less aggressive response can be imposed by lowering the value of T_t). The combined approach seems to be less sensitive to the characteristics of the process. The two techniques considered for the PI(D) controller in automatic reset configuration do not show particular differences. With respect to them, a significant improvement in the performance does not actually emerges by applying the VSPID or the preload technique, despite they present a tuning parameter.

4
Set-point Weighting

4.1 Introduction

One of the main difficulties in the tuning of the PID parameters is to address at the same time different control specifications. In particular, as already mentioned in Section 1.6.2, achieving a high load disturbance rejection performance generally results in an aggressive tuning that provides a too oscillatory set-point step response. This is actually true especially when the apparent dead time of the process is small (with respect to the dominant time constant). If both specifications are of concern in a given application, a good solution is to use a set-point weight, *i.e.*, a two-degree-of-freedom control scheme (Araki, 1988), where both a feedback and a feedforward action are exploited. In this way, the set-point response task can be addressed without modifying the load disturbance rejection performance.

In this chapter, different methodologies for the design of the feedforward part are described. Specifically, the use of a fixed value for the set-point weight is compared to the use of a variable set-point weight, where the latter choice is made in order to avoid the increase of the rise time due to the smoothing of the reference signal applied to the closed-loop system.

4.2 Constant Set-point Weight Design

In the following analysis, the PID control law considered is

$$u(t) = K_p \left(\beta r(t) - y(t) + \frac{1}{T_i} \int_0^t e(\tau)d\tau - T_d \frac{dy(t)}{dt} \right). \quad (4.1)$$

Note that, for the sake of simplicity, the derivative filter is not considered here. It will be specified in the different considered methodologies. The adoption of this control law corresponds to the adoption of a feedforward/feedback control strategy that is depicted in the block diagram of Figure 1.11, where

$$C(s) = K_p \left(1 + \frac{1}{T_i s} + T_d s\right) \tag{4.2}$$

and

$$F(s) = \frac{1 + \beta T_i s}{1 + T_i s + T_i T_d s^2}, \tag{4.3}$$

as it can be easily derived from Expression (1.32). Intuitively, given a set of parameters K_p, T_i and T_d, the adoption of a set-point weight $\beta < 1$ allows the reduction of the overshoot in the set-point response, since the effect of the proportional action is reduced (note again that this is achieved without affecting the load disturbance rejection performance). From another point of view, the overshoot is reduced by smoothing the set-point signal by means of the filter F.

However, this is countered by the fact that a slower response (in terms of rise time) is obtained. Actually, parameter β has a clear physical meaning and its tuning can be easily done by taking into account the fact that reducing its value produces a more sluggish response with less overshoot.

This fact is clarified by the following example. Consider the process

$$P(s) = \frac{1}{10s + 1} e^{-4s}. \tag{4.4}$$

A PID controller tuned according to the Ziegler–Nichols rules with $K_p = 3$, $T_i = 8$ and $T_d = 2$, and with different set-point weights, namely $\beta = 1$, $\beta = 0.5$ and $\beta = 0$ has been applied. The resulting process outputs in the three cases are plotted in Figure 4.1. It is evident that the effect of decreasing the set-point weight value is indeed to make the response less oscillatory. In order to understand better the role of the set-point filter F, its output in the three experiments (*i.e.*, the actual signal applied to the closed-loop system) is shown in Figure 4.2. Finally, it is worth stressing that the reduction of the overshoot obviously implies that the control effort is reduced as well, as it appears from the resulting control signals plotted in Figure 4.3.

In any case, in order to avoid time-consuming experiments, tuning rules have been devised to determine explicitly a suitable value of β starting from a model of the process (see, (Hang et al., 1991; Åström and Hägglund, 1995; Åström et al., 1998; Leva and Colombo, 1999; Panagopoulos et al., 2002)). For example, if a dominant pole design method is applied (see Section 2.3), the set-point weight can be chosen in order to define a zero that cancels the additional pole whose position is not imposed (Åström and Hägglund, 1995). Alternatively, the set-point weight can be determined so that the maximum of the magnitude of the Bode diagram of the transfer function between the set-point and the process variable is equal to one (Panagopoulos et al., 2002). In case this is not achieved even with $\beta = 0$ the adoption of an additional first-order filter is suggested. In fact, the absence of resonance peaks prevents the occurrence of large overshoots.

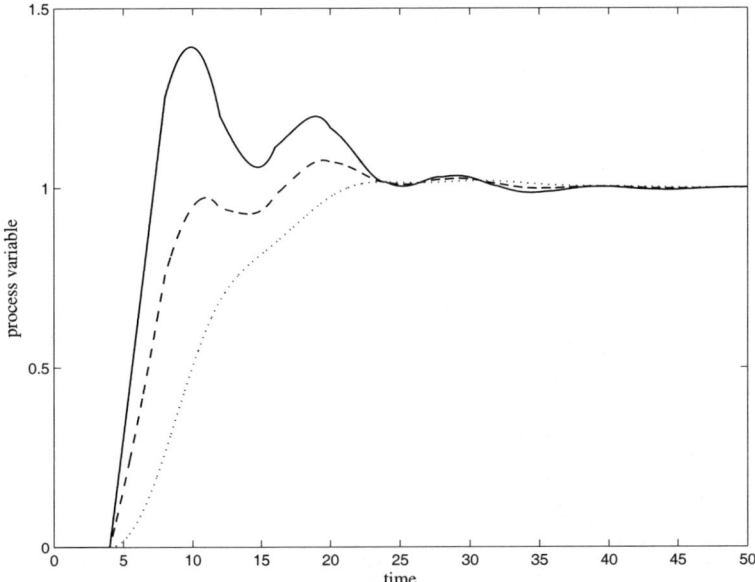

Fig. 4.1. Process variable for different values of the set-point weight. Solid line: $\beta = 1$; dashed line: $\beta = 0.5$; dotted line: $\beta = 0$.

These considerations demonstrate how the set-point weight is indeed a very useful parameter because it allows the solution of the tuning problem when both the set-point following and the load disturbance rejection performances are of concern with a modest increase of the design effort.

4.3 Variable Set-point Weighting

4.3.1 Methodology

In the previous sections, it has been pointed out that when the normalised dead time Θ of the process (*i.e.*, the ratio between the apparent dead time of the process and the dominant time constant) is small, the tuning that provides a good load disturbance rejection performance also gives a set-point response that exhibits a large overshoot. The use of a (constant) set-point weighting allows the reduction of the overshoot but this is countered by an increasing of the rise time. With the aim of providing a smaller overshoot without decreasing the set-point response speed, the use of a variable set-point response is proposed in (Hang and Cao, 1996).
The PID controller adopted in this case is the one described in Expression (4.1), where the derivative action is filtered by means of a first-order filter whose time constant is $T_d/10$ and where the value of the set-point weight

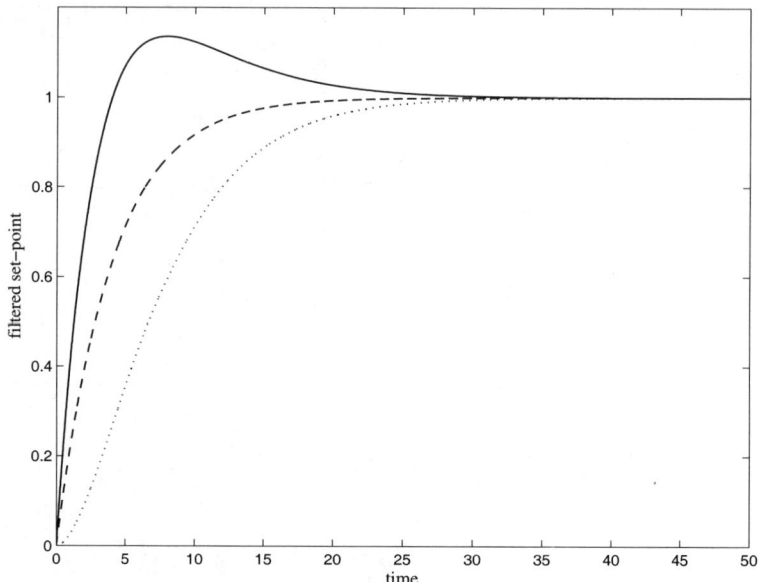

Fig. 4.2. Filtered set-point for different values of the set-point weight. Solid line: $\beta = 1$; dashed line: $\beta = 0.5$; dotted line: $\beta = 0$.

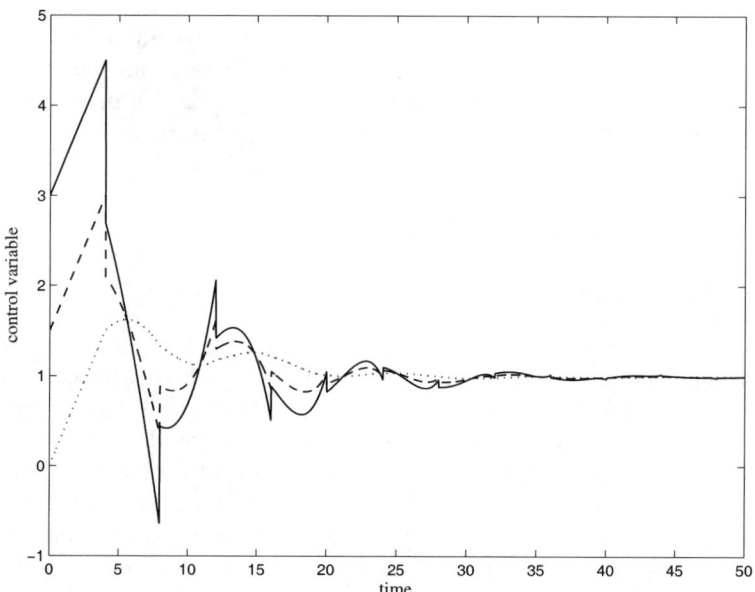

Fig. 4.3. Control variables for different values of the set-point weight. Solid line: $\beta = 1$; dashed line: $\beta = 0.5$; dotted line: $\beta = 0$.

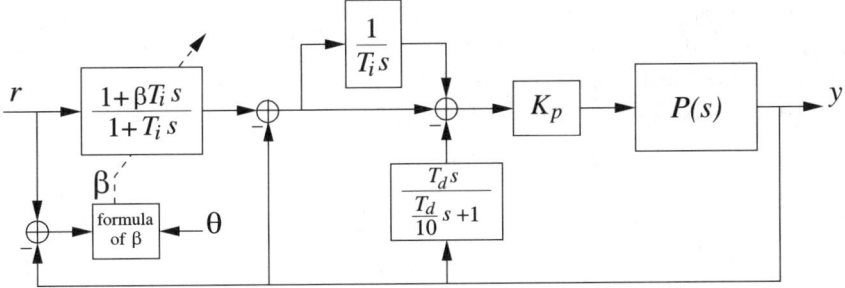

Fig. 4.4. Block diagram of the control scheme with variable set-point weighting

β varies during the transient, depending on the control error and on the normalised dead time. The corresponding block diagram of the control scheme is shown in Figure 4.4. Note that, with respect to Expression (4.3), here the resulting set-point filter is of first order because the derivative action is applied to the process variable instead of to the control error as in Expression (4.2) (the control scheme is indeed equivalent).

The variable set-point weighting is actually provided by switching among three constant values at selected time instants during the transient response. The rationale of this method is to use a (relatively) high value of the weight at the beginning of the transient response in order to speed up the response. Then, the weight is lowered to avoid the overshoot and eventually it is raised again to eliminate the undershoot. Note that it has been found by many simulations that the use of three constant values of the set-point weight (*i.e.*, of two switching times) is sufficient to obtain satisfactory results by employing a simple tuning rule for the overall controller. Actually, good results cannot be achieved in general with just one switching instant and the use of more than two switching instants does not provide a significant improvement.

An automatic tuning procedure has been devised for the overall control scheme (Hang and Cao, 1996). First, a relay-feedback experiment is performed (see Chapter 7). From this experiment, the ultimate gain K_u and the ultimate period T_u are estimated. Then, by assuming that the process is described by a second-order-plus-dead-time transfer function with coincident poles, *i.e.*, by the transfer function

$$P(s) = \frac{K}{(Ts+1)^2} e^{-Ls}, \qquad (4.5)$$

the time constant T and the dead time L are calculated as

$$T = \frac{T_u}{2\pi} \sqrt{K_u K - 1} \qquad (4.6)$$

and

$$L = \frac{T_u}{2\pi} \left(\pi - 2 \arctan \frac{2\pi T}{T_u} \right) = \frac{T_u}{2\pi} \left(\pi - 2 \arctan(\sqrt{K_u K - 1}) \right). \qquad (4.7)$$

It is worth noting that the process gain K has to be known in advance but this is not a serious drawback as it can be determined by considering steady-state values of the control variable and of the process variable.
Then, the apparent dead time $\hat{L}$ of the process is determined as

$$\hat{L} = L + 0.28T \qquad (4.8)$$

and the major time constant $\hat{T}$ is determined as

$$\hat{T} = 2.72T. \qquad (4.9)$$

At this point, by considering Expressions (4.7)–(4.9), the normalised dead time Θ of the process can be calculated as

$$\Theta = \frac{\hat{L}}{\hat{T}} = \frac{\pi - 2\arctan(\sqrt{K_u K - 1})}{2.72\sqrt{K_u K - 1}} + 0.1. \qquad (4.10)$$

The PID parameters are then determined by applying the following refined Ziegler–Nichols tuning rule:

$$K_p = 0.6 K_u, \qquad (4.11)$$

$$T_i = (-0.22\Theta + 0.53)T_u, \qquad (4.12)$$

$$T_d = \frac{T_i}{4}. \qquad (4.13)$$

Suitable values for the set-point weight and for the switching time instants are found empirically by means of a large number of simulations. Denote by β_1 the first set-point weight value (to be applied before the first switching time instant), by β_s the second weight value (to be applied between the first and the second switching time instant) and by β_m its final value (to be applied after the second switching time instant). They are determined as:

$$\beta_1 = 1, \qquad (4.14)$$

$$\beta_s = 0.2, \qquad (4.15)$$

$$\beta_m = 0.4\Theta^2 - 0.05\Theta + 0.58. \qquad (4.16)$$

The switching time instants are determined based on the current control error. In particular, denoting as e_0 the control error at the beginning of the transient, the first change of the set-point weight value is applied when the control error is

$$e_s = 0.85 e_0, \qquad (4.17)$$

while the second change is performed when the control error is

$$e_m = (-2.77\Theta^2 + 3.11\Theta - 0.25) e_0. \qquad (4.18)$$

4.3.2 Simulation Results

In order to verify the effectiveness of the variable set-point weighting technique, the following processes are considered:

$$P_1(s) = \frac{1}{(1+sT)^2}e^{-sL}; \quad T = 1, 10; \quad L = 0.1, 0.4, 0.8; \quad (4.19)$$

$$P_2(s) = \frac{1}{(1+s)^3}; \quad (4.20)$$

$$P_3(s) = \frac{1}{(1+s)(1+0.5s)(1+0.25s)(1+0.125s)}e^{-sL}; L = 0, 0.1; \quad (4.21)$$

$$P_4(s) = \frac{(1-0.5s)}{(1+s)^3}. \quad (4.22)$$

For each process, the automatic tuning method has been applied. The resulting PID parameters are shown in Table 4.1. Note that no saturation limits have been considered in order to avoid biassing the results. As illustrative examples, the resulting set-point step responses for some of the processes considered are plotted in Figures 4.5–4.9. The results obtained are compared with that obtained by considering the case of no set-point weighting, (*i.e.*, with $\beta = 1$) and with a constant set-point weight. In this latter case, its value β^* has been determined in order to minimise the integrated absolute error, defined as

$$IAE = \int_0^\infty |r(t) - y(t)|dt. \quad (4.23)$$

It can be seen that the use of a variable set-point weighting provides a smaller overshoot, with respect to the case of no set-point weighting, without impairing significantly the rise time (note again that the process variable is actually

Table 4.1. PID parameters for the variable set-point weighting technique

Process			K_p	T_i	T_d
$P_1(s),$	$T = 1,$	$L = 0.1$	10.84	0.76	0.19
$P_1(s),$	$T = 1,$	$L = 0.4$	3.21	1.40	0.35
$P_1(s),$	$T = 1,$	$L = 0.8$	1.76	1.88	0.47
$P_1(s),$	$T = 10,$	$L = 0.1$	91.4	2.68	0.67
$P_1(s),$	$T = 10,$	$L = 0.4$	26.8	4.85	1.21
$P_1(s),$	$T = 10,$	$L = 0.8$	13.5	6.83	1.71
$P_2(s)$			4.41	1.82	0.46
$P_3(s),$	$L = 0$		3.72	1.10	0.28
$P_3(s),$	$L = 0.1$		2.88	1.24	0.31
$P_4(s)$			1.78	2.41	0.60

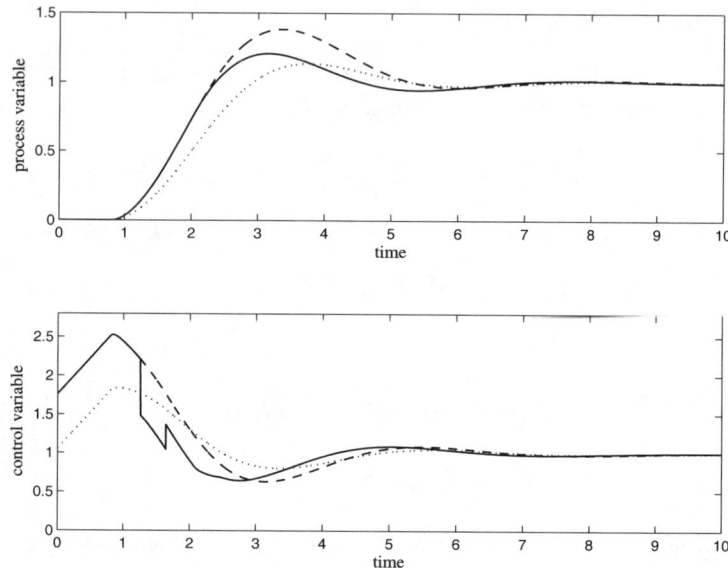

Fig. 4.5. Process and control variable for $P_1(s)$ with $T = 1$ and $L = 0.8$. Solid line: variable set-point weighting; dashed line: no set-point weighting; dotted line: fixed set-point weight $\beta = \beta^* = 0.6$.

the same in the two cases at the beginning of the transient response, since for the variable set-point weighting technique it is $\beta = 1$ until the reduction of the control error is of 15%).

In order to understand better the significance of the methodology, the resulting filtered set-point for process $P_2(s)$ and $P_4(s)$ for the cases again of variable set-point weighting, of no set-point weighting and of set-point weighting with $\beta = \beta^*$ are shown in Figures 4.10 and 4.11 respectively.

It appears that the set-point is suitably shaped in order to provide a fast response with a small overshoot at the same time. Indeed, at the beginning of the transient response the set-point signal is not filtered so that a fast response is achieved. Then, the set-point is abruptly lowered so that the response is significantly damped. Finally, the set-point is filtered as in the case of a fixed set-point weighting.

It is worth noting at this point that the tuning rule provided for the PID control might be to aggressive in practical cases. Indeed, oscillatory responses result for experiments made with the laboratory setups presented in Sections A.1 and A.2 (note that the dynamics in both cases is actually nonlinear). Thus, a detuning of the controller should be considered in general.

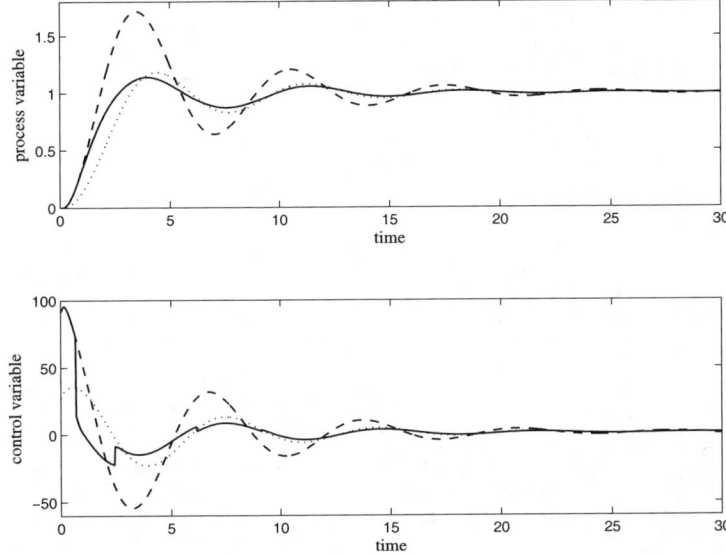

Fig. 4.6. Process and control variable for $P_1(s)$ with $T = 10$ and $L = 0.1$. Solid line: variable set-point weighting; dashed line: no set-point weighting; dotted line: fixed set-point weight $\beta = \beta^* = 0.3$.

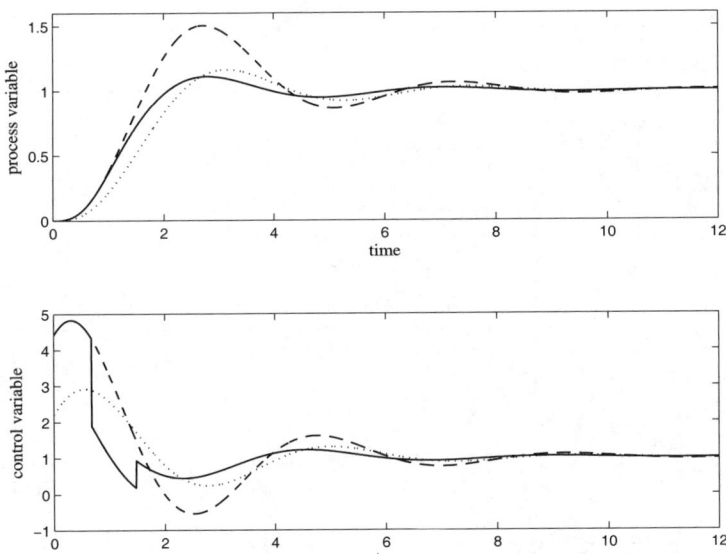

Fig. 4.7. Process and control variable for $P_2(s)$. Solid line: variable set-point weighting; dashed line: no set-point weighting; dotted line: fixed set-point weight $\beta = \beta^* = 0.5$.

70   4 Set-point Weighting

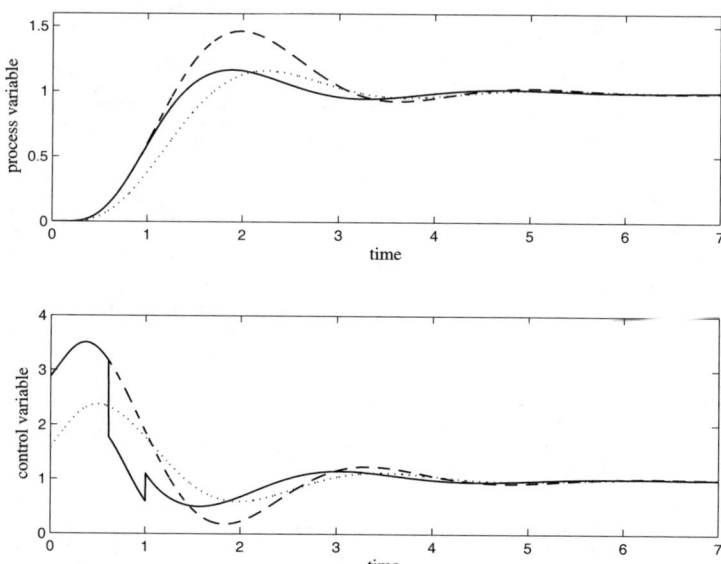

Fig. 4.8. Process and control variable for $P_3(s)$ with $L = 0.1$. Solid line: variable set-point weighting; dashed line: no set-point weighting; dotted line: fixed set-point weight $\beta = \beta^* = 0.55$.

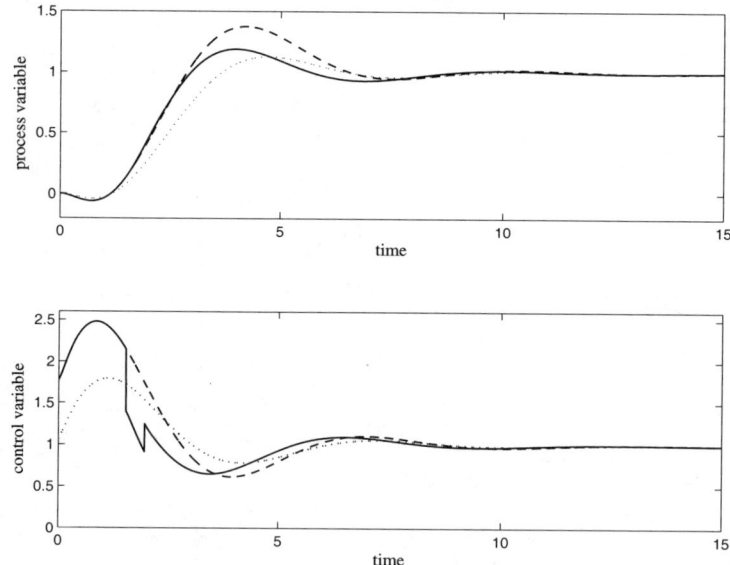

Fig. 4.9. Process and control variable for $P_4(s)$. Solid line: variable set-point weighting; dashed line: no set-point weighting; dotted line: fixed set-point weight $\beta = \beta^* = 0.6$.

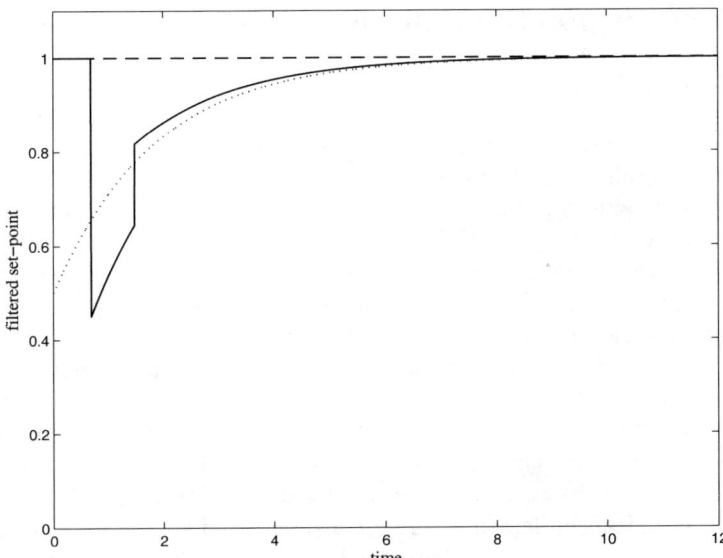

Fig. 4.10. Filtered set-point for $P_2(s)$. Solid line: variable set-point weighting; dashed line: no set-point weighting; dotted line: fixed set-point weight $\beta = \beta^* = 0.5$.

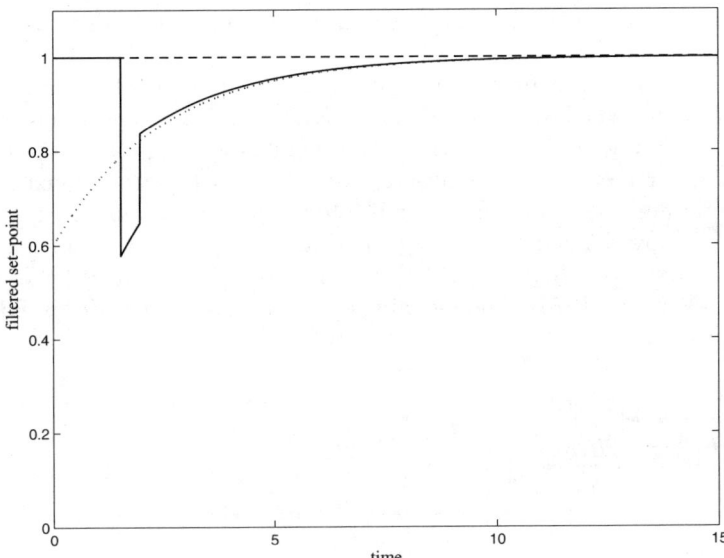

Fig. 4.11. Filtered set-point for $P_4(s)$. Solid line: variable set-point weighting; dashed line: no set-point weighting; dotted line: fixed set-point weight $\beta = \beta^* = 0.6$.

4.4 Fuzzy Set-point Weighting

4.4.1 Methodology

Fuzzy logic has been successfully applied widely in the control field in order to exploit a linguistic model of the process to be controlled (usually given by the operator) (Tzafestas, 1994). For an excellent introduction to fuzzy control see, for example, (Driankov et al., 1993).

The use of a fuzzy inference system can be adopted to determine the value of the set-point weight during the transient response in order to decrease the rise time and at the same to reduce the overshoot (and therefore to decrease the settling time as well) (Visioli, 1999). Consider the PID control law (4.1), where in this case the filter to make the corresponding transfer function proper is applied to the overall control law (see (1.23)). Then, a fuzzy inference system is adopted to determine the value of the weight $\beta(t)$ depending on the current value of the system error $e(t)$ and its time derivative $\dot{e}(t)$ (denoted equivalently by Δe if the digital implementation is considered). The idea, put succinctly, is simply that β has to be increased when the convergence of the process output $y(t)$ to $r(t)$ has to be speeded up, and decreased when the divergence trend of $y(t)$ from $r(t)$ has to be slowed down. For the sake of simplicity, the methodology is implemented in such a way that the output $f(t)$ of the fuzzy module is added to a constant parameter w, resulting in the coefficient $\beta(t)$ that multiplies the set-point. The overall control scheme is shown in Figure 4.12.

The two inputs of the fuzzy inference system, the control error e and its derivative $\dot{e}$, are scaled by two coefficients, K_{in1} and K_{in2} respectively, in order to match the range $[-1, 1]$ on which the membership functions are defined. Five triangular membership functions (see Figure 4.13) are defined for each input, while nine triangular membership functions (see Figure 4.14) over the range $[-1, 1]$ are defined for the output that is scaled by a coefficient K_{out}. The rule matrix (see Figure 4.15) is based on the Macvicar–Whelan matrix (Macvicar-Whelan, 1976). The meaning of the linguistic variables is explained in Table 4.2.

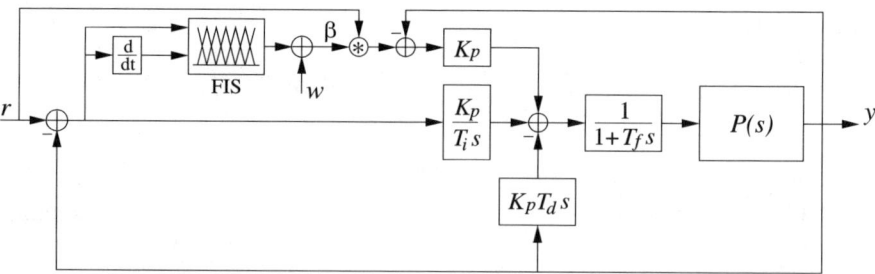

Fig. 4.12. Overall control scheme for the fuzzy set-point weighting methodology

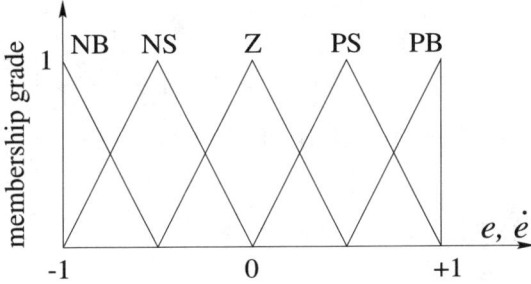

Fig. 4.13. Membership functions for the inputs of the fuzzy inference system

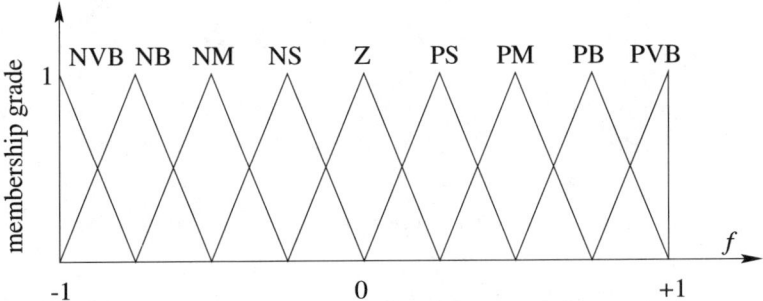

Fig. 4.14. Membership functions for the output of the fuzzy inference system

4.4.2 Tuning Procedure

The fuzzy set-point weighting technique involves the tuning of many parameters, namely, the four PID parameters K_p, T_i, T_d and the filter time constant T_f, the three scaling coefficients of the fuzzy inference system K_{in1}, K_{in2} and K_{out}, and the basic set-point weight value w. Further, the peak values of the membership functions can be modified, as well as the rules, in order to improve the performance.

It is suggested that the PID parameters are selected by adopting the Ziegler–Nichols tuning rules based on the step response, in order to obtain a satisfactory load disturbance rejection performance (the filter time constant can be selected so that the filter dynamics is negligible). Then, the fuzzy inference system parameters (including the basic set-point weight w) can be tuned by adopting the typical procedure for fuzzy controllers (Zheng, 1992). In particular, at the beginning the value of K_{in1} can simply be chosen as the inverse of the amplitude of the step of the set-point. For the other parameters, a practical procedure is to set $w = 1$ and then keep increasing the value of K_{out} (starting from $K_{out} = 0$), accordingly modifying the value of K_{in2} in order to normalise the input $\dot{e}$, as long as the performance improves. Then, this procedure can be iterated by decreasing the value of w until no better results

	Δe				
	NB	NS	Z	PS	PB
NB	NVB	NB	NM	NS	Z
NS	NB	NM	NS	Z	PS
e Z	NM	NS	Z	PS	PM
PS	NS	Z	PS	PM	PB
PB	Z	PS	PM	PB	PVB

Fig. 4.15. Table of the rules of the fuzzy inference system

Table 4.2. Meaning of the linguistic variables in the fuzzy inference system of the fuzzy set-point weighting methodology

NVB	Negative Very Big
NB	Negative Big
NM	Negative Medium
NS	Negative Small
Z	Zero
PS	Positive Small
PM	Positive Medium
PB	Positive Big
PVB	Positive Very Big

are achieved. At the end, the peak values of the membership functions have to be tuned, especially to limit oscillations of the system output, increasing the action of the fuzzy module when the output of the system is close to the set-point but its derivative is still high. Finally, the rules may also be modified to improve the response, although often in practical cases this is not necessary. Note that this procedure can be automated (Visioli, 2000), although it can be very time-consuming. An alternative effective method for the tuning of the fuzzy inference system is by using genetic algorithms (Mitchell, 1998), which have been successfully adopted in the optimisation of fuzzy controllers (see for example (Homaifar and McCormick, 1995)). The idea is to search

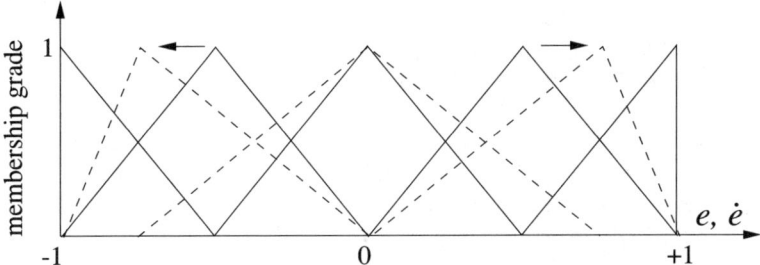

Fig. 4.16. Example of a vertex translation for the two inputs of the fuzzy inference system

the optimal values of the parameters of the fuzzy module with respect to a determined objective function. In the case of the fuzzification of the set-point weight, having fixed the value of K_{in1} equal to the inverse of the amplitude of the step of the set-point, the values of K_{in2}, K_{out}, w and the position of the membership functions are searched in order to minimise the value of the integrated absolute error (4.23). Specifically, the optimal position of all the peaks of the membership functions is searched, although the extreme and the central ones are kept fixed, for the sake of simplicity. The vertexes of the bases are accordingly translated, as shown in Figure 4.16. In order to limit the search space, the symmetry between the membership functions around the zero is imposed, both for the two inputs and the output. The procedure consists of evaluating a series of step responses in order to permit to the genetic algorithm to converge to the optimal solution (in a stochastic sense). Note that, if a model of process is available, this can be done off-line with the use of a proper simulation environment. Note also that by choosing other objective functions, different design specifications can be satisfied; for example, a step response with zero overshoot can be obtained.

4.4.3 Simulation Results

Simulation results are provided in order to understand better the fuzzy set-point weighting technique and to evaluate its effectiveness. As for the variable set-point weighting methodology, saturation limits on the control variable have not been considered. First, the following two processes are considered:

$$P_a(s) = \frac{1}{s^2 + 2\xi s + 1}; \xi = 0.2, \ 0.8; \tag{4.24}$$

$$P_b(s) = \frac{1}{s(1+s)}. \tag{4.25}$$

It is evident that by applying an analytical design method to $P_a(s)$ an arbitrary fast response can be achieved in principle. Thus, the application of the fuzzy set-point weighting technique in this case is mainly done to verify its potentiality to fully recover the set-point following performance with the PID parameters tuned for load disturbance rejection purpose (here it is $K_p = 3.34$, $T_i = 0.95$, $T_d = 0.24$ for the case where $\xi = 0.2$ and $K_p = 9.01$, $T_i = 0.80$, $T_d = 0.20$ for the case where $\xi = 0.8$). The result obtained after having applied the genetic-based tuning procedure is plotted in Figure 4.17 for the case $\xi = 0.2$. As in Section 4.3.2, it is compared with the case of no set-point weight (i.e., $\beta = 1$) and with the case of the optimal (fixed) set-point weight β^* that minimises the integrated absolute error (4.23). Results for the case of $\xi = 0.8$ are very similar and are not shown for the sake of brevity. It can be seen that the fuzzy set-point weighting technique provides a very fast response (with virtually no overshoot) despite the unappropriate tuning of the PID parameters for the set-point following point of view. Such a high performance is far from what can be achieved with the use a fixed (although optimal) set-point weight. Obviously, this is countered by a resulting high control effort (the control variable in the case of a fixed set-point weight cannot be evaluated in Figure 4.17 due to the scaling imposed by the control variable in the case of the fuzzy set-point weighting).

The responses for the non self-regulating process $P_b(s)$ are plotted in Figure 4.18 ($K_p = 1.22$, $T_i = 1.96$, $T_d = 0.49$). The same considerations made before apply also in this case.

Then, the fuzzy set-point weighting methodology has been applied to the same processes (4.19)–(4.22) of Section 4.3.2. The adopted PID parameters are shown in Table 4.3. Results related to some of the considered processes are plotted in Figures 4.19–4.23. It can be seen that the fuzzy set-point weighting method allows a decrease in many cases in both the rise time and the overshoot at the same time (obviously, the control effort increases as well). In order to understand better the methodology, the (varying) fuzzified set-point weight for process $P_2(s)$ and process $P_4(s)$ is shown in Figure 4.24 and in Figure 4.25 respectively. It appears that it is the significant variation of the set-point weight that allows a high performance in the set-point following task to be obtained.

Finally, it is worth noting that the fuzzy set-point weighting technique is effective also if a different tuning rule is employed for the PID parameters (Visioli, 1999) (see Section 4.4.4), in case a saturation limit for the control variable is considered and in case a load disturbance occurs (Visioli, 2001b). Further, in (Visioli, 2001b) it has also been shown that this method outperforms other fuzzy logic based methods in which the PID parameters are modified during the transient response.

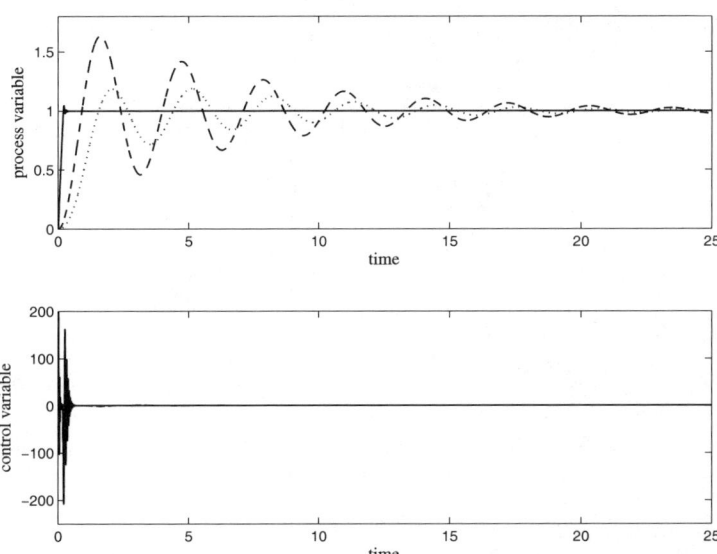

Fig. 4.17. Process variable and control variable for process $P_a(s)$ with $\xi = 0.2$. Solid line: fuzzy set-point weighting; dashed line: no set-point weighting; dotted line: fixed set-point weight $\beta = \beta^* = 0.2$.

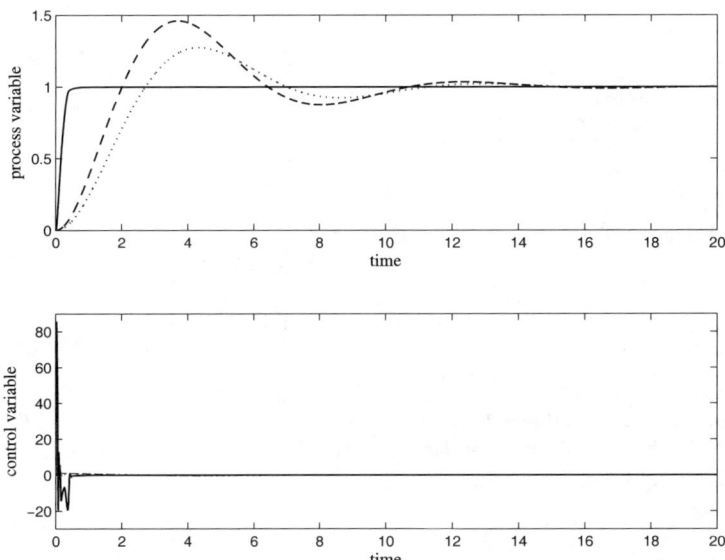

Fig. 4.18. Process variable and control variable for process $P_b(s)$. Solid line: fuzzy set-point weighting; dashed line: no set-point weighting; dotted line: fixed set-point weight $\beta = \beta^* = 0.5$.

4 Set-point Weighting

Table 4.3. PID parameters for the fuzzy set-point weighting technique

Process			K_p	T_i	T_d
$P_1(s)$,	$T=1$,	$L=0.1$	12.6	0.70	0.17
$P_1(s)$,	$T=1$,	$L=0.4$	3.39	1.46	0.36
$P_1(s)$,	$T=1$,	$L=0.8$	1.92	2.12	0.53
$P_1(s)$,	$T=10$,	$L=0.1$	120	2.24	0.56
$P_1(s)$,	$T=10$,	$L=0.4$	30	4.48	1.12
$P_1(s)$,	$T=10$,	$L=0.8$	15.6	6.24	1.56
$P_2(s)$			4.80	1.80	0.45
$P_3(s)$,	$L=0$		4.05	1.10	0.27
$P_3(s)$,	$L=0.1$		3.03	1.30	0.32
$P_4(s)$			1.92	2.64	0.66

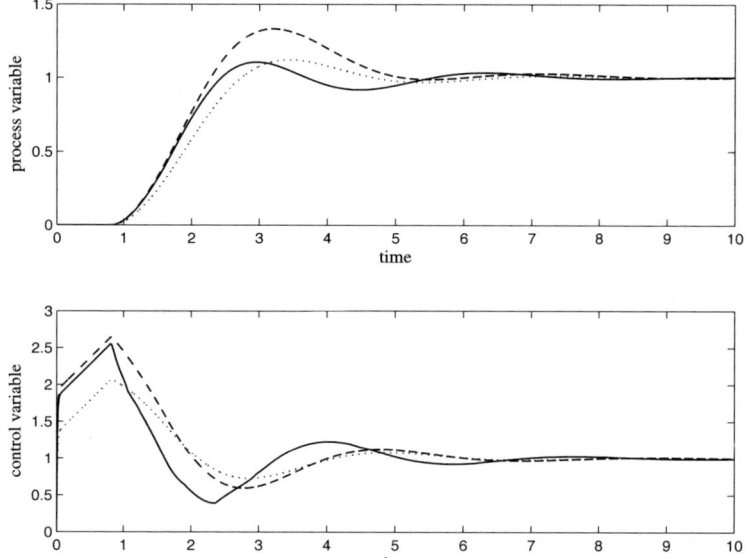

Fig. 4.19. Process variable and control variable for process $P_1(s)$ with $T=1$ and $L=0.8$. Solid line: fuzzy set-point weighting; dashed line: no set-point weighting; dotted line: fixed set-point weight $\beta = \beta^* = 0.7$.

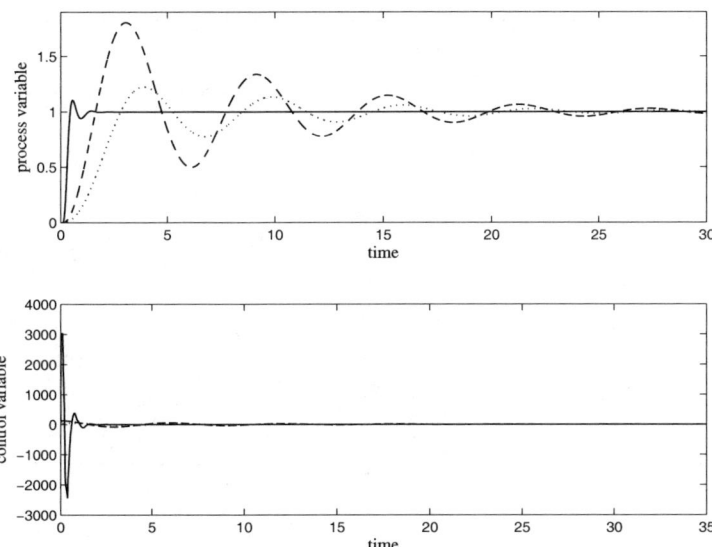

Fig. 4.20. Process variable and control variable for process $P_1(s)$ with $T = 10$ and $L = 0.1$. Solid line: fuzzy set-point weighting; dashed line: no set-point weighting; dotted line: fixed set-point weight $\beta = \beta^* = 0.2$.

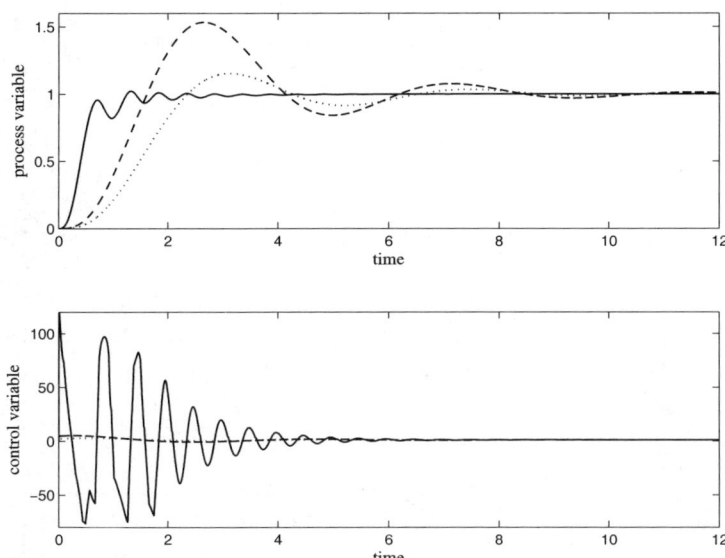

Fig. 4.21. Process variable and control variable for process $P_2(s)$. Solid line: fuzzy set-point weighting; dashed line: no set-point weighting; dotted line: fixed set-point weight $\beta = \beta^* = 0.45$.

80   4 Set-point Weighting

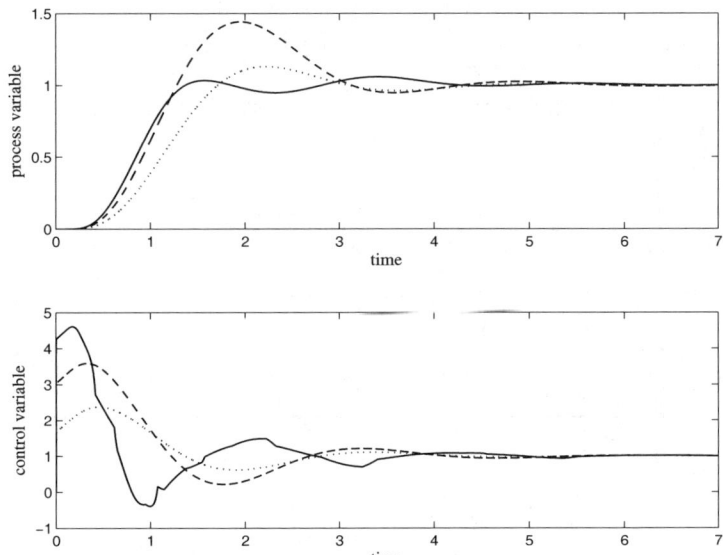

Fig. 4.22. Process variable and control variable for process $P_3(s)$ with $L = 0.1$. Solid line: fuzzy set-point weighting; dashed line: no set-point weighting; dotted line: fixed set-point weight $\beta = \beta^* = 0.55$.

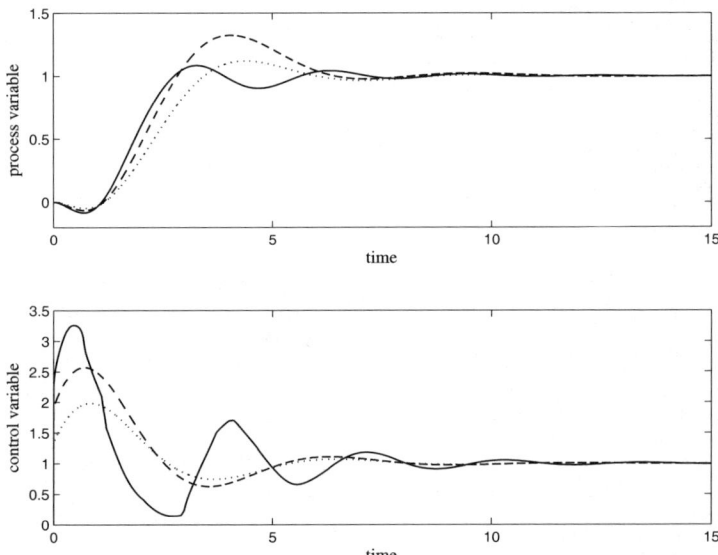

Fig. 4.23. Process variable and control variable for process $P_4(s)$. Solid line: fuzzy set-point weighting; dashed line: no set-point weighting; dotted line: fixed set-point weight $\beta = \beta^* = 0.7$.

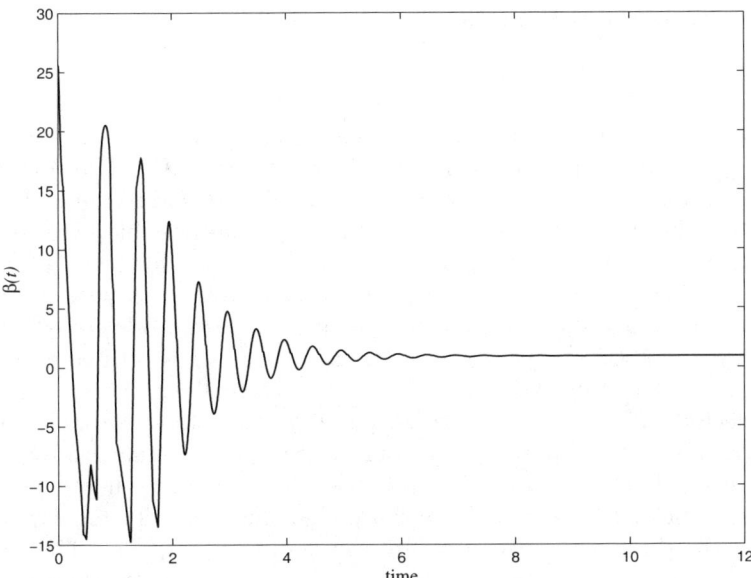

Fig. 4.24. Fuzzified set-point weight for process $P_2(s)$

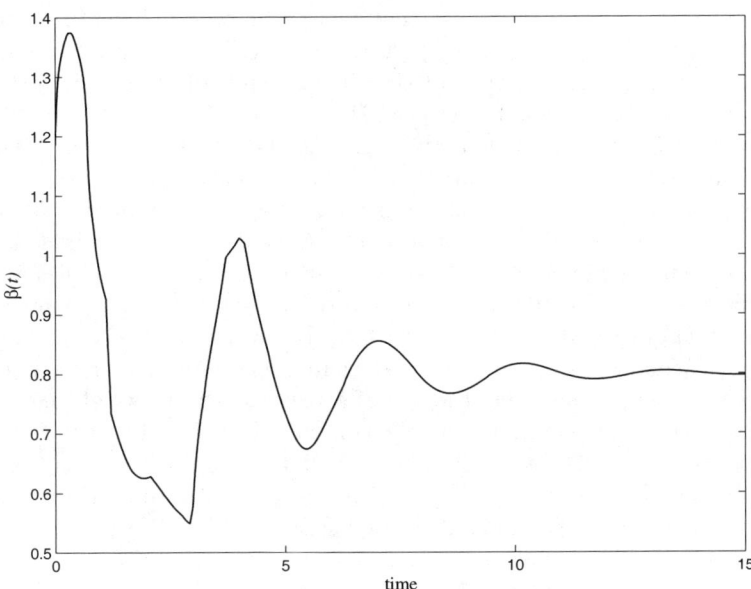

Fig. 4.25. Fuzzified set-point weight for process $P_4(s)$

4.4.4 Experimental Results

Level Control

The fuzzy set-point weighting methodology has been tested in a level control task with the tank apparatus described in Section A.1. The control task requires an output transition from 1.5 V to 3 V. In this context, the adoption of the Ziegler–Nichols tuning rule has not been possible because of the resulting high value of the proportional gain (due to the low value of the normalised dead time) which is not suitable with the present saturation limits of the actuator. Thus, the following parameters have been selected after a few trials: $K_p = 0.008$, $T_i = 0.3$, $T_d = 0.124$, $T_f = 0.1$. It is worth stressing that, although these are certainly not the most convenient parameters, they have been selected in order to verify the capability of the fuzzy set-point weighting methodology to recover the set-point following performance after a rough tuning of the PID controller. The parameters of the fuzzy inference system have been selected by applying a genetic algorithm after having estimated a model of the process by evaluating an open-loop step response. The resulting function $f(e, \Delta e)$ (which is actually implemented as a look-up table in the PC-based controller) is plotted in Figure 4.26. It is worth stressing that in this case a very high value of the basic set-point weight w results (indeed, it is $w = 31.3$). This is due to the fact that the PID parameters adopted provide a fairly sluggish response that has to be speeded up (note the low value of the proportional gain K_p). Actually, this is a remarkable feature of the methodology, since it turns out that it is capable of improving the performance "independently" of the selected PID gains.

The process variable obtained by applying the fuzzy set-point weighting methodology is shown in Figure 4.27. It is compared with the case of no set-point weight and with the use of a fixed set-point weight $\beta = 31.3$ (note that the selection of a different value for the fixed set-point weight does not affect the results significantly). The corresponding control variables are plotted in Figure 4.28. Finally, the variation of the set-point weight during the transient response is shown in Figure 4.29. It can be seen that set-point following performance is much improved from every point of view (rise time, overshoot, settling time) with the use of a fuzzy set-point weighting. Indeed, this is achieved without increasing the control effort. Finally, it is worth noting that the noisy set-point weight signal that results in the fuzzy set-point weighting technique does not imply that the control signal is noisy as well because of the use of the output-filtered form for the PID control which is actually essential in this case.

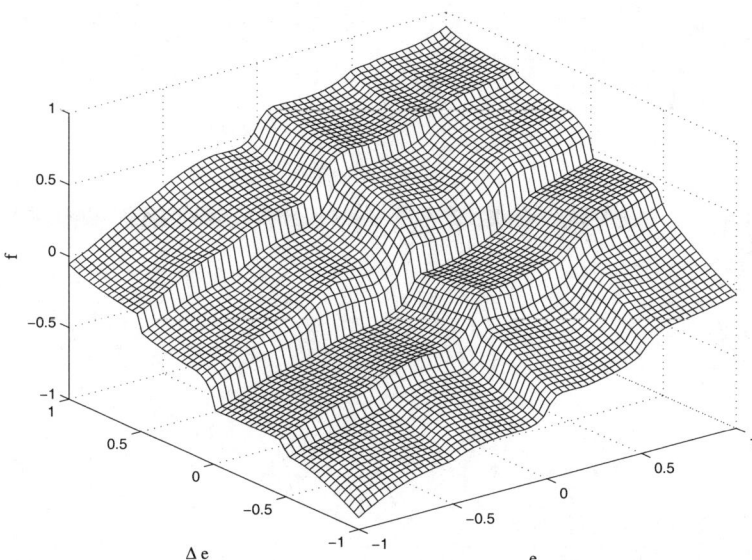

Fig. 4.26. Function implemented by the fuzzy inference scheme for the level control experiment

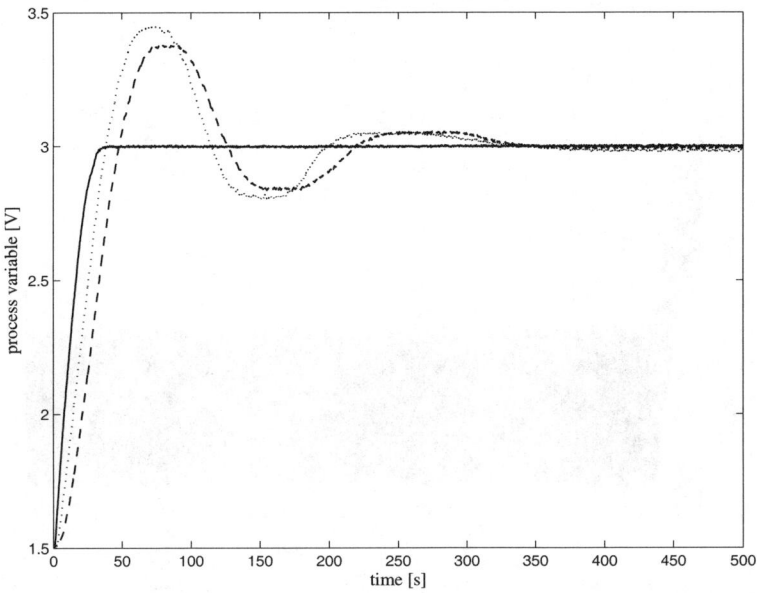

Fig. 4.27. Process variable for the level control experiment. Solid line: fuzzy set-point weighting; dashed line: no set-point weighting; dotted line: fixed set-point weight $\beta = 31.3$.

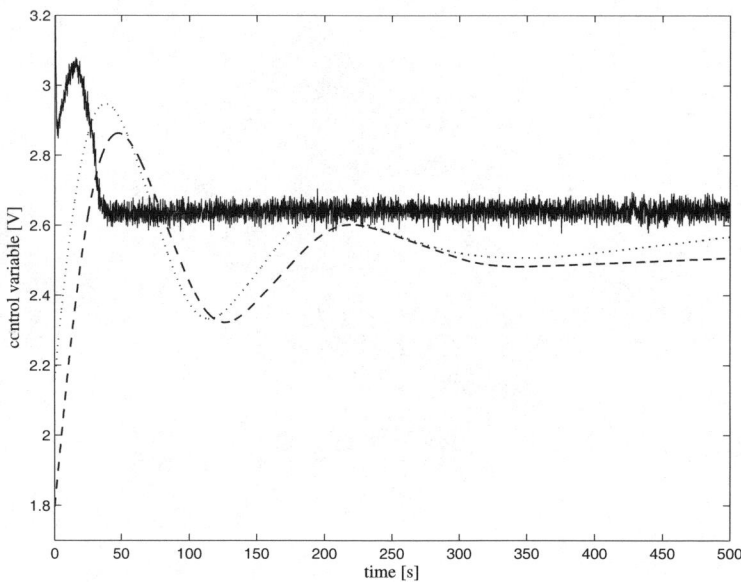

Fig. 4.28. Control variable for the level control experiment. Solid line: fuzzy set-point weighting; dashed line: no set-point weighting; dotted line: fixed set-point weight $\beta = 31.3$.

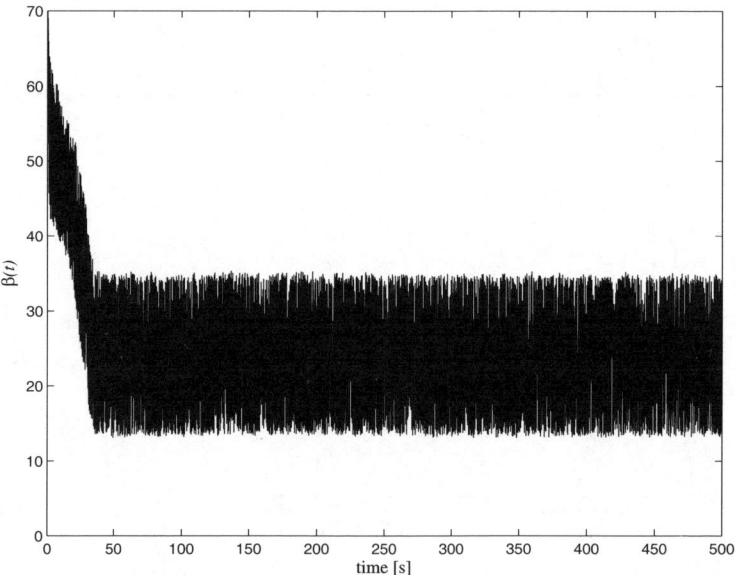

Fig. 4.29. Set-point weight for the level control experiment with fuzzy set-point weighting

Temperature Control

The effectiveness of the fuzzy set-point weighting technique is evaluated also in the context of temperature control with the experimental setup described in Section A.2. The control task is to perform an output transition from the initial condition (determined by the room temperature) to a new steady-state value of 2 V. As for the level control example, also in this case the PID parameters determined by applying the Ziegler–Nichols rule have been modified in order to cope with the saturation limits of the actuator. The adopted PID parameters are $K_p = 0.50$, $T_i = 300$, $T_d = 14$, $T_f = 0.7$. Then, a genetic algorithm has been employed for the design of the fuzzy inference mechanism (see the resulting function in Figure 4.30) and to determine the basic set-point weight $w = 6.46$. Note that, as for the level control experiment, $w > 1$. Similar considerations can be made in this case, by taking into account the requirement that the response has to be speeded up.

The resulting process variable is plotted in Figure 4.31. It is compared with the case of no set-point weighting and with the case of a fixed set-point weight $\beta = 0.5$ (note that the use of fixed set-point weight $\beta = 6.46$ results in an excessive overshoot). The corresponding control variables are shown in Figure 4.32. As expected, the fuzzy set-point weighting method causes a greater control effort and a more noisy control signal. This is clearly explained by evaluating the fuzzified weight signal plotted in Figure 4.33. In any case it is evident that the use of a fuzzy set-point weighting allows to improve significantly the set-point following performance achieved by the PID controller.

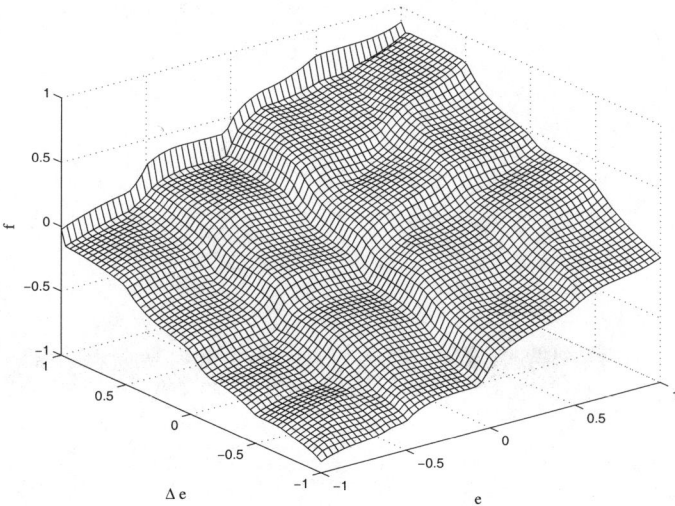

Fig. 4.30. Function implemented by the fuzzy inference scheme for the temperature control experiment

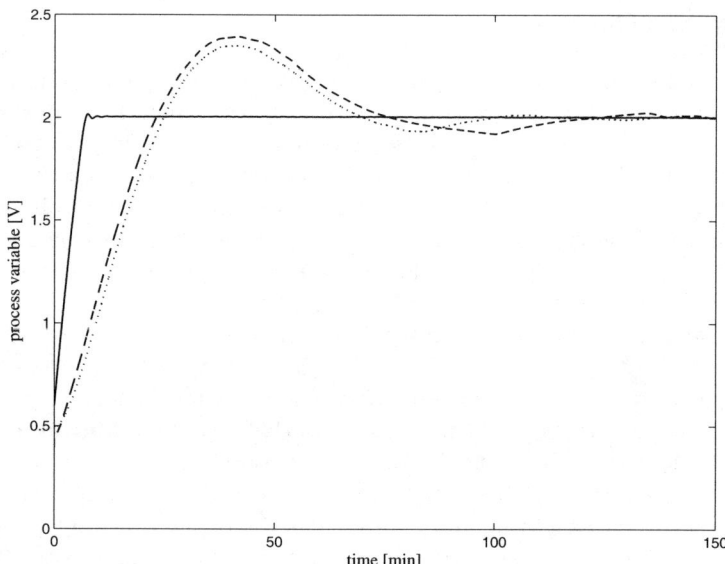

Fig. 4.31. Process variable for the level control experiment. Solid line: fuzzy set-point weighting; dashed line: no set-point weighting; dotted line: fixed set-point weight $\beta = 0.5$.

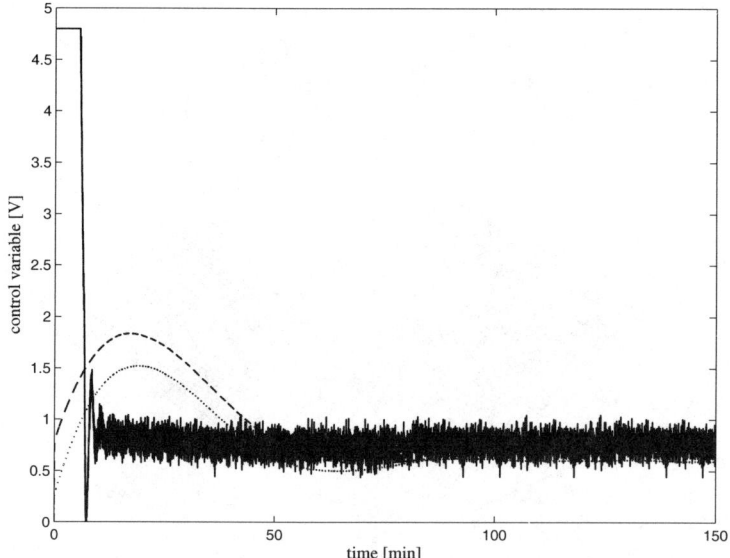

Fig. 4.32. Control variable for the level control experiment. Solid line: fuzzy set-point weighting; dashed line: no set-point weighting; dotted line: fixed set-point weight $\beta = 0.5$.

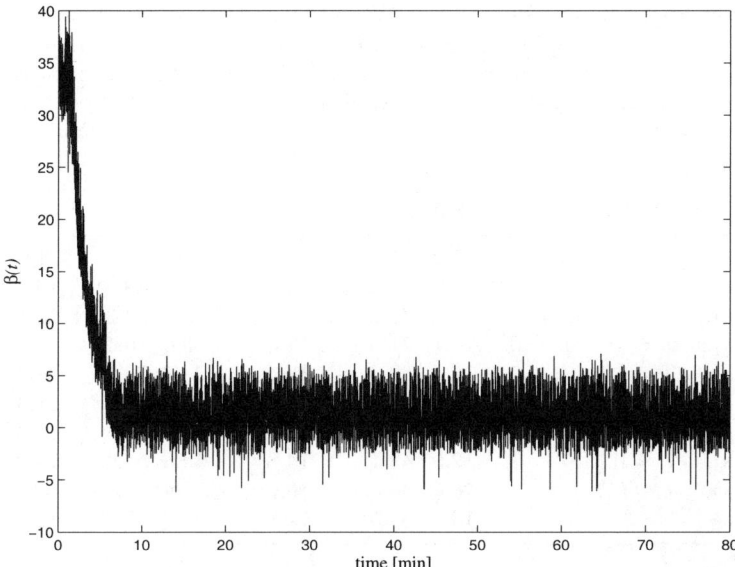

Fig. 4.33. Set-point weight for the level control experiment with fuzzy set-point weighting

4.5 Discussion

The use of the different methods for the set-point weight is discussed hereafter. For the processes (4.19)–(4.22) considered previously in the simulation results sections, a comparison between the different considered methods from the point of view of the achieved overshoot, rise time, 5% settling time and integrated absolute error is reported in Tables 4.4–4.7.

Clearly, the use of a fixed set-point weight is the simplest method to adopt and actually allows in general to reduce the overshoot more than the other methods (by paying the price of a further increased rise time). As already mentioned, the variable set-point weighting technique allows the maintenance of almost the same rise time of the case without set-point weight by reducing the overshoot at the same time. It can be remarked that an automatic tuning procedure is given so that the design extra effort required by the user is not significant. However, it is not clear how to do this if the attained overshoot is still excessive in a given application. Thus, the method is appropriate when the overshoot limit is not of major concern.

The best performance is generally achieved by the fuzzy set-point weighting methodology. This is done however at the expense of an increased complexity of the overall controller (although the tuning of the PID controller can be done with less effort since the fuzzified set-point weight allows to recover in any case the set-point following performance) and of an increased control

4 Set-point Weighting

Table 4.4. Resulting overshoot [%] for the considered methodologies. VSW: variable set-point weighting; RZN: refined Ziegler–Nichols tuning with no set-point weighting; RZN-OSPW: refined Ziegler–Nichols tuning with optimal (fixed) set-point weighting; FSW: fuzzy set-point weighting; ZN: Ziegler–Nichols tuning with no set-point weighting; ZN-OSPW: Ziegler–Nichols tuning with optimal (fixed) set-point weighting.

Process			VSW	RZN	RZN-OSPW	FSW	ZN	ZN-OSPW
$P_1(s)$,	$T=1$,	$L=0.1$	17.0	62.2	19.2	9.32	73.8	21.5
$P_1(s)$,	$T=1$,	$L=0.4$	19.7	49.2	15.7	5.64	49.3	14.4
$P_1(s)$,	$T=1$,	$L=0.8$	20.6	38.2	13.5	10.6	33.3	12.1
$P_1(s)$,	$T=10$,	$L=0.1$	14.1	71.5	18.3	10.6	80.4	22.6
$P_1(s)$,	$T=10$,	$L=0.4$	17.4	69.6	18.4	2.77	75.7	28.0
$P_1(s)$,	$T=10$,	$L=0.8$	17.3	64.4	18.1	2.65	72.6	21.3
$P_2(s)$			11.1	50.4	16.3	2.26	53.3	15.3
$P_3(s)$,	$L=0$		12.9	48.1	14.8	5.86	51.0	16.2
$P_3(s)$,	$L=0.1$		16.6	46.3	16.0	5.94	44.1	13.1
$P_4(s)$			18.9	37.5	12.8	8.50	32.4	12.1

Table 4.5. Resulting rise time for the considered methodologies. VSW: variable set-point weighting; RZN: refined Ziegler–Nichols tuning with no set-point weighting; RZN-OSPW: refined Ziegler–Nichols tuning with optimal (fixed) set-point weighting; FSW: fuzzy set-point weighting; ZN: Ziegler–Nichols tuning with no set-point weighting; ZN-OSPW: Ziegler–Nichols tuning with optimal (fixed) set-point weighting.

Process			VSW	RZN	RZN-OSPW	FSW	ZN	ZN-OSPW
$P_1(s)$,	$T=1$,	$L=0.1$	0.42	0.34	0.56	0.21	0.31	0.53
$P_1(s)$,	$T=1$,	$L=0.4$	0.71	0.68	1.01	0.68	0.66	1.00
$P_1(s)$,	$T=1$,	$L=0.8$	1.03	1.03	1.41	1.08	0.99	1.28
$P_1(s)$,	$T=10$,	$L=0.1$	1.69	1.16	2.03	0.21	0.99	1.72
$P_1(s)$,	$T=10$,	$L=0.4$	2.83	2.12	3.72	0.92	1.99	3.11
$P_1(s)$,	$T=10$,	$L=0.8$	3.81	3.04	5.16	1.77	2.76	4.77
$P_2(s)$			1.23	0.98	1.39	0.41	0.95	1.40
$P_3(s)$,	$L=0$		0.69	0.59	0.85	0.52	0.57	0.81
$P_3(s)$,	$L=0.1$		0.72	0.66	0.91	0.69	0.66	0.92
$P_4(s)$			1.31	1.28	1.78	1.20	1.25	1.64

effort. Note that, however, the trade-off between aggressiveness and control effort can be easily handled by modifying the value of the parameter K_{out}. In fact, the presence of parameter K_{out} is useful for excluding the fuzzification of the set-point weight from the overall control scheme if the operator prefers to avoid its use (for example when the normalised dead time of the process has a high value). Actually, by setting $K_{out} = 0$, a classical control scheme with a fixed set-point weighting results. Note that this solution naturally arises from

Table 4.6. Resulting 5% settling time for the considered methodologies. VSW: variable set-point weighting; RZN: refined Ziegler–Nichols tuning with no set-point weighting; RZN-OSPW: refined Ziegler–Nichols tuning with optimal (fixed) set-point weighting; FSW: fuzzy set-point weighting; ZN: Ziegler–Nichols tuning with no set-point weighting; ZN-OSPW: Ziegler–Nichols tuning with optimal (fixed) set-point weighting.

Process			VSW	RZN	RZN-OSPW	FSW	ZN	ZN-OSPW
$P_1(s)$,	$T=1$,	$L=0.1$	2.46	3.52	2.69	0.99	4.19	3.37
$P_1(s)$,	$T=1$,	$L=0.4$	2.97	4.74	3.44	1.85	4.62	3.34
$P_1(s)$,	$T=1$,	$L=0.8$	5.77	4.98	4.76	5.02	4.67	4.23
$P_1(s)$,	$T=10$,	$L=0.1$	12.0	18.2	12.2	1.03	21.9	16.5
$P_1(s)$,	$T=10$,	$L=0.4$	16.7	28.1	22.3	1.75	32.7	27.1
$P_1(s)$,	$T=10$,	$L=0.8$	22.2	31.7	24.3	3.40	37.8	30.3
$P_2(s)$			5.04	7.73	5.93	1.65	7.77	5.93
$P_3(s)$,	$L=0$		2.20	3.56	2.55	2.66	3.55	3.43
$P_3(s)$,	$L=0.1$		3.40	3.95	2.91	3.64	3.66	2.80
$P_4(s)$			7.62	7.61	5.91	5.26	5.95	5.44

Table 4.7. Resulting integrated absolute error for the considered methodologies. VSW: variable set-point weighting; RZN: refined Ziegler–Nichols tuning with no set-point weighting; RZN-OSPW: refined Ziegler–Nichols tuning with optimal (fixed) set-point weighting; FSW: fuzzy set-point weighting; ZN: Ziegler–Nichols tuning with no set-point weighting; ZN-OSPW: Ziegler–Nichols tuning with optimal (fixed) set-point weighting.

Process			VSW	RZN	RZN-OSPW	FSW	ZN	ZN-OSPW
$P_1(s)$,	$T=1$,	$L=0.1$	0.58	1.06	0.58	0.34	1.26	0.87
$P_1(s)$,	$T=1$,	$L=0.4$	1.23	1.77	1.12	1.08	1.74	1.48
$P_1(s)$,	$T=1$,	$L=0.8$	1.99	2.43	1.76	1.95	2.26	2.08
$P_1(s)$,	$T=10$,	$L=0.1$	1.96	4.62	2.18	0.34	5.46	3.31
$P_1(s)$,	$T=10$,	$L=0.4$	3.77	7.86	3.80	1.27	8.79	5.95
$P_1(s)$,	$T=10$,	$L=0.8$	5.20	9.87	5.11	2.49	11.0	7.65
$P_2(s)$			1.19	2.22	1.36	0.51	2.27	1.84
$P_3(s)$,	$L=0$		0.79	1.30	0.82	0.71	1.32	1.11
$P_3(s)$,	$L=0.1$		1.02	1.50	1.00	0.96	1.44	1.26
$P_4(s)$			2.49	3.08	2.21	2.20	2.83	2.61

the genetic tuning procedure just with the increasing of the normalised dead time of the process.

It is worth stressing at this point that techniques based on the modification of the set-point are already available in industrial single-station controllers. For example, in some Yokogawa temperature controllers, the so-called Fuzzy Overshoot Suppressor is implemented. It consists of lowering the set-point value by means of a fuzzy inference system, namely of substituting the set-point signal with a sub-set-point signal during the transient when the occurrence of an excessive overshoot is estimated. This fact is depicted in Figure 4.34. The performance obtained by this method have been compared with those obtained by the variable set-point weighting method (Hang and Cao, 1996) and by the fuzzy set-point weighting method (Visioli and Veronesi, 1999). Results demonstrate that the varying just the set-point weight instead of the set-point signal is preferable, although it has to be highlighted that a fair comparison is difficult to perform since the exact algorithm of the industrial controller is not available.

Finally, it has to be noted that the algorithms previously described differ from the classical gain scheduling approach. Actually, the purpose of a gain scheduling is to address process nonlinearities by adopting different sets of (fixed) PID parameters in different operating regions. Conversely, the use of a time-varying set-point weight aims at improving the transient response for a process with linear dynamics. Further, in the gain scheduling approach the set of PID parameters to be adopted depends of the absolute value of an auxiliary variable (for example the set-point or the process variable) that is representative of the current operating region. On the contrary, in the described techniques the current value of the set-point weight is determined by the current control error (and its first time derivative) (Hang and Cao, 1996).

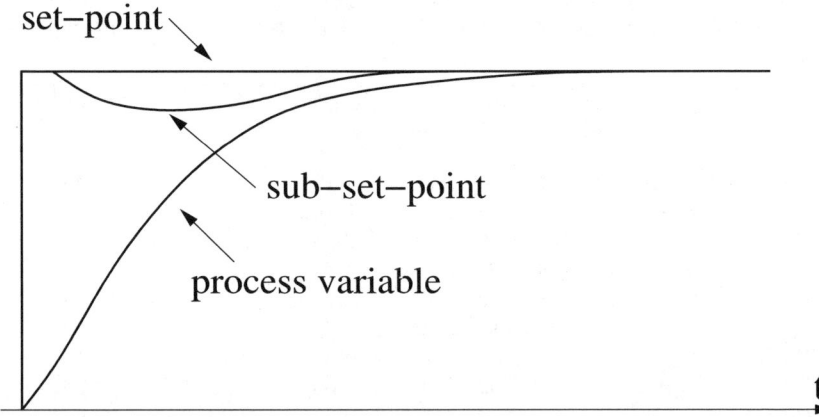

Fig. 4.34. Sketch of a result of the implementation of the Fuzzy Overshoot Suppressor

4.6 Conclusions

In this chapter the role of the set-point weighting in the PID control law has been underlined and it has been shown how it can be effectively adopted in order to decouple the problem of providing good performance at the same time both in the set-point following and in the load disturbance rejection task. For processes with small normalised dead time, the adoption of a time varying (control error dependent) set-point weight can represent a valid choice to avoid a large overshoot without increasing the rise time. In this context, two methodologies, namely the variable set-point weighting and the fuzzy set-point weighting technique, have been thoroughly analysed in order to provide a clear characterisation of them and to understand their applicability in practical situations. It has been shown that the use of such techniques indeed represents a valuable solution for the implementation of a high-performance controller by retaining at the same time the overall ease of use of a PID controller.

5
Use of a Feedforward Action

5.1 Introduction

The main purpose of using feedback is to compensate for external disturbances and for model uncertainties. Actually, when a sufficiently accurate model of the process is available (and the process dynamics does not change significantly during the process operations), control performance can be improved in general by conveniently employing an additional feedforward (open-loop) control law. Different methodologies for the design and the implementation of a feedforward control law, to be adopted in conjunction with the feedback action provided by a PID controller, are described in this chapter. It is shown how the problem can be approached from different points of view. In particular, regarding the set-point following task, two kinds of approaches are presented: the design of a causal feedforward action and of a noncausal feedforward action. In the first case a nonlinear control law is described and its advantages with respect to the standard methodology are outlined. In the second case, to be employed when desired process output transitions are known in advance, strategies based on input-output inversion are explained both in the continuous-time and in the discrete-time framework. Finally, a brief review of the use of feedforward for disturbance compensation is also provided.

5.2 Linear Causal Feedforward Action

The standard methodology for the implementation of a feedforward action for the improvement of set-point following task is that shown in Figure 5.1 (note that this scheme can be made equivalent to the one of Figure 1.11 by a proper modification of the block diagram), where $M(s)$ is a reference model that gives the desired response of a set-point change and $G(s)$ is chosen as

$$G(s) = \frac{M(s)}{\tilde{P}(s)}. \tag{5.1}$$

94   5 Use of a Feedforward Action

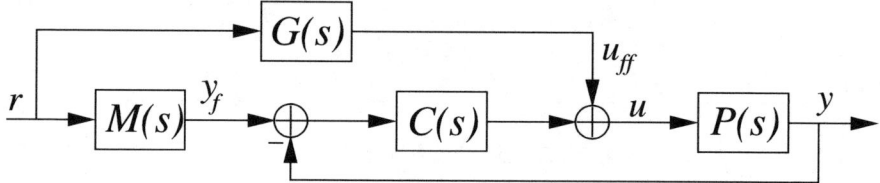

Fig. 5.1. Block diagram for the standard implementation of feedforward action for set-point following task

where $\tilde{P}(s)$ is the minimum-phase part of the process transfer function $P(s)$. Note that this is actually a general scheme and can be implemented with any feedback controller C, although the following analysis will assume the adoption of a PID controller. Obviously, the effectiveness of feedforward control heavily depends on the accuracy of the estimated process model (see the remarkable result presented in (Devasia, 2002)). In any case, even if a perfect model is available, the design of $M(s)$ is a crucial issue, as it represents the desired performance. It has to contain the nonminimum-phase (i.e., the non invertible) part of $P(s)$ and also it should take into account actuator limits. The following example illustrates this issue. Consider the process

$$P(s) = \frac{1}{10s+1} e^{-5s}, \qquad (5.2)$$

and an ideal output-filtered PID controller (1.23) with $K_p = 2.4$, $T_i = 10$, $T_d = 2.5$ and $T_f = 0.1$. Then, the transfer function $M(s)$ is designed as

$$M(s) = \frac{1}{2s+1} e^{-5s}, \qquad (5.3)$$

in order to speed up the closed-loop control system response with respect to the open-loop system. Thus, the transfer function $G(s)$ results to be

$$G(s) = \frac{10s+1}{2s+1}. \qquad (5.4)$$

If no saturation limits are considered, the unit set-point step response is plotted in Figure 5.2. Note that the output of the PID controller, i.e., the control error, is always zero. Conversely, if a saturation limit $u_{sat} = 2$ is applied to the control variable, the result is that shown in Figure 5.3 (note the different time scaling). It can be seen that the feedforward action is not exploited due to the saturation limits and a (possibly unexpected) worse performance result (note that there is no integrator windup effect). Indeed, the design of the reference model has to take into account the saturation limits of the actuator, since a too fast response cannot be imposed. In other words, even if a nonminimum-phase dynamics is absent, the performance achievable is obviously limited by the physical constraints of the actuator and the overall design cannot leave this aspect out of consideration.

5.2 Linear Causal Feedforward Action   95

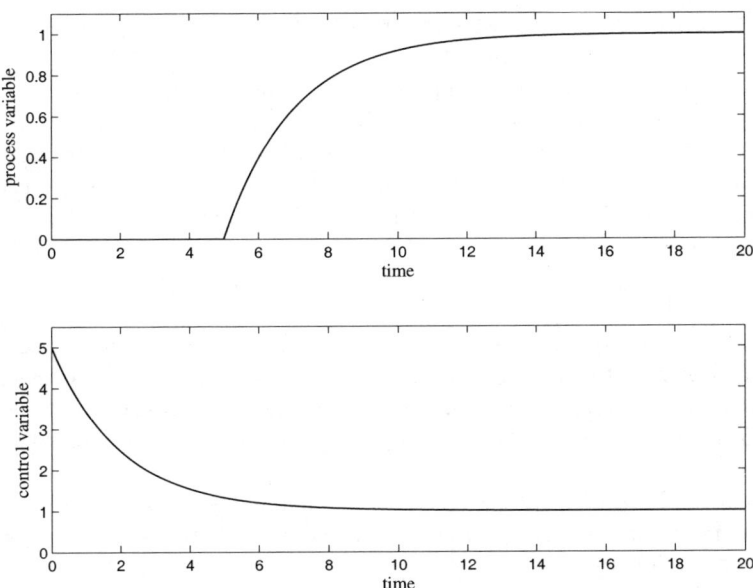

Fig. 5.2. Example of use of the standard implementation of feedforward action with no saturation limits

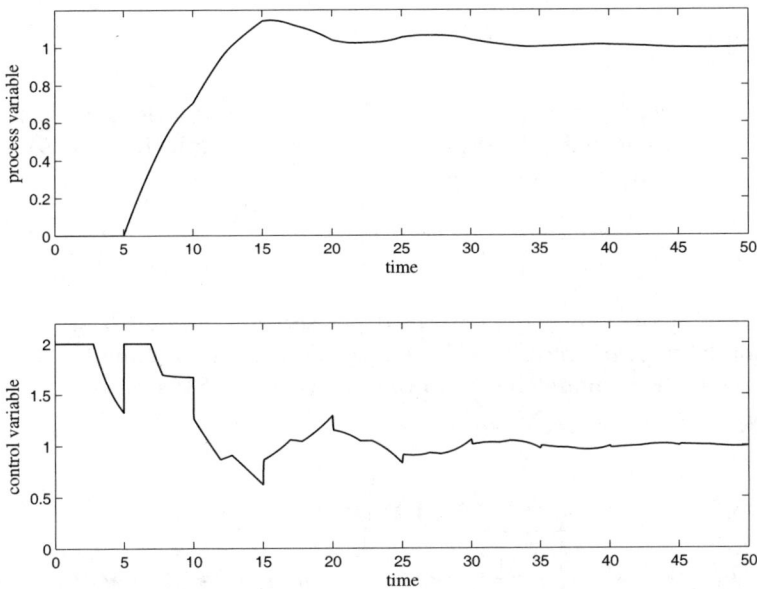

Fig. 5.3. Example of use of the standard implementation of feedforward action with saturation limits

5.3 Nonlinear Causal Feedforward Action

A significant improvement in the set-point following performances can be obtained by employing a nonlinear feedforward action, as shown in (Wallen, 2000; Wallen and Åström, 2002), where a technique inspired by the bang-bang control strategy (Lewis, 1996) is devised to achieve a fast response to set-point changes. With the same aim of fully exploiting the capabilities of the actuator, namely, in order to take into account the actuator nonlinearity (without impairing the ease of use of the overall control system), the following methodology has been proposed in (Visioli, 2004).

Assume that it is required to design a control scheme based on a PID controller plus a feedforward term aiming at achieving a transition of the process output y from the value y_0 to the value y_1 in a predefined time interval of duration τ. In the following, for the sake of clarity and without loss of generality, it will be assumed $y_0 = 0$ and $y_1 > 0$.

The devised PID plus feedforward control scheme is shown in Figure 5.4, and implements the following design technique. First, the process is described by a FOPDT model, i.e.:

$$P(s) = \frac{K}{Ts+1} e^{-Ls}. \tag{5.5}$$

Based on this model, the output u_{ff} of the feedforward block FF is defined as follows:

$$u_{ff}(t) = \begin{cases} \bar{u}_{ff} & \text{if } t < \tau \\ \dfrac{y_1}{K} & \text{if } t \geq \tau \end{cases} \tag{5.6}$$

where the value of $\bar{u}_{ff}$ is determined, after trivial calculations, in such a way that the process output y (which is necessarily zero until time $t = L$) is y_1 at time $t = \tau + L$. It produces the result:

$$\bar{u}_{ff} = \frac{y_1/K}{1 - e^{-\tau/T}}. \tag{5.7}$$

In this way, if the process is described perfectly by Model (5.5), an output transition in the time interval $[L, \tau + L]$ occurs. Then, at time $t = \tau + L$ the output settles at value y_1 thanks to the constant value assumed by $u_{ff}(t)$ for $t \geq \tau$.

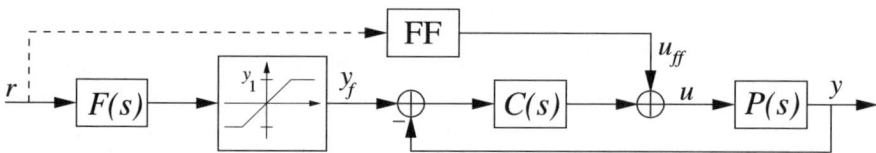

Fig. 5.4. Block diagram of the PID plus nonlinear feedforward action control scheme

Then, a suitable reference signal y_f has to be applied to the closed-loop system. It is desired that y_f be equal to the desired process output that would be obtained in the case where the process is modelled perfectly by Expression (5.5). Thus, the step reference signal r of amplitude y_1 has to be filtered by the system

$$F(s) = \frac{K\bar{u}_{ff}}{\frac{y_1}{Ts+1}} e^{-Ls} \tag{5.8}$$

and then saturated at the level y_1.

It is worth stressing at this point that this method exploits the fact that a process output transition is required instead of tracking a general reference signal. In the latter case the typical control scheme of Figure 5.1 has to be adopted. The presence of many set-point changes can be instead easily handled by the PID plus nonlinear feedforward control system. Indeed, in case a new value of the set-point is selected during a previously determined transient response, it is sufficient to determine the feedforward action for the new value and to sum it to the one that is currently applied. Analogously, the reference signal determined for the latest set-point change has to be summed to the one related to the previous one.

The overall control scheme design involves the selection of the transition time τ and of the PID parameters. The choice of a sensible value of τ can be made by the user either directly or through a (possibly) more intuitive reasoning. For example, the user might select a ratio between the bandwidth of the open-loop system and that of the closed-loop one, from which the value of τ can be determined easily. Obviously, decreasing the value of τ means that the value of $\bar{u}_{ff}$ (and therefore of the overall manipulated variable) increases, and too low a value of τ might imply that the determined control variable cannot be applied due to the saturation of the actuator. Thus, alternatively, the operator might first select the value of $\bar{u}_{ff}$ depending on the desired control effort (defined typically as a percentage of the maximum limit of the manipulated variable) and determine consequently the value of τ. In this way the potentiality of the actuator can be fully exploited and the problems associated with the use of the standard control scheme of Figure 5.1 are avoided. In any case, the design parameter τ has a clear physical meaning, as it handles the trade-off between performance, robustness and control activity (Kristiansson and Lennartson, 2001; Morari and Zafiriou, 1989). Indeed, it has the same role of the time constant of the reference model $M(s)$ in the classic technique. It can be therefore exploited to satisfy the specific requirements of a given application.

The tuning of the PID controller should take into account the robustness issue, since the feedforward action is based on a simple FOPDT model of the plant and the compensation of the (unavoidable) modelling errors is left to the feedback control law. To this respect, it is very useful to consider the analysis made in (Wallen, 2000), where it is shown that the deviations due to the modelling errors between the desired and the actual output can be treated as the effect of a load disturbance $d = G_d u_{ff}$ where

$$G_d(s) = \frac{P(s) - F(s)}{P(s)}(1 + P(s)C(s)). \qquad (5.9)$$

Thus, by considering that the process output results to be the superposition of the effects of the feedforward action and of the load disturbances (*i.e.*, of true load disturbances and the "fictitious" one d due to the modelling errors) and by considering also that in the nominal case the set-point following performances are determined only by the feedforward action, it is sensible to tune the PID controller by taking into account its load disturbance rejection performances.

5.3.1 Simulation Results

The following simulation results are given in order to understand better the significance of the technique presented and to compare it with the classic method of Section 5.2. In all the considered cases it is assumed that $y_1 = 1$, namely, a unit step is applied to the set-point signal at time $t = 0$ (starting from null initial conditions). An ideal output-filtered PID controller (1.23) is adopted and no actuator saturation limits are considered (this case is addressed in Section 5.3.2 where experimental results are shown).
As a first example, consider the process

$$P_1(s) = \frac{1}{s+1}e^{-0.5s}, \qquad (5.10)$$

and a PID controller with $K_p = 2.4$, $T_i = 1$ and $T_d = 0.25$ ($T_f = 0.01$). The selected transition time is $\tau = 0.5$ and therefore, by applying Equation (5.6), it results $\bar{u}_{ff} = 2.54$, *i.e.*, the feedforward signal $u_{ff}(t)$ is equal to 2.54 for $0 \leq t < 0.5$ and equal to one for $t \geq 0.5$ s. This feedforward signal and the obtained process output are plotted in Figure 5.5 (solid line). Note that in this case, being the process modelled perfectly, we have that the output of the PID controller is zero for $t \geq 0$ s, *i.e.*, the reference signal y_f for the closed-loop system is equal to the obtained process output.
For the sake of comparison, the classic control scheme of Figure 5.1 has been applied to the same process (and the same PID controller). By choosing

$$M(s) = \frac{1}{0.5s+1}e^{-0.5s}$$

we obtain (see (5.1))

$$G(s) = \frac{s+1}{0.5s+1}$$

The resulting feedforward signal and process output are also shown in Figure 5.5 (dashed line). Also in this case the output of the PID controller is zero for

$t \geq 0$ s as there are no model uncertainties.

It appears that, with the new approach, the settling time has been shortened about five times, as is obvious because in the classic scheme 0.5 is the time constant of the overall system (which is of first order), instead of the actual transition time. In order to evaluate better the significance of the results, the conventional control scheme has been implemented by setting the value of the time constant of the reference model to $\tau/5$, i.e.:

$$M(s) = \frac{1}{0.1s+1} e^{-0.5s}$$

and, consequently,

$$G(s) = \frac{s+1}{0.1s+1}$$

Results are again shown in Figure 5.5 (dash-dot line). It appears that a value of the settling time comparable with the one obtained with the new methodology has been achieved in this case by much increasing the control effort.

Evidently, the much better performance obtained with the adopted nonlinear feedforward action (despite no extra design effort required from the user) is due to the fact that the control signal is kept at a constant level for the time necessary to achieve the desired process output transition and this cannot be obtained with a linear feedforward action.

As a second example, consider the process

$$P_2(s) = \frac{1}{(s+1)^4} e^{-0.5s}. \tag{5.11}$$

By applying the area method (see Chapter 7), a FOPDT model (5.5) of the process is estimated, resulting in $K = 1$, $T = 2.12$, and $L = 2.38$. Based on this model, the PID parameters can be selected as $K_p = 1.07$, $T_i = 4.76$, $T_d = 1.19$ and $T_f = 0.01$. By fixing $\tau = 2$, the resulting value of the constant feedforward action (see (5.6)) is $\bar{u}_{ff} = 1.637$. Results related to the nonlinear feedforward approach are reported in Figure 5.6. As for the process $P_1(s)$, a comparison with the standard approach has been performed. Thus,

$$M(s) = \frac{1}{0.4s+1} e^{-2.38s}$$

and, consequently,

$$G(s) = \frac{2.12s+1}{0.4s+1}$$

Results related to this case are shown in Figure 5.7.

As for process $P_1(s)$ (5.10), it turns out that the approach based on the use of a nonlinear feedforward action provides a performance similar to that obtained with the classic one but with a much less control effort. The same kind of experiment is repeated with $\tau = 10$ for the PID plus nonlinear feedforward scheme ($\bar{u}_{ff} = 1.01$) and it is compared with the classic linear approach where

$$M(s) = \frac{1}{2s+1} e^{-2.38s}.$$

Results are shown in Figures 5.8 and 5.9 respectively.

The role played by parameter τ of the nonlinear feedforward control method and by the selected time constant of the reference model in the linear feedforward control method in handling the trade-off between performance, robustness and control activity becomes apparent.

From the presented results, it is evident that the robustness of the two approaches with respect to modelling uncertainties is basically the same. Further, it can be seen that, as expected, the two approaches result in a more and more similar performance as the desired process output transition time is increased.

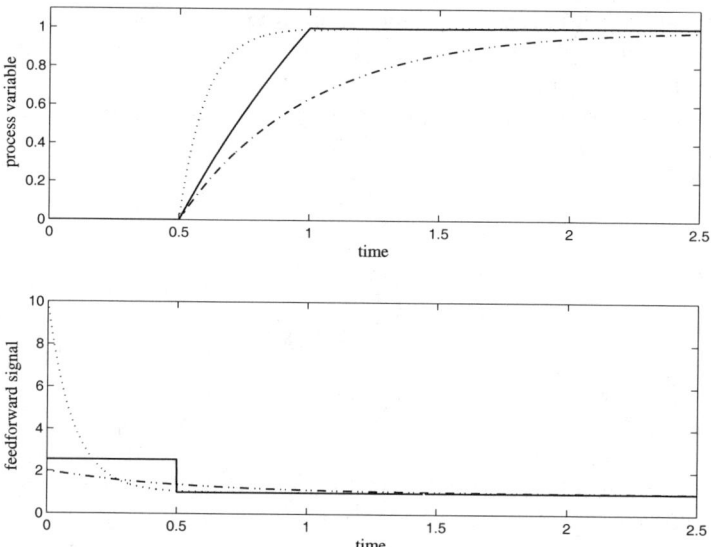

Fig. 5.5. Simulation results for the process $P_1(s)$. Solid line: PID plus nonlinear feedforward action; dash-dot line: PID plus linear feedforward action with a FOPDT reference model with a time constant of 0.5; dotted line: PID plus linear feedforward action with a FOPDT reference model with a time constant of 0.1.

5.3 Nonlinear Causal Feedforward Action   101

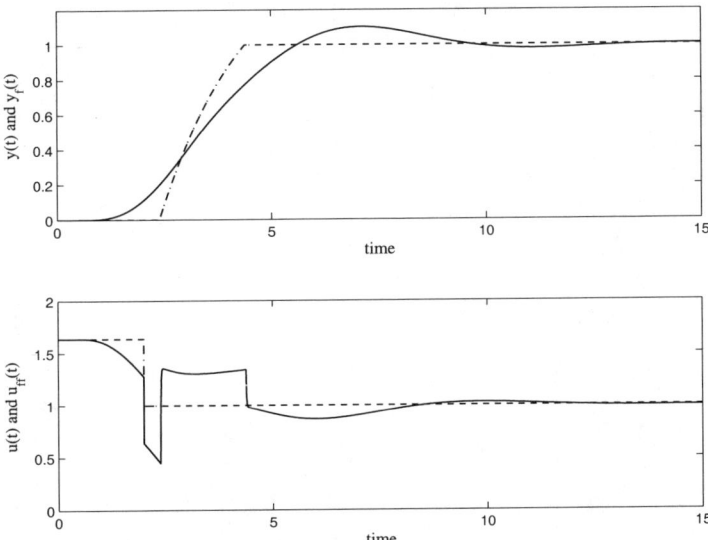

Fig. 5.6. Process output, reference signal, manipulated variable and feedforward signal with the PID plus nonlinear feedforward action scheme for process $P_2(s)$ ($\tau = 2$). Solid line: $y(t)$ and $u(t)$; dashed line: $y_f(t)$ and $u_{ff}(t)$.

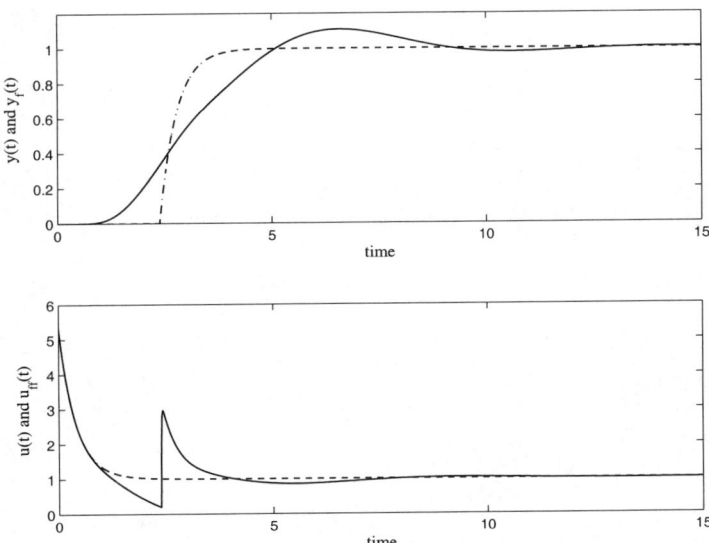

Fig. 5.7. Process output, reference signal, manipulated variable and feedforward signal with the classic linear feedforward action scheme for process $P_2(s)$ with a FOPDT reference model with a time constant of 0.4. Solid line: $y(t)$ and $u(t)$; dashed line: $y_f(t)$ and $u_{ff}(t)$.

102   5 Use of a Feedforward Action

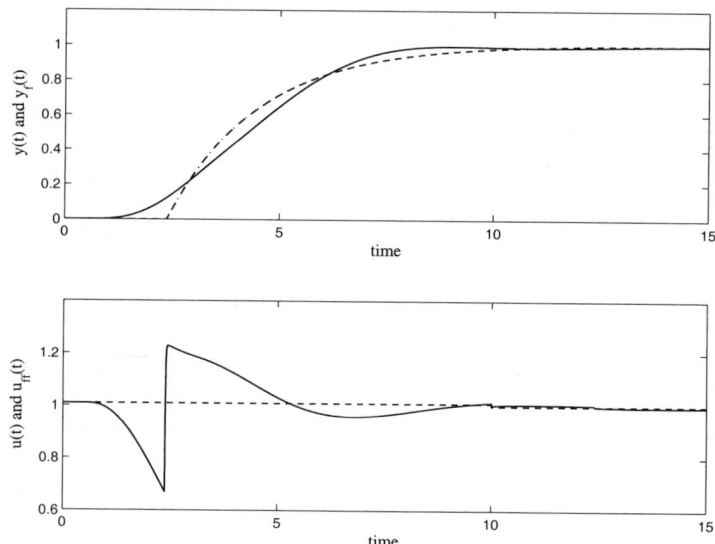

Fig. 5.8. Process output, reference signal, manipulated variable and feedforward signal with the PID plus nonlinear feedforward action scheme for process $P_2(s)$ ($\tau = 10$). Solid line: $y(t)$ and $u(t)$; dashed line: $y_f(t)$ and $u_{ff}(t)$.

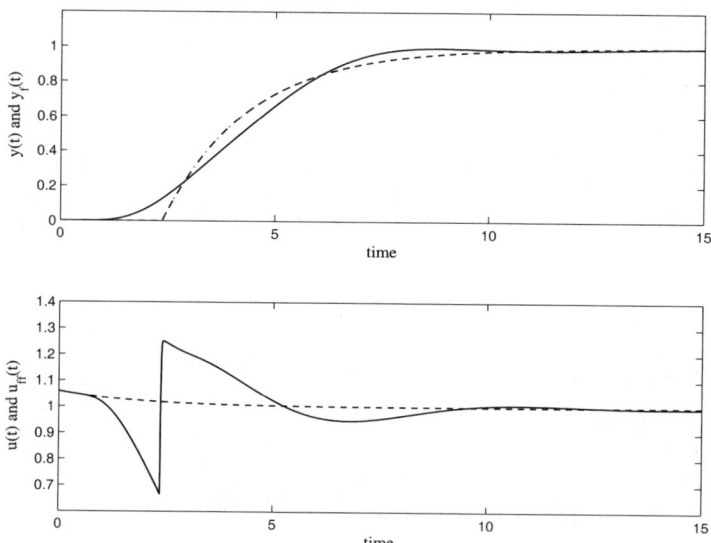

Fig. 5.9. Process output, reference signal, manipulated variable and feedforward signal with the classic linear feedforward action scheme for process $P_2(s)$ with a FOPDT reference model with a time constant of 2. Solid line: $y(t)$ and $u(t)$; dashed line: $y_f(t)$ and $u_{ff}(t)$.

5.3.2 Experimental Results

Level Control

The PID plus nonlinear feedforward control scheme has been tested in a level control problem by means of the double-tank apparatus described in Section A.1. In particular, a single tank has been considered and a transition from the initial level $y_0 = 2$ V to the final level $y_1 = 4$ V has been required. By taking into account that the maximum value of the manipulated variable is $u_{max} = 5$ V, the maximum absolute value for the feedforward signal $u_{ff}(t)$ (denoted u_{ff}^M) has been fixed to 4.8 V. This choice is justified by the need of giving to the feedback PID controller the capability to compensate for the model uncertainties. The FOPDT model of the tank in the operating range has been estimated by applying the area method to an open-loop step response. It results in

$$P(s) = \frac{1.2}{20s+1} e^{-1.5s}, \tag{5.12}$$

i.e., it is $K = 1.2$, $T = 20$ s and $L = 1.5$ s. A PI controller is then tuned by fixing $K_p = 7.41$ and $T_i = 20$. The transition time τ has then been determined by considering the selected value of u_{ff}^M. From Equation (5.6), it can be derived

$$\tau = -T \log \left(-\frac{y_1 - y_0}{K \bar{u}_{ff}} + 1 \right), \tag{5.13}$$

where the value of $\bar{u}_{ff}$, by taking into account the initial output value, has to be calculated as

$$\bar{u}_{ff} = u_{ff}^M - \frac{y_0}{K}. \tag{5.14}$$

By considering the values of the parameters, it results $\tau = 15.18$ s. The nonlinear feedforward approach has been compared to the case of no feedforward action and to the linear feedforward approach. In this latter case the reference model has been selected as

$$M(s) = \frac{1}{\frac{\tau}{5}s + 1} e^{-1.5s}, \tag{5.15}$$

so that the feedforward block transfer function becomes

$$G(s) = \frac{1}{1.2} \cdot \frac{20s+1}{3.06s+1}. \tag{5.16}$$

The resulting process variable for the control systems considered is plotted in Figure 5.10. The corresponding (PID plus feedforward) controller output is reported in Figure 5.11. It appears that the resulting controller output when no feedforward action is present and when a linear feedforward action is employed is much greater than the actual saturation limit of the actuator

104   5 Use of a Feedforward Action

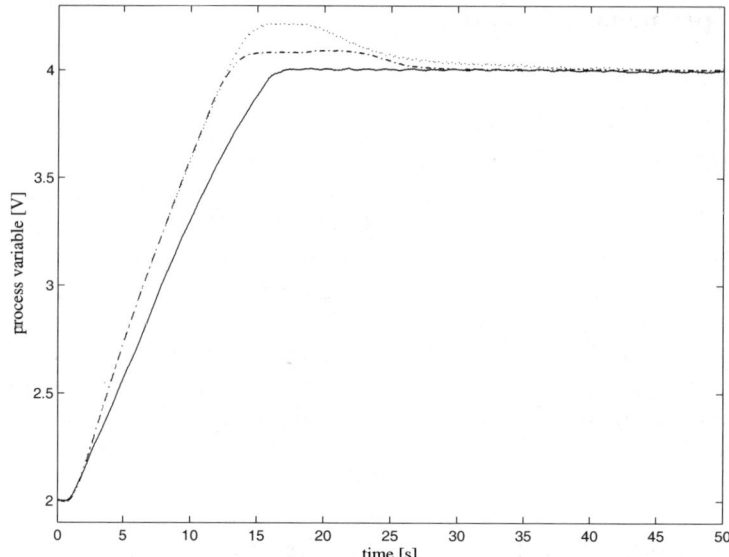

Fig. 5.10. Process variable for the level control task. Solid line: PID plus nonlinear feedforward action; dash-dot line: PID with no feedforward action; dotted line: PID plus linear feedforward action.

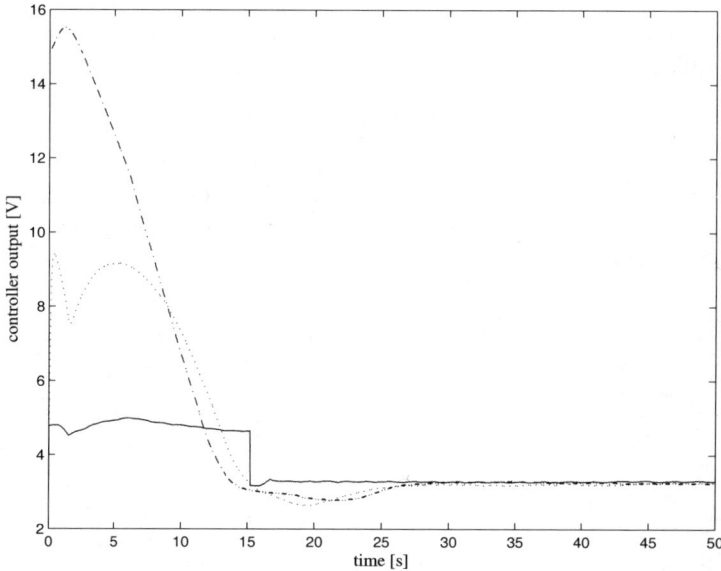

Fig. 5.11. Controller output for the level control task. Solid line: PID plus nonlinear feedforward action; dash-dot line: PID with no feedforward action; dotted line: PID plus linear feedforward action.

in a significant time interval and therefore it is not exploited. Conversely, the controller output for the case of the nonlinear feedforward action is, as expected, about 4.8 V at the beginning of the transient (variations up to 5 V are due to the model uncertainties). This implies that a slightly higher rise time occurs but the overall performance is much more satisfactory (there is no overshoot and the settling time is smaller).

Another experiment has been performed by considering the other tank present in the equipment and by adding an artificial dead time of 10 s. In this case the control requirement is to accomplish an output transition from $y_0 = 2$ V to $y_1 = 3$ V. The estimated process model is

$$P(s) = \frac{1.98}{29s + 1} e^{-11s}, \quad (5.17)$$

and the PI controller parameters are selected as $K_p = 1.2$ and $T_i = 33$. By applying a reasoning analogous to the previous case, the transition time is chosen as $\tau = 6$ s. Then, the model reference for the classic linear feedforward approach is selected as

$$M(s) = \frac{1}{1.2s + 1} e^{-11s}, \quad (5.18)$$

so that the feedforward block transfer function results to be

$$G(s) = \frac{1}{1.98} \cdot \frac{29s + 1}{1.2s + 1}. \quad (5.19)$$

The resulting process variable and the corresponding process input (note that the saturation limit is $u_{max} = 5$ V) for the three considered control schemes are reported in Figures 5.12 and 5.13 respectively.

From the presented experimental results, critical issues associated with the classical feedforward design appear. In particular, in the presented cases, the linear approach is not worthy to being applied with respect to the standard PID control, since the decreasing of the rise time is paid by a larger overshoot and by an increased amplitude of the manipulated variable. Conversely, the nonlinear feedforward approach outperforms the other ones.

Actually, it is evident that a higher saturation level would be necessary for the classical feedforward scheme in order to provide a performance similar to that achieved by adopting the technique proposed in (Visioli, 2004), and this highlights its main advantage, namely that the use of a nonlinear (piecewise constant) feedforward action allows to obtain in general a low rise time and a low overshoot in the set-point step response despite a less maximum value of the control variable is required.

106   5 Use of a Feedforward Action

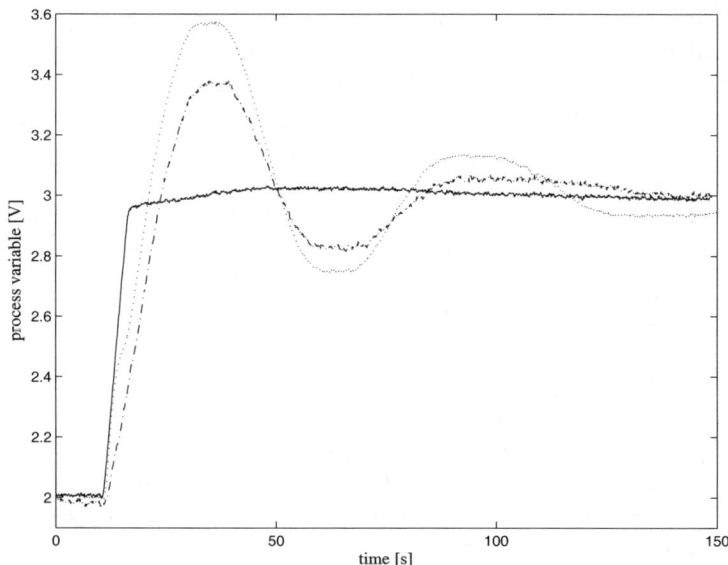

Fig. 5.12. Process variable for the level control task with additional dead time. Solid line: PID plus nonlinear feedforward action; dash-dot line: PID with no feedforward action; dotted line: PID plus linear feedforward action.

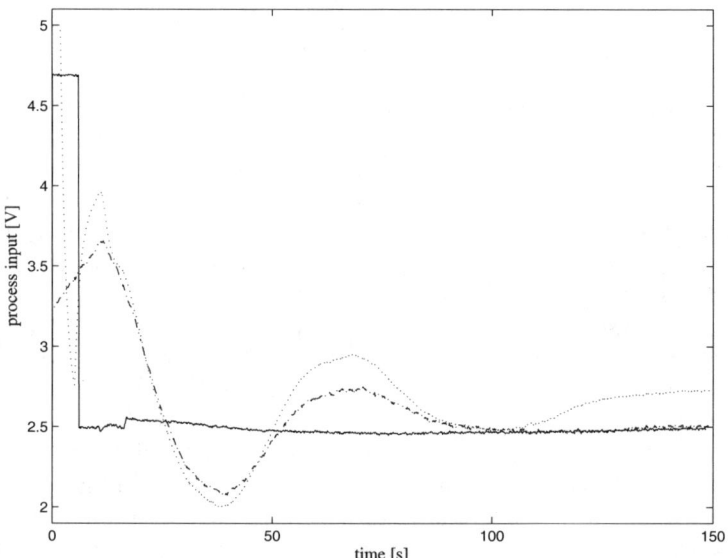

Fig. 5.13. Controller output for the level control task with additional dead time. Solid line: PID plus nonlinear feedforward action; dash-dot line: PID with no feedforward action; dotted line: PID plus linear feedforward action.

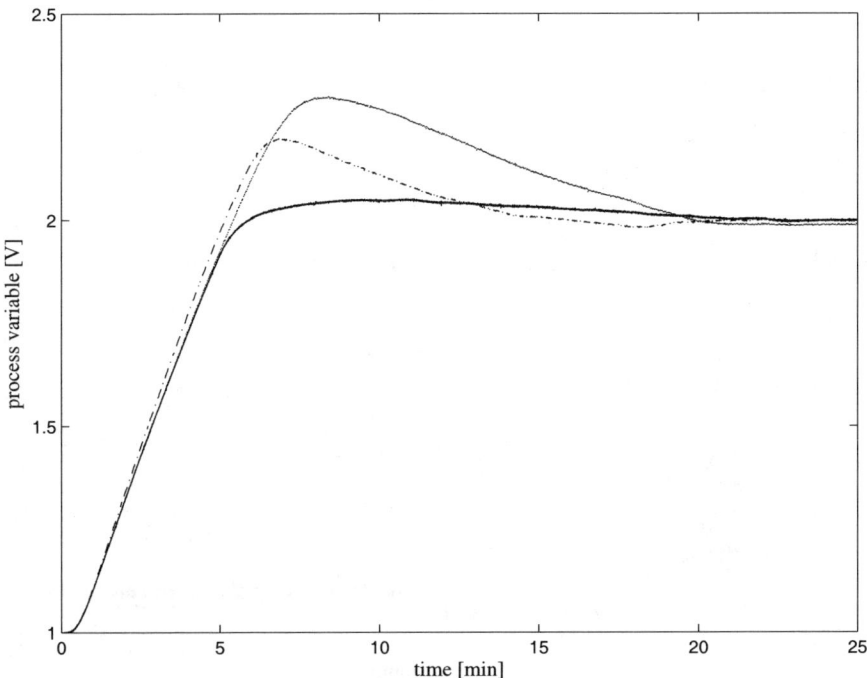

Fig. 5.14. Process variable for the temperature control task. Solid line: PID plus nonlinear feedforward action; dash-dot line: PID with no feedforward action; dotted line: PID plus linear feedforward action.

Temperature Control

The PID plus nonlinear feedforward control scheme has been tested also in a temperature control problem by means of the experimental setup described in Section A.2. A transition from the initial level $y_0 = 1$ V to the final level $y_1 = 2$ V has been required. As in the level control case, by taking into account that the maximum value of the manipulated variable is $u_{max} = 5$ V, the maximum absolute value for the feedforward signal $u_{ff}(t)$ has been fixed to $u_{ff}^M = 4.8$ V. The FOPDT model of the oven in the operating range has been estimated by applying the area method to an open-loop step response. It becomes

$$P(s) = \frac{1.33}{1400s + 1}e^{-30s}, \qquad (5.20)$$

i.e., it is $K = 1.2$, $T = 1400$ s and $L = 30$ s. A PID controller is then tuned by fixing $K_p = 2.0$ and $T_i = 300$. The transition time τ has then been determined by following the same reasoning as in the level control example. It results $\tau = 287.7$ s. Based on this result, the model reference for the classic linear feedforward approach is selected as

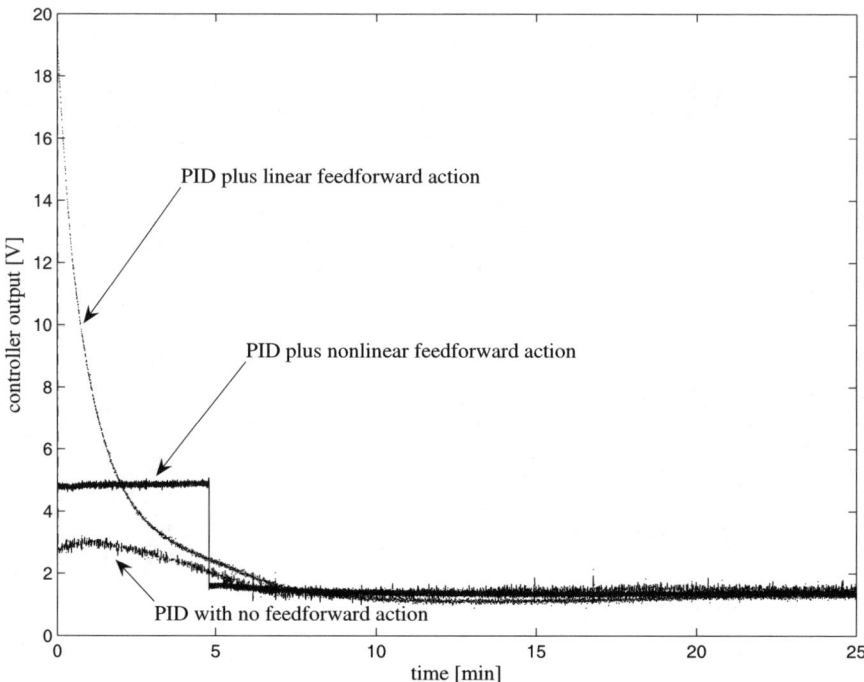

Fig. 5.15. Controller output for the temperature control task. Solid line: PID plus nonlinear feedforward action; dash-dot line: PID with no feedforward action; dotted line: PID plus linear feedforward action.

$$M(s) = \frac{1}{57.5s + 1} e^{-30s}, \qquad (5.21)$$

so that the feedforward block transfer function becomes

$$G(s) = \frac{1}{1.33} \cdot \frac{1400s + 1}{57.5s + 1}. \qquad (5.22)$$

The resulting process variable and the corresponding controller output (note that the saturation limit is $u_{max} = 5$ V) for the three control schemes considered are shown in Figures 5.14 and 5.15 respectively.

It appears that in this case the standard linear feedforward action provides a better performance than pure PID control, but the nonlinear feedforward action allows the significant decrease of the overshoot with respect to both of them. Indeed, considerations similar to those made for the level control task apply also in this case.

5.4 Noncausal Feedforward Action: Continuous-time Case

5.4.1 Generalities

In the previous sections, it has been shown that the set-point following performance of a feedback control system can be significantly improved by the application of a properly designed (causal) feedforward action. From a different point of view, when the desired output trajectory is known in advance, a feedforward action determined by means of a stable inversion technique can be applied (Zou and Devasia, 1999). Roughly speaking, the approach consists in selecting a desired output function that meets the control requirements and then determining, by inverting the system dynamics, the input function that causes that selected output signal. It is worth noting that the concept of dynamic input-output inversion (Hunt et al., 1996; Devasia et al., 1996) has been already proven to be effective in different areas of the automatic control field, such as motion control (Piazzi and Visioli, 2000; Perez and Devasia, 2003), flight control (Hunt and Meyer, 1997), robust control (Piazzi and Visioli, 2001c; Piazzi and Visioli, 2001a).

In the context of PID control, the input-output inversion technique can be exploited to determine a suitable command signal to be applied to the closed-loop control system, instead of the typical step signal, in order to achieve a high performance (i.e., low rise time and low overshoot at the same time) when the process output is required to assume a new value. Indeed, assume that the process variable is required to achieve a steady-state value y_1 starting from a steady-state value y_0. If a causal feedforward action is adopted, the control scheme of Figure 1.11, which comprises the feedforward approaches described until now, is based on the causal filtering of a step signal (of amplitude $y_1 - y_0$) by means of the system described by the transfer function $F(s)$. The resulting signal is then applied to the closed-loop system. Conversely, if an inversion approach is exploited, the scheme shown in Figure 5.16 is employed. In this case a step signal is not employed, but the knowledge in advance of y_1 is adopted by a command signal generator block to calculate a suitable command signal r to be applied to the closed-loop PID control system.

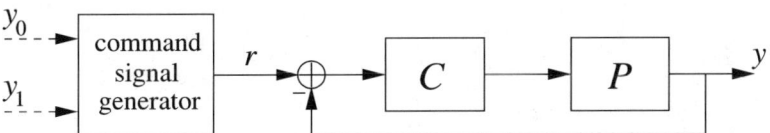

Fig. 5.16. Control scheme based on input-output inversion

5.4.2 Methodology

Modelling

The design methodology based on input-output inversion proposed in (Piazzi and Visioli, 2006) is based on a theoretical framework that might appear to be somewhat complicated. However, the theoretical development can be made transparent to the user and therefore the use of the technique does not impair the ease of use that is an essential requirement in the context of PID control. The fundamental passages are describe hereafter in some detail in order to understand better the underlying concepts of the overall methodology.
As a first step of the devised method, the process to be controlled (assumed to be self-regulating) is modelled as a FOPDT transfer function, *i.e.*:

$$P(s; K, T, L) = \frac{K}{Ts+1} e^{-Ls}, \qquad (5.23)$$

but then, in order to have a rational transfer function, the dead-time term is approximated by means of a second-order Padè approximation. In this way, the approximated process transfer function results to be:

$$\tilde{P}(s; K, T, L) = \frac{K}{Ts+1} \frac{1 - Ls/6 + L^2 s^2/12}{1 + Ls/6 + L^2 s^2/12}. \qquad (5.24)$$

Note that if the process is non self-regulating, it can be modelled as an integrator-plus-dead-time (IPDT) transfer function, *i.e.*:

$$\tilde{P}(s; K, L) = \frac{K}{s} \frac{1 - Ls/6 + L^2 s^2/12}{1 + Ls/6 + L^2 s^2/12}. \qquad (5.25)$$

Then the methodology is basically the same for the FOPDT case and details for this case are omitted hereafter (an example is presented in Section 5.4.3).

PID Controller Design

An output filtered PID controller in ideal form (1.23) is employed as a feedback controller. For the sake of clarity, its transfer function is recalled here:

$$C(s; K_p, T_i, T_d, T_f) = K_p \left(1 + \frac{1}{T_i s} + T_d s\right) \frac{1}{T_f s + 1}. \qquad (5.26)$$

The tuning of the parameters can be done according to any of the many methods proposed in the literature or even by a trial-and-error procedure. However, since the purpose of the overall procedure is the attainment of a high performance in the set-point following task, disregarding of the controller gains, it is sensible to select the PID parameters aiming only at obtaining a good load rejection performance.

Output Function Design

At this point, a desired output function that defines the transition from a setpoint value y_0 to another y_1 (to be performed in the time interval $[0, \tau]$) has to be selected. Without loss of generality and for the sake of clarity assume $y_0 = 0$. A sensible choice is to adopt a so-called "transition" polynomial (Piazzi and Visioli, 2001b), i.e., a polynomial function that satisfies boundary conditions and that is parameterised by the transition time τ. It is formally defined as

$$y_d(t) = c_{2k+1} t^{2k+1} + c_{2k} t^{2k} + \cdots + c_1 t + c_0 \tag{5.27}$$

The polynomial coefficients can be uniquely found by solving the following linear system, in which boundary conditions at the endpoints of interval $[0, \tau]$ are imposed:

$$\begin{cases} y_d(0) = 0; \quad y_d(\tau) = y_1 \\ y_d^{(1)}(0) = 0; \quad y_d^{(1)}(\tau) = 0 \\ \vdots \\ y_d^{(k)}(0) = 0; \quad y_d^{(k)}(\tau) = 0 \end{cases} \tag{5.28}$$

The results can be expressed in closed-form as follows ($t \in [0, \tau]$):

$$y_d(t; \tau) = y_1 \frac{(2k+1)!}{k!} \sum_{i=k+1}^{2k+1} \frac{(-1)^{i-k-1}}{i(i-k-1)!(2k+1-i)!} \left(\frac{t}{\tau}\right)^i \tag{5.29}$$

Expression (5.29) represents a monotonic function with neither undershooting nor overshooting and its use is therefore very appealing in a practical context. The order of the polynomial can be selected by imposing the order of continuity of the command input that results from the input-output inversion procedure (Piazzi and Visioli, 2001b). Specifically, since the plant is modelled as a FOPTD transfer function (see (5.23)), its relative degree is equal to one. Taking into account that the relative degree of the PID controller is zero, the relative degree of the overall closed-loop system is one. Thus, a third order polynomial ($k = 1$) suffices if a continuous command input function is required, i.e.:

$$y_d(t; \tau) = y_1 \left(-\frac{2}{\tau^3} t^3 + \frac{3}{\tau^2} t^2 \right) \quad t \in [0, \tau]. \tag{5.30}$$

Outside the interval $[0, \tau]$ the function $y(t; \tau)$ is equal to 0 for $t < 0$ and equal to y_1 for $t > \tau$.

Stable Input–Output Inversion Algorithm

Once the closed-loop system is designed and the desired output function is selected, the problem of finding the command signal $r(t; K, T, L, K_p, T_i, T_d, T_f, \tau)$

that provides the desired output function has to be solved. For the sake of clarity of notation, the dependence of the functions and of the resulting coefficients from the parameters K, T, L, K_p, T_i, T_d, T_f is omitted in the following analysis. The closed-loop transfer function be denoted as

$$H(s) := \frac{C(s)\tilde{P}(s)}{1+C(s)\tilde{P}(s)} = K_1 \frac{b(s)}{a(s)} \qquad (5.31)$$

where $b(s)$ and $a(s)$ are monic polynomials. As $H(s)$ is nonminimum phase, the straightforward inversion of the dynamics, namely, the calculation of $Y_d(s)/H(s)$, where $Y_d(s)$ is the Laplace transform of $y_d(t)$, would produce an unbounded command input function, which cannot be obviously adopted in practice. In other words, a stable dynamic inversion procedure is necessary, that is a bounded input function has to be found in order to produce the desired output (Piazzi and Visioli, 2005).

The numerator of the transfer function (5.31) can be rewritten as follows:

$$b(s) = b_-(s)b_+(s)$$

where $b_-(s)$ and $b_+(s)$ denote the polynomials associated to the zeros with negative real part (*i.e.*, those of the PID controller) and positive real part (*i.e.*, those of the Padè approximation) respectively. From (5.25) we have

$$b_+(s) = (s - z_R^+)^2 + z_I^{+^2} \qquad (5.32)$$

where $Z_R^+ = 3/L$, $Z_I^+ = \sqrt{3}/L$ correspond to the complex zeros $z_R^+ \pm j z_I^+ \in \mathbb{C}_+$. From (5.26) three cases can be distinguished (depending on the selected PID parameters):

$$b_-(s) = (s - z_1^-)(s - z_2^-) \qquad (5.33)$$

$$b_-(s) = (s - z^-)^2 \qquad (5.34)$$

$$b_-(s) = (s - z_R^-)^2 + z_I^{-^2} \qquad (5.35)$$

corresponding to real distinct zeros (5.33), real coincident zeros (5.34), and complex zeros (5.35) respectively. Now, consider the inverse system of (5.31) whose transfer function can be written as:

$$H(s)^{-1} = \gamma_0 + \gamma_1 s + H_0(s)$$

where γ_0 and γ_1 are suitable constants and $H_0(s)$, a strictly proper rational function, represents the zero dynamics. This can be uniquely decomposed according to

$$H_0(s) = H_0^-(s) + H_0^+(s) = \frac{c(s)}{b_-(s)} + \frac{d(s)}{b_+(s)}$$

5.4 Noncausal Feedforward Action: Continuous-time Case

where $c(s) = c_1 s + c_0$ and $d(s) = d_1 s + d_0$ are first-order polynomials with coefficients depending on K, T, L, K_p, T_i, T_d and T_f. The modes associated to $b^-(s)$ and $b^+(s)$ be denoted by $m_i^-(t)$, $i = 1, 2$, and by $m_i^+(t)$, $i = 1, 2$ respectively. More specifically, the unstable zero modes are given by

$$m_1^+(t) = e^{z_R^+ t} \cos z_I^+ t \qquad m_2^+(t) = e^{z_R^+ t} \sin z_I^+ t \qquad (5.36)$$

while the stable zero ones are given according to the cases (5.33), (5.34), and (5.35) by:

$$m_1^-(t) = e^{z_1^- t} \qquad m_2^-(t) = e^{z_2^- t} \qquad (5.37)$$

$$m_1^-(t) = e^{z^- t} \qquad m_2^-(t) = t e^{z^- t} \qquad (5.38)$$

$$m_1^-(t) = e^{z_R^- t} \cos z_I^- t \qquad m_2^-(t) = e^{z_R^- t} \sin z_I^- t \qquad (5.39)$$

With $\mathcal{L}$ the Laplace transform operator, define:

$$\eta_0^-(t) := \mathcal{L}^{-1}[H_0^-(s)]$$

and

$$\eta_0^+(t) := \mathcal{L}^{-1}[H_0^+(s)].$$

The following propositions and the following theorem represent the solution to the stable dynamic inversion problem.

Proposition 5.1.

$$\int_0^t \eta_0^+(t-v) y_d(v; \tau) dv = H_0^+(0) y_d(t; \tau) + \frac{1}{\tau^3} \left(p_1^+(\tau) m_1^+(t) + p_2^+(\tau) m_2^+(t) \right) + \frac{1}{\tau^3} T_0^+(t; \tau) \qquad (5.40)$$

where

$$T_0^+(t, \tau) = \begin{cases} s_0^+(t) + s_1^+(t) \tau & \text{if } t \in [0, \tau] \\ q_1^+(\tau) m_1^+(t-\tau) + q_2^+(\tau) m_2^+(t-\tau) & \text{if } t > \tau \end{cases} \qquad (5.41)$$

and $p_i^+(\tau)$, $q_i^+(\tau)$, $i = 1, 2$ are suitable τ-polynomials and $s_i^+(t)$, $i = 0, 1$ are suitable t-polynomials.

Proposition 5.2.

$$\int_0^t \eta_0^-(t-v) y_d(v; \tau) dv = H_0^-(0) y_d(t; \tau) + \frac{1}{\tau^3} \left(p_1^-(\tau) m_1^-(t) + p_2^-(\tau) m_2^-(t) \right) + \frac{1}{\tau^3} T_0^-(t, \tau) \qquad (5.42)$$

where

$$T_0^-(t, \tau) = \begin{cases} s_0^-(t) + s_1^-(t) \tau & \text{if } t \in [0, \tau] \\ q_1^-(\tau) m_1^-(t-\tau) + q_2^-(\tau) m_2^-(t-\tau) & \text{if } t > \tau \end{cases} \qquad (5.43)$$

and $p_i^-(\tau)$, $q_i^-(\tau)$, $i = 1, 2$ are suitable τ-polynomials and $s_i^-(t)$, $i = 0, 1$ are suitable t-polynomials.

Theorem 5.3. *The function* $r(t;\tau)$ *defined as*

$$r(t;\tau) = -\frac{1}{\tau^3} \left(p_1^+(\tau) m_1^+(t) + p_2^+(\tau) m_2^+(t) \right. \\ \left. - q_1^+(\tau) m_1^+(t-\tau) - q_2^+(\tau) m_2^+(t-\tau) \right) \quad \text{if } t < 0 \tag{5.44}$$

$$r(t;\tau) = \gamma_1 \dot{y}_d(t;\tau) + \gamma_0 y_d(t;\tau) + H_0(0) y_d(t;\tau) + \\ \frac{1}{\tau^3} \left(s_0^+(t) + s_0^-(t) + s_1^+(t)\tau + s_1^-(t)\tau - q_1^+(\tau) m_1^+(t-\tau) - \right. \\ \left. q_2^+(\tau) m_2^+(t-\tau) + p_1^-(\tau) m_1^-(t) + p_2^-(\tau) m_2^-(t) \right) \quad \text{if } t \in [0,\tau] \tag{5.45}$$

$$r(t;\tau) = \gamma_0 + H_0(0) + \frac{1}{\tau^3} \left(p_1^-(\tau) m_1^-(t) + p_2^-(\tau) m_2^-(t) \right. \\ \left. + q_1^-(\tau) m_1^-(t-\tau) + q_2^-(\tau) m_2^-(t-\tau) \right) \quad \text{if } t > \tau. \tag{5.46}$$

is bounded over $(-\infty, +\infty)$ *and* $r(t;\tau)$ *causes the desired output* $y_d(t;\tau)$.

Proofs of the above propositions and of the above theorem can be found in (Piazzi and Visioli, 2005).

Summarising, the determined function $r(t; K, T, L, K_p, T_i, T_d, T_f, \tau)$ exactly solves the stable inversion problem for FOPDT processes (in which the deadtime term has been substituted by a Padè approximation) controlled by a PID controller (5.26) and for a family of output functions, which depend on the free transition time τ.

Actually, from a practical point of view, since the synthesised function (5.44)–(5.46) is defined over the interval $(-\infty, +\infty)$, it is necessary to adopt a truncated function $r_a(t;\tau)$, resulting therefore in an approximate generation of the desired output $y_d(t;\tau)$. In particular, a preactuation time t_s and a postactuation time t_f can be selected so that $r_a(t;\tau) = 0$ for $t < t_s$ and $r_a(t;\tau) = y_1$ for $t > t_p$. By taking into account that the preactuation and postactuation inputs (*i.e.*, the input defined for $t < 0$ and $t > \tau$ respectively) converge exponentially to zero at time $t \to -\infty$ and to y_1 at time $t \to +\infty$, an arbitrarily precise approximation can be accomplished (Piazzi and Visioli, 2005). Practically, the method suggested in (Perez and Devasia, 2003) can be adopted. It consists of selecting

$$t_s = -\frac{10}{D_{rhp}} \tag{5.47}$$

and

$$t_p = \frac{10}{D_{lhp}} \tag{5.48}$$

where D_{rhp} and D_{lhp} are the minimum distance of the right- and left-half plane poles respectively from the imaginary axis of the complex plane. Hence, the approximate command signal to be actually used is

$$r_a(t;\tau) := \begin{cases} 0 & \text{for } t < t_s \\ r(t;\tau) & \text{for } t_s \leq t \leq t_f \\ y_1 & \text{for } t > t_f. \end{cases} \quad (5.49)$$

It is worth highlighting that the preactuation time depends only on the (apparent) dead time of the process, as this determines the unstable zeros of the closed-loop systems by means of the Padè approximation. Conversely, the postactuation time depends on the tuning of the PID parameters because the stable zeros of the closed-loop systems are those of the controller.

Discussion

The presented stable input-output inversion procedure can be performed by means of a symbolic computation, *i.e.*, a closed-form expression of the command input function $r(t; K, T, L, K_p, T_i, T_d, T_f, \tau)$ results. Indeed, the actual command signal to be applied for a given plant and a given controller is determined by substituting the actual value of the parameters into the resulting closed-form expression and this actually motivates its strong appeal in the context of PID control. In this framework, the choice of using a second-order Padè approximation is motivated, from one side, by keeping the expression of $r(; K, T, L, K_p, T_i, T_d, T_f, \tau)$ as simple as possible and, from the other side, by providing an approximation as good as possible, since the basic rationale of this method is to apply a model-based feedforward control action.

In any case, it is worth noting that the presented inversion procedure is based on a general one (Piazzi and Visioli, 2005), where $H(s)$ can be the rational transfer function of any (stable) system, provided that there are not purely imaginary zeros. Thus, as already mentioned, the proposed approach can be straightforwardly applied also to integral (and unstable) processes $\tilde{P}(s)$, as it is based on the inversion of the dynamics of the closed-loop system $H(s)$. Analogously, the same method can be trivially extended to PI, P and PD control.

It appears also that the devised method can be extended also to high-order processes. Thus, a more accurate model of the process, if available, can be fully exploited. However, in this case, the inversion procedure has to be performed on purpose. Conversely, if a FOPDT (or a IPDT) model is employed, the determined general closed-form expression of $r(t; K, T, L, K_p, T_i, T_d, T_f, \tau)$ can be used.

Once the PID controller has been tuned, the only free design parameter is the transition time τ. Its role is basically the same of the transition time in the causal nonlinear feedforward method described in Section 5.3, namely, it allows to handle the trade-off between performance, robustness and control activity. It can be selected therefore by applying an analogous reasoning. However, since a closed-form expression of the control variable can be easily derived, the transition time can be also determined by solving an optimisation problem where its value has to be minimised subject to actuator constraints.

5.4.3 Simulation Results

In the following examples the process output has to perform a transition from 0 to $y_1 = 1$. The methodology is evaluated with processes with different dynamics.

FOPDT Process

Consider the following FOPDT process:

$$P_1(s) = \frac{1}{10s + 1} e^{-6s}. \tag{5.50}$$

To prove the effectiveness of the method with different PID tunings, three sets of PID parameters have been considered, namely, the one given by the Ziegler–Nichols step response PID formula ($K_p = 2$, $T_i = 12$, $T_d = 3$), the one given by the Ziegler–Nichols step response PI formula ($K_p = 1.5$, $T_i = 18$, $T_d = 0$), and the one that results from the minimization of the ISTE integral criterion for the load disturbance rejection (Zhuang and Atherton, 1993) ($K_p = 2.41$, $T_i = 7.33$, $T_d = 2.74$). In the first and in the third case it has been set $T_f = 0.01$ (for a PI controller the filter is not necessary). To give a clear idea of the different performance achieved with the considered tuning rules, the set-point step response is plotted in Figure 5.17. Note the high initial value of the control variable due to the poor filtering of the derivative action. Saturation limits have not been applied in order to avoid to bias the results. Then, the noncausal feedforward approach has been applied by always fixing the transition time to $\tau = 10$. The resulting value of the preactuation and postactuation times in the three cases are $t_s = -20$ and $t_p = 60$, $t_p = 180$, $t_p = 54.8$ respectively (note that, for convenience, the time axis has been properly shifted in order to have $t_s = 0$). The determined command functions are reported in Figure 5.18 and the corresponding process outputs and control variables are plotted in Figure 5.19. It appears that the inversion-based methodology is able to provide low rise times and low overshoots at the same time but, most of all, is able to provide basically the same response despite a very different PI(D) tuning, as it is evidenced by the very different step responses they provide. Note that this is achieved with a much lower control effort with respect to the classic case since the step signal is substituted by a smoother signal.

High-order Process

As a second example the following high-order process is considered:

$$P_2(s) = \frac{1}{(s+1)^8}. \tag{5.51}$$

By applying the area method (see Chapter 7), a FOPDT transfer function has been estimated, resulting in $K = 1$, $T = 3.04$ and $L = 4.97$. With respect

5.4 Noncausal Feedforward Action: Continuous-time Case   117

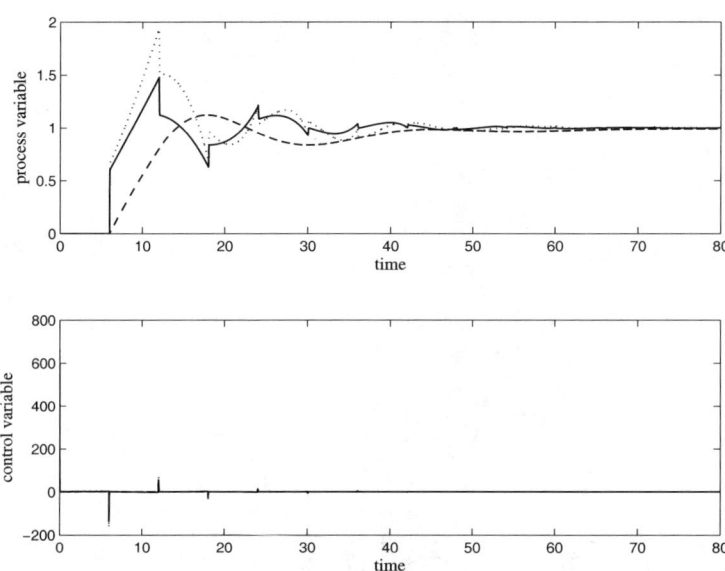

Fig. 5.17. Set-point step response for process $P_1(s)$. Solid line: Ziegler–Nichols PID tuning; dashed line: Ziegler–Nichols PI tuning; dotted line: ISTE criterion tuning.

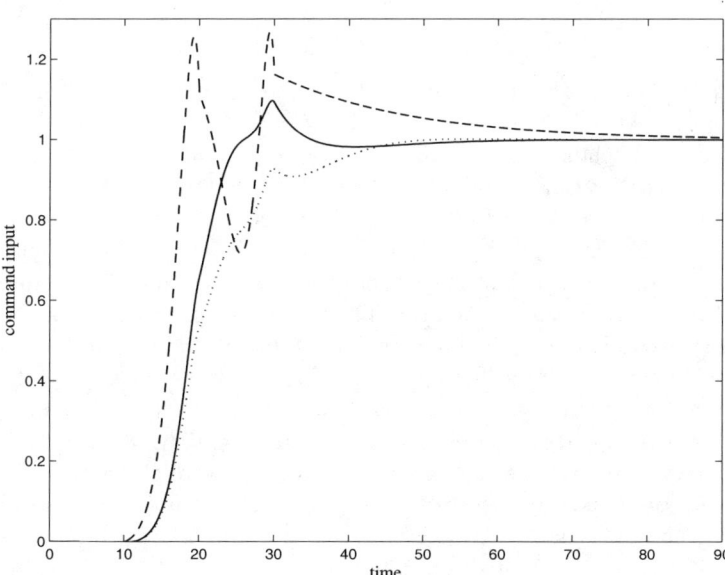

Fig. 5.18. Inversion-based command input for process $P_1(s)$. Solid line: Ziegler–Nichols PID tuning; dashed line: Ziegler–Nichols PI tuning; dotted line: ISTE criterion tuning.

118   5 Use of a Feedforward Action

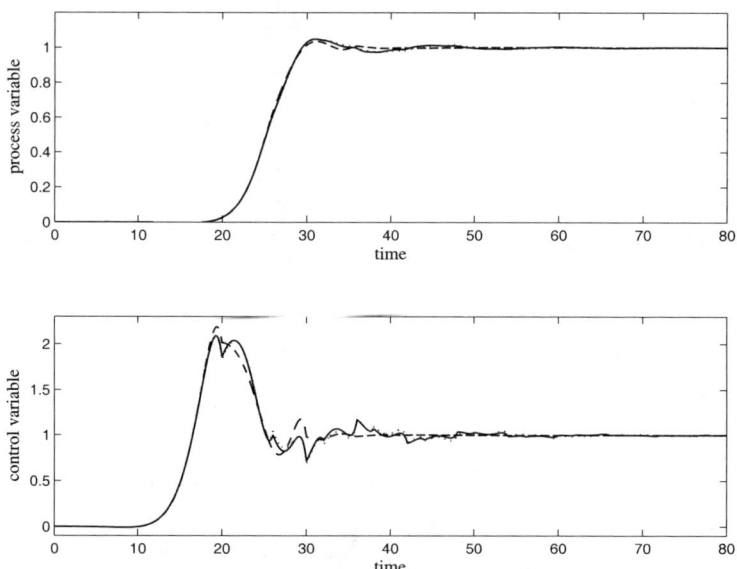

Fig. 5.19. Response with the noncausal feedforward approach for process $P_1(s)$. Solid line: Ziegler–Nichols PID tuning; dashed line: Ziegler–Nichols PI tuning; dotted line: ISTE criterion tuning.

to these parameters, the same tuning formulae as for the FOPDT example has been adopted, resulting in $K_p = 0.73$, $T_i = 9.93$, $T_d = 2.48$ ($T_f = 0.01$) for the Ziegler–Nichols PID tuning, $K_p = 0.55$, $T_i = 14.90$, $T_d = 0$ for the Ziegler–Nichols PI tuning, and $K_p = 1.06$, $T_i = 4.26$, $T_d = 2.48$ ($T_f = 0.01$) for the minimization of the ISTE integral criterion. Set-point step responses are plotted in Figure 5.20. Note that the control variable is shown just for the beginning of the transient response, otherwise it cannot be evaluated because of the scaling. In any case, the different performance achieved without the feedforward action clearly emerges. The noncausal feedforward action has then been determined by selecting a transition time equal to $\tau = 20$ for all the adopted tuning rules. The resulting values of the preactuation and postactuation times in the three cases are $t_s = -16.57$ and $t_p = 51.23$, $t_p = 149$, $t_p = 49.6$ respectively. The command functions determined by applying the input-output inversion procedure are reported in Figure 5.21 and the corresponding control system responses are plotted in Figure 5.22. The same considerations of the previous example can be done also on this case. Further, the robustness of the method with respect to model uncertainties appears. Indeed, the feedback controller reduces the effects of the model uncertainties in the frequency range of the command input and therefore the application of such command input to the closed-loop system is effective.

5.4 Noncausal Feedforward Action: Continuous-time Case

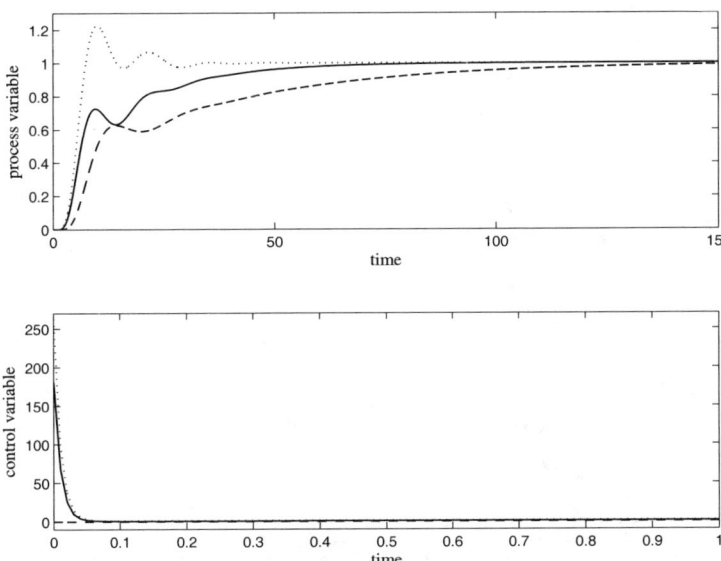

Fig. 5.20. Set-point step response for process $P_2(s)$. Solid line: Ziegler–Nichols PID tuning; dashed line: Ziegler–Nichols PI tuning; dotted line: ISTE criterion tuning.

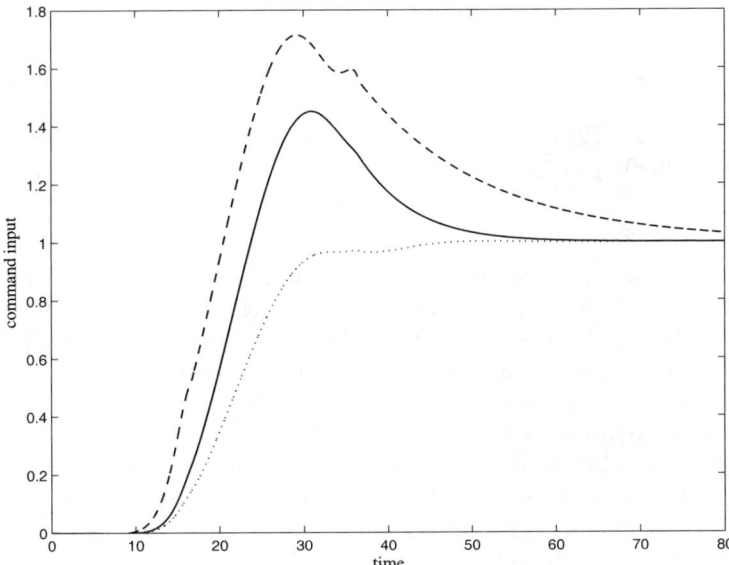

Fig. 5.21. Inversion-based command input for process $P_2(s)$. Solid line: Ziegler–Nichols PID tuning; dashed line: Ziegler–Nichols PI tuning; dotted line: ISTE criterion tuning.

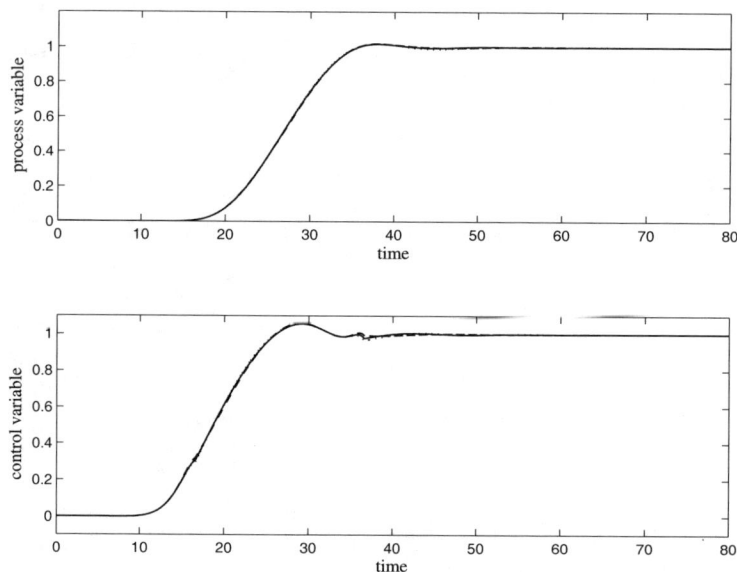

Fig. 5.22. Response with the noncausal feedforward approach for process $P_2(s)$. Solid line: Ziegler–Nichols PID tuning; dashed line: Ziegler–Nichols PI tuning; dotted line: ISTE criterion tuning.

IPDT Process

As an example for IPDT processes, the following process is considered (Wang and Cluett, 2000):

$$P_3(s) = \frac{0.0506}{s}e^{-6s}. \tag{5.52}$$

Two sets of PID parameters have been adopted, namely, the one based on the minimisation of the integrated square error for set-point responses (Visioli, 2001a) (which is actually a PD controller with $K_p = 3.394$ and $T_d = 2.94$) and the one proposed in (Wang and Cluett, 2000) ($K_p = 2.0123$, $T_i = 31.2030$, $T_d = 1.5674$). In both cases it has been set again $T_f = 0.01$. The classic set-point step responses are plotted in Figure 5.23.

The noncausal feedforward action has been determined by setting $\tau = 10$. The resulting value of the preactuation and postactuation times are $t_s = -20.01$s and $t_p = 29.41$ and $t_p = 295.48$ for the PD and PID case respectively. The determined command functions are reported in Figure 5.24 and the corresponding control system responses are plotted in Figure 5.25.

It appears that the same considerations that have been done for self-regulating processes can be done also for integrating processes. Indeed, almost the same performance is obtained despite the different tuning strategies that have been employed.

5.4 Noncausal Feedforward Action: Continuous-time Case   121

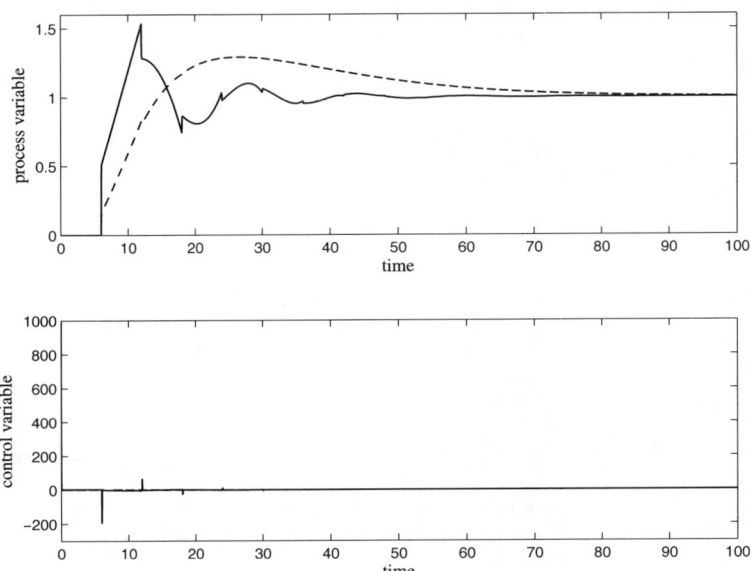

Fig. 5.23. Set-point step response for process $P_3(s)$. Solid line: PD control; dashed line: PID control.

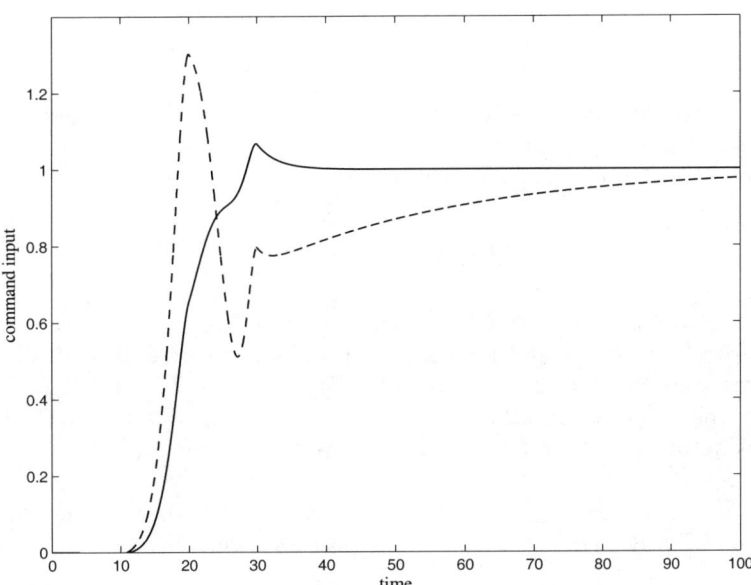

Fig. 5.24. Inversion-based command input for process $P_3(s)$. Solid line: PD control; dashed line: PID control.

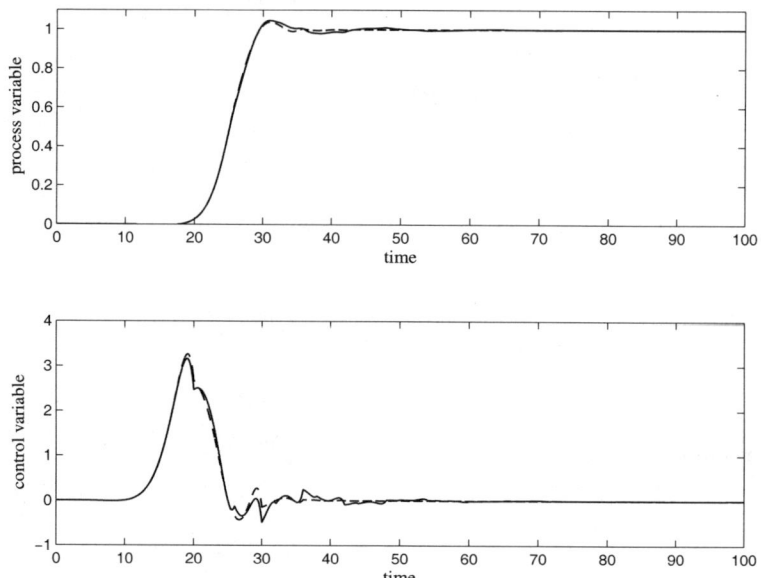

Fig. 5.25. Response with the noncausal feedforward approach for process $P_3(s)$. Solid line: PD control; dashed line: PID control.

Unstable Process

The presented noncausal feedforward technique can be straightforwardly applied also to unstable processes.
Consider for example the process model by the following transfer function:

$$P_4(s) = \frac{1}{s-1}e^{-0.2s}. \tag{5.53}$$

Also in this case, two sets of PID parameters have been selected, the one that results from the minimisation of the integrated square error of the load disturbance response (Visioli, 2001a) ($K_p = 6.85$, $T_i = 0.36$, $T_d = 0.12$, $T_f = 0.01$) and the one proposed in (Ho and Xu, 1998), i.e., $K_p = 3.46$ and $T_i = 1.47$, which is actually a PI controller. The desired transition time has been fixed to $\tau = 1$.

Results are shown in Figures 5.26–5.28. It can be seen that the methodology retains its effectiveness also for unstable processes and therefore its generality appears.

5.4 Noncausal Feedforward Action: Continuous-time Case

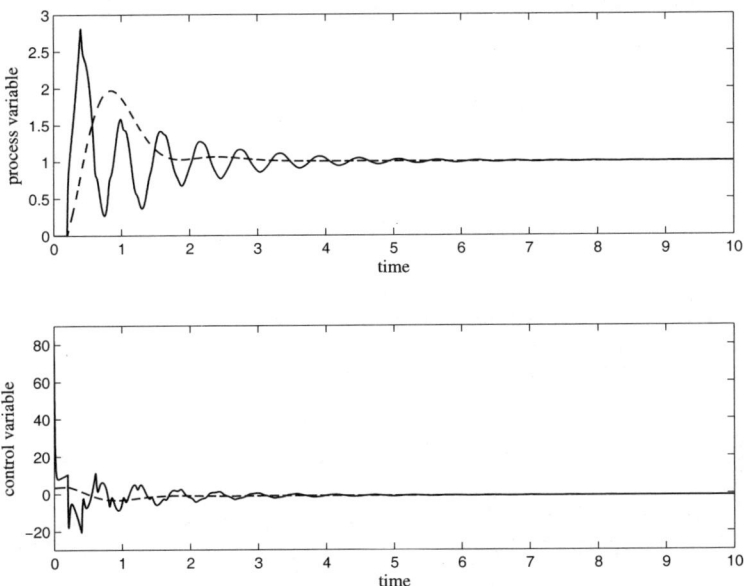

Fig. 5.26. Set-point step response for process $P_4(s)$. Solid line: PID control; dashed line: PI control.

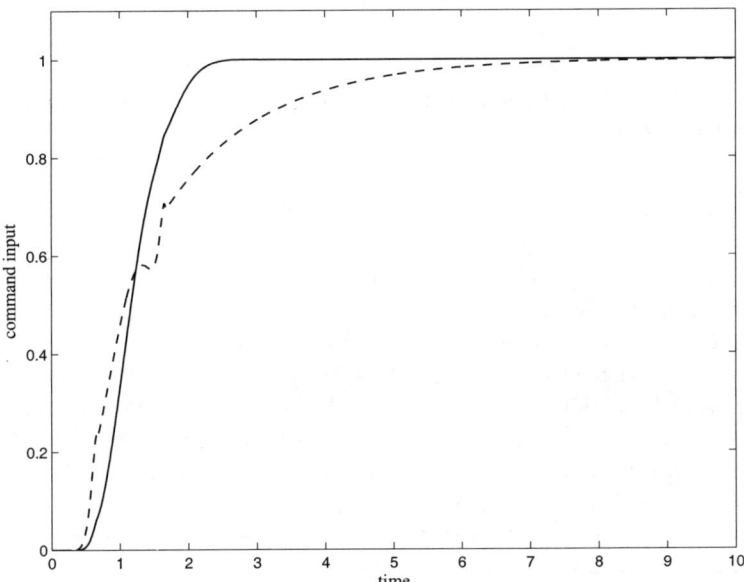

Fig. 5.27. Inversion-based command input for process $P_4(s)$. Solid line: PID control; dashed line: PI control.

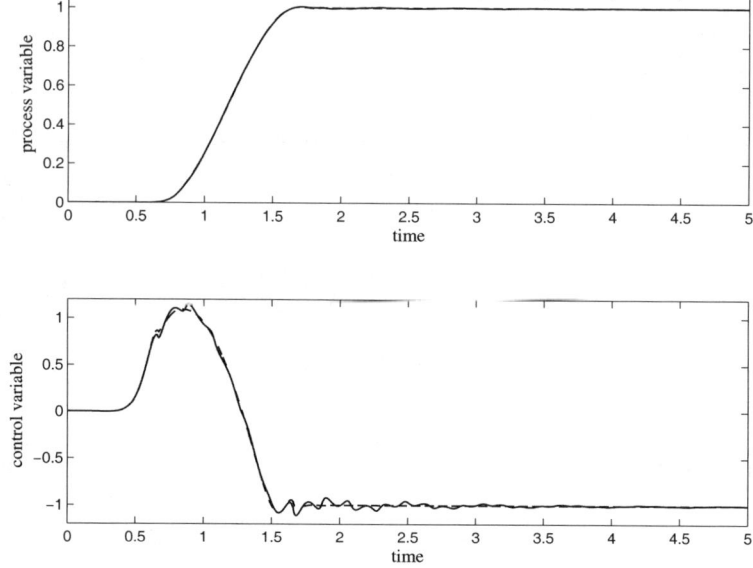

Fig. 5.28. Response with the noncausal feedforward approach for process $P_4(s)$. Solid line: PD control; dashed line: PID control.

Role of the Parameter τ

In order to clarify better the role of the transition time τ, the following result is presented. Consider the process

$$P_5(s) = \frac{1}{(s+1)^4}. \tag{5.54}$$

The PID controller is tuned by means of a genetic algorithm in order to achieve the minimum integrated absolute error for a load step disturbance. It results $K_p = 3.50$, $T_i = 1.77$, and $T_d = 1.47$ ($T_f = 0.01$). Then, the area method has been adopted to estimate a FOPDT transfer function: it results $K = 1$, $T = 2.12$ and $L = 1.89$. The process variable and the control variable resulting from the application of the noncausal feedforward approach with a value of the transition time τ that ranges from 1 to 10 are reported in Figure 5.29 and 5.30 respectively. It can be easily seen that when the transition time increases, the obtained process output is more similar to the desired one (expressed by the polynomial form (5.30)) and the control effort decreases.

Finally, it is worth stressing that the robustness of the (continuous-time) noncausal approach with respect to the identification method adopted for the estimation of the FOPDT (or IPDT) transfer function can be verified in (Visioli and Piazzi, 2005).

5.4 Noncausal Feedforward Action: Continuous-time Case 125

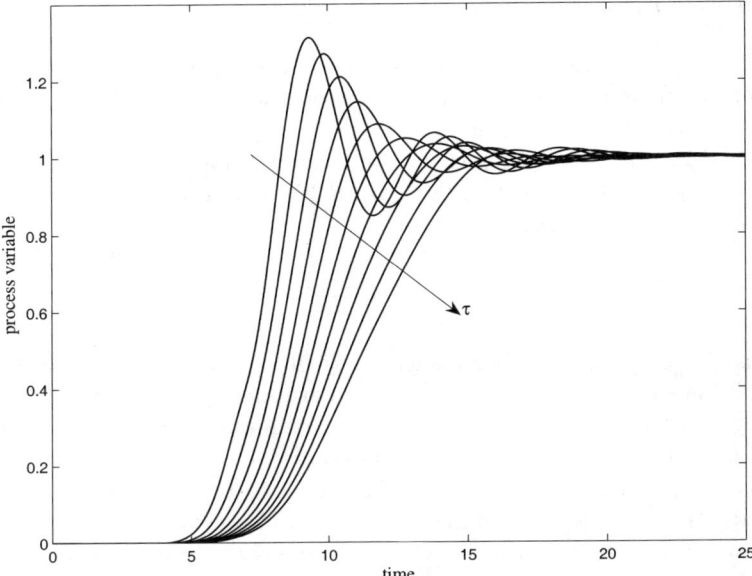

Fig. 5.29. Process variable for process $P_5(s)$ with different value of the transition time τ

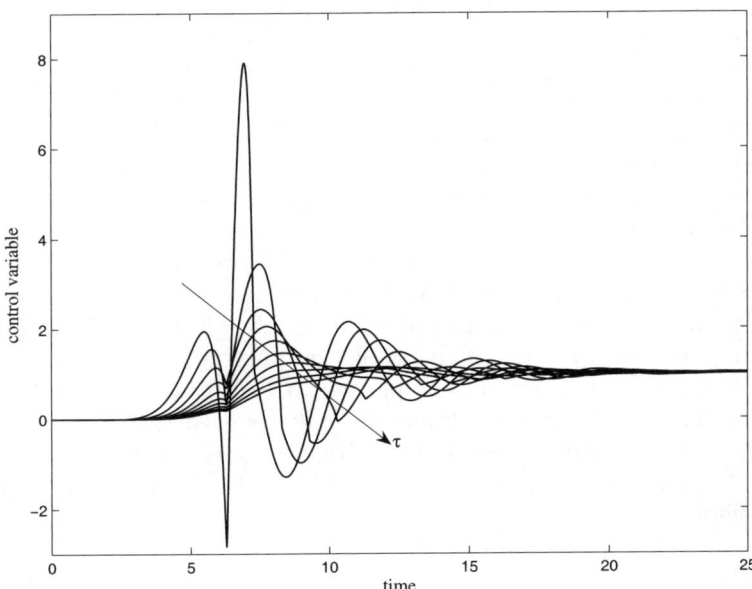

Fig. 5.30. Control variable for process $P_5(s)$ with different value of the transition time τ

5.4.4 Experimental Results

The effectiveness of the (continuous-time) noncausal feedforward approach has been verified by applying it to a level control task by means of the laboratory equipment described in Section A.1.

In particular, a single tank has been considered and a transition from the initial level $y_0 = 2$ V to the final level $y_1 = 3$ V has been required. The model of the system has been estimated as

$$P(s) = \frac{1.93}{26s + 1} e^{-s}. \tag{5.55}$$

Then, a PI controller has been selected with $K_p = 2.5$ and $T_i = 8.3$. The noncausal feedforward action has been determined by selecting a transition time of $\tau = 10$ s and $\tau = 20$ s. The resulting command input $r(t)$ is plotted in both cases in Figure 5.31. The obtained process variable and controller output in the two experiments are plotted in Figures 5.32 and 5.33, respectively. The achieved performance is compared with that obtained by applying a step to the set-point signal. Note that the actual process input saturates at 5 V.

It can be seen that, as expected, with the smallest value of the transition time the rise time is the smallest (almost the same of the step response), but the control effort is the highest (actually higher than that obtained with the step: this can be explained by evaluating the determined command input that exceed the desired new steady-state value during the transient).

The noncausal approach allows in any case the reduction the overshoot. The fact that almost the same overshoot is achieved with both values of the desired transition time is probably due to the nonlinear dynamics that has been neglected in the design phase.

The methodology has also been applied by adding (via software) an additional dead time of 10 s to the process input (the same model (5.55) has been adopted by changing the dead time from 1 s to 11 s). The PI parameters have been selected as $K_p = 1.24$, $T_i = 31$. Again, the two values of $\tau = 10$ s and $\tau = 20$ s for the transition time have been selected.

The calculated command input signals are reported in Figure 5.34. The corresponding process variables and controller outputs, compared again with the case of a set-point step signal, are plotted in Figure 5.35 and 5.36 respectively. In this case the inversion-based approach outperforms the standard one, by avoiding significant oscillations, despite the rise time is not increased for $\tau = 10$ s and only slightly increased for $\tau = 20$ s (in this latter case less control effort is required).

5.4 Noncausal Feedforward Action: Continuous-time Case

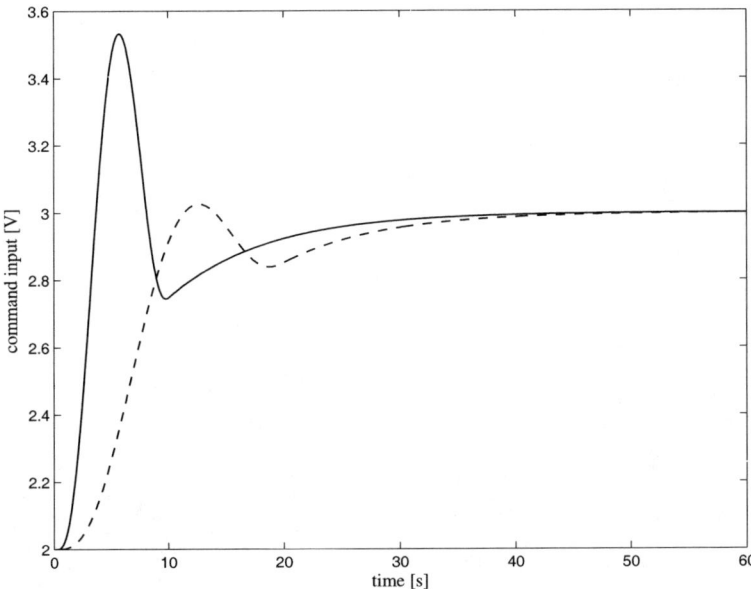

Fig. 5.31. Inversion-based command input for the level control task. Solid line: $\tau = 10$; dashed line: $\tau = 20$.

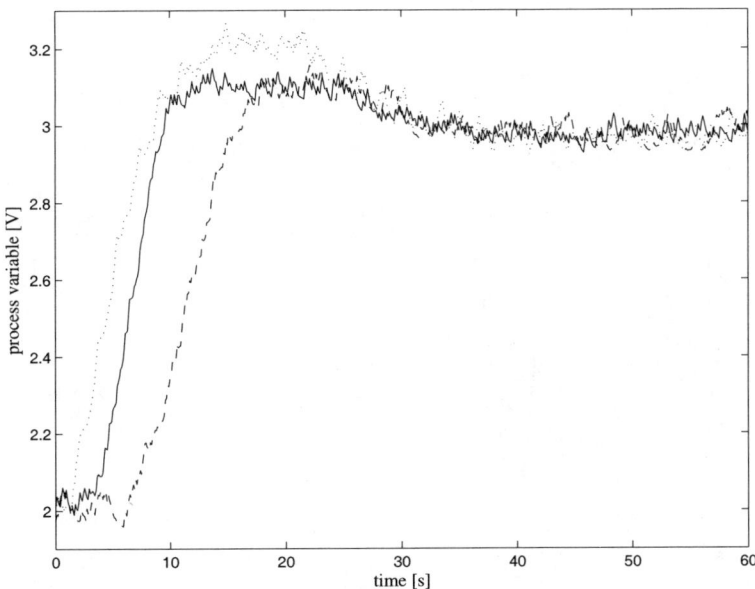

Fig. 5.32. Process variable for the level control task. Solid line: noncausal approach with $\tau = 10$; dashed line: noncausal approach with $\tau = 20$; dotted line: step response.

128   5 Use of a Feedforward Action

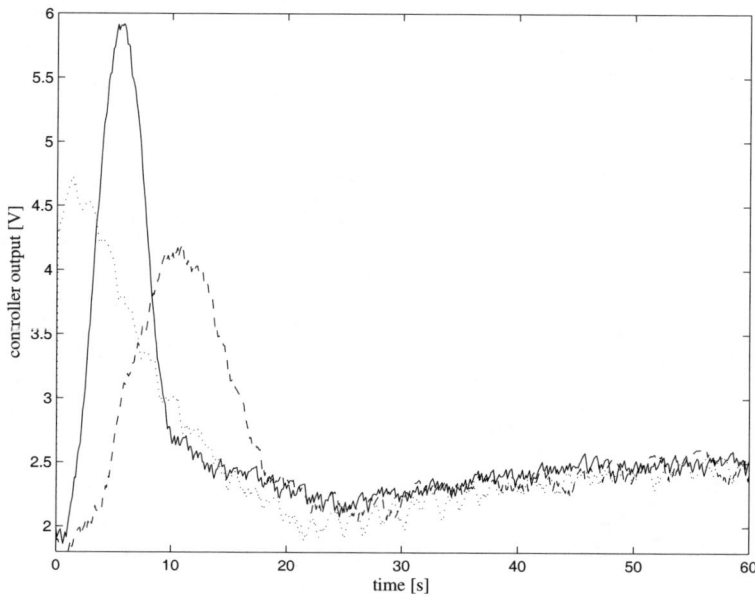

Fig. 5.33. Controller output for the level control task. Solid line: noncausal approach with $\tau = 10$; dashed line: noncausal approach with $\tau = 20$; dotted line: step response.

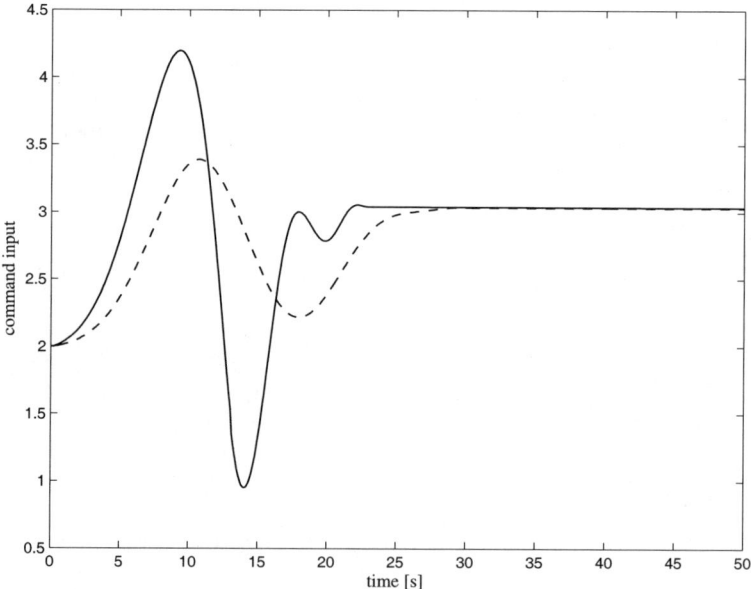

Fig. 5.34. Inversion-based command input for the level control task with additional dead time. Solid line: $\tau = 10$; dashed line: $\tau = 20$.

5.4 Noncausal Feedforward Action: Continuous-time Case 129

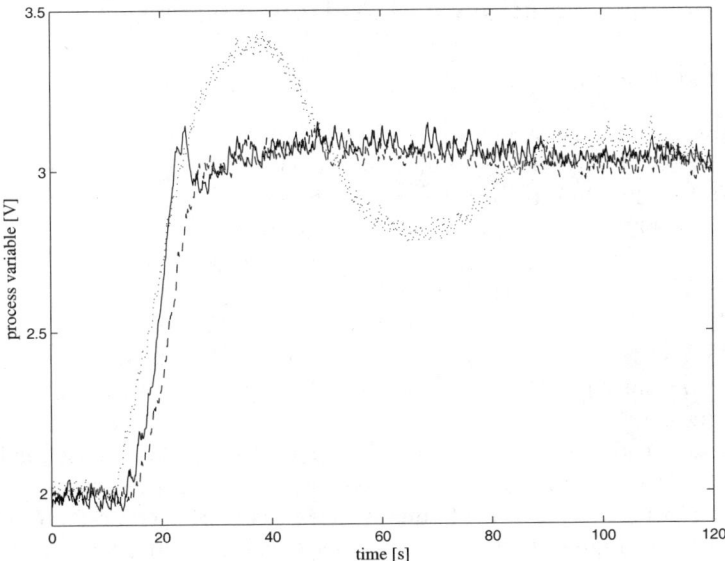

Fig. 5.35. Process variable for the level control task with additional dead time. Solid line: noncausal approach with $\tau = 10$; dashed line: noncausal approach with $\tau = 20$; dotted line: step response.

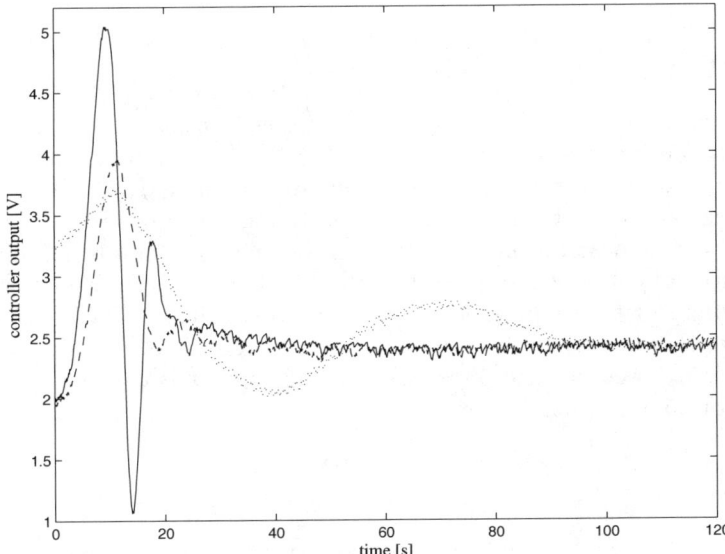

Fig. 5.36. Controller output for the level control task with additional dead time. Solid line: noncausal approach with $\tau = 10$; dashed line: noncausal approach with $\tau = 20$; dotted line: step response.

5.5 Noncausal Feedforward Action: Discrete-time Case

5.5.1 Methodology

A noncausal feedforward action can be designed also in a different context, namely, by inverting the dynamics of the closed-loop system after having identified it in the discrete-time framework by means of a step response (Visioli and Piazzi, 2003).

Consider again the scheme shown in Figure 5.16. Actually, the controller C can be of any type (provided that the closed-loop system is asimptotically stable), but for the sake of simplicity it is assumed that it is a PID controller. As for the method described in Section 5.4, the aim is to find the command function $r(t)$ that produces a desired system output transition from y_0 to y_1, starting from time $t = 0$, but here no *a priori* knowledge on the process model is assumed. Despite the process and the controller are defined in the continuous-time domain, sampled data are considered in the following analysis (actually, nowadays the use of microprocessors is the common practice in industrial environments). It is assumed that the sampling time T has been chosen suitably by any standard technique (Åström and Wittenmark, 1997). An identification experiment can be easily performed by applying a step signal to the input of the closed-loop system. A closed-loop system model can then be obtained by considering the truncated response ($t \in \{T, 2T, \ldots, NT\}$):

$$y(t) = y_0 + g_{t/T} r(0) + \sum_{i=1}^{\frac{t}{T}-1} g_i \left[r(t - iT) - r(t - (i+1)T) \right] \qquad (5.56)$$

where $g_i := g(iT)$, $i = 1, \ldots, N$ are the sampled output values in response to a unit-step input (see Figure 5.37) and $r(t)$ is the system input. For the sake of simplicity and without loss of generality, assume $y_0 = 0$. The number N of parameters has to be taken sufficiently high in order to allow a sufficiently accurate description of the system, but not too high to minimise the computational effort of the control strategy. From a practical point of view, the sampling of the step response in order to obtain parameters g_i should stop when the process output remains close to its steady-state value for a sufficiently long time. For the presented methodology, it is convenient to write Expression (5.56) in matrix form:

$$Y = GR \qquad (5.57)$$

where

$$Y = \begin{bmatrix} y(T) \\ y(2T) \\ y(3T) \\ \vdots \\ y(NT) \end{bmatrix}$$

5.5 Noncausal Feedforward Action: Discrete-time Case

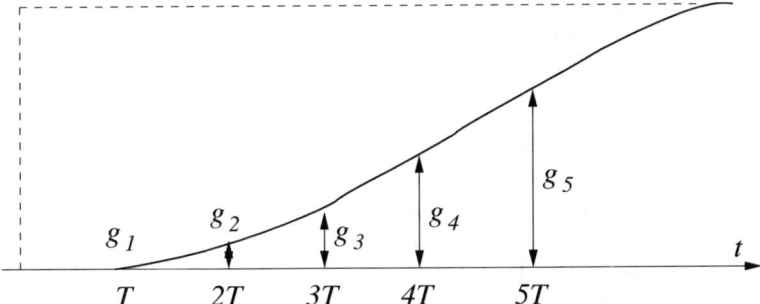

Fig. 5.37. Model coefficients based on step response

$$G = \begin{bmatrix} g_1 & 0 & 0 & \cdots & 0 \\ -g_1 + g_2 & g_1 & 0 & \cdots & 0 \\ -g_2 + g_3 & -g_1 + g_2 & g_1 & \cdots & 0 \\ \vdots & \vdots & \vdots & \ddots & 0 \\ -g_{N-1} + g_N & -g_{N-2} + g_{N-1} & -g_{N-3} + g_{N-2} & \cdots & g_1 \end{bmatrix}$$

and

$$R = \begin{bmatrix} r(0) \\ r(T) \\ r(2T) \\ \vdots \\ r((N-1)T) \end{bmatrix}$$

It is worth noting that in many cases it might not be necessary to perform an *ad hoc* identification experiment (*i.e.*, to stop the normal process operations) in order to apply the presented methodology. In fact, as the model is obtained by evaluating a standard closed-loop step response, data taken from an output transition performed during routine process operations can be adopted. Obviously, it is important that the collected data be representative of a true step response (and therefore operations such as filtering and detrending might be necessary (Leva *et al.*, 2001)) and if an unmeasured load disturbance occurs during the transient response, they should not be adopted. In this context, it can be useful to adopt the method proposed in (Hägglund and Åström, 2000) to detect load disturbances (see Section 8.4.1).

The desired output function is chosen again as a transition polynomial (5.29). In contrast with the continuous-time case, here its order can be chosen arbitrarily. Indeed, the order of the polynomial can be selected in order to handle the trade-off between the need to decrease the rise time and the need to decrease the control effort, taking into account that the rise time decreases and the control effort increases when the order of the polynomial increases. In general, a good choice in this context is to select $k = 2$, *i.e.*, the desired output function is:

$$y_d(t;\tau) = \begin{cases} y_1 \left(\dfrac{6}{\tau^5}t^5 - \dfrac{15}{\tau^4}t^4 + \dfrac{10}{\tau^3}t^3 \right) & \text{if } 0 \leq t \leq \tau \\ y_1 & \text{if } t > \tau \end{cases} \quad . \tag{5.58}$$

Regarding the choice of the value of the transition time τ, the same considerations done in the continuous-time case can be applied also in this case.

Once the desired output function has been selected, i.e., the array Y_d has been constructed, then the corresponding closed-loop system input $r(t)$ that causes $y_d(t;\tau)$ can be easily determined by simply inverting the system using Expression (5.57). In order for matrix G to be invertible by a standard numeric algorithm, it should be well-conditioned, for example there must not be a row (or a column) where all the elements are very small with respect to the elements of other rows (or columns). This happens when the process has a true dead time or an apparent dead time (i.e., when the process is of high order), which causes some of the first sampled output values g_i of the step response to be null or almost null. Thus, denote by k the number of the first rows of G in which all the elements are less than a selected threshold ε. Then, matrix $\hat{G}$ can be obtained by removing the first k rows and the last k columns from G. Subsequently, by evaluating $y_d(t;\tau)$ at the first $N-k$ sampling time intervals, the array $Y_d = [y_d(T;\tau) \quad y_d(2T;\tau) \quad \cdots \quad y_d((N-k)T;\tau)]^T$ can be easily constructed. The first $N-k$ values of the command reference input are then determined by applying the following expression:

$$\hat{R} = [r(T) \quad r(2T) \quad \cdots \quad r((N-k)T)]^T = \hat{G}^{-1}Y_d. \tag{5.59}$$

In this way, the input function can be calculated by simply determining the inverse of a matrix, which can be performed by using different algorithms (see for example (Press et al., 1995)).

Note that if the sampling time T and the value of N have been selected appropriately, as well as the value of τ, then the last element of the array $\hat{R}$ actually corresponds to the steady-state value of the input and therefore the value of $r((N-k)T)$ can be applied to the closed-loop system for $t > (N-k)T$. Note also that, since the first k rows and the last k columns have been removed from matrix G, the output function obtained is delayed by kT with respect to the desired one. Actually, the dead time is removed in the model of the closed-loop system transfer function adopted in the dynamic inversion.

In the presence of measurement noise, as is always the case in practical applications, the method can be successfully applied provided that the step response function employed for the identification of the closed-loop system model (5.56) is appropriately filtered. Since the required filtering can be performed off-line, a zero-phase noncausal filter can be applied in order to avoid a phase distortion. Further, the presence of the noise has to be considered when matrix $\hat{G}$ is constructed from G. Actually, due to the noise measurements, it is sensible to redefine parameter ε as a noise band NB (Åström et al., 1993), i.e., a threshold value that determines, as before, whether the sampled value g_i has to be discarded. Specifically, if $|g_i| < NB$, then g_i is considered to be zero in the construction of matrix $\hat{G}$. The value of NB can be easily selected by monitoring that process output for a sufficiently long time when the process is at steady-state.

Finally, it is worth stressing that, as in the continuous-time case, the inversion-based design of the feedforward action is independent of the PID design. It is therefore convenient to tune the controller in order to guarantee good load disturbance performance, if this is of concern. In fact, even if this implies that the predicted closed-loop step response is unsatisfactory, the feedforward action is capable to provide an output transition with low rise time and low overshoot.

5.5.2 Simulation Results

Some simulation examples are presented in order to illustrate the noncausal feedforward methodology in the discrete-time case and to evaluate its effectiveness. For the sake of clarity, measurement noise is taken into account only in the last example. For all the examples presented the PID controller has an output-filtered ideal structure (with the derivative action applied to the process variable) described by the following expression:

$$U(s) = K_p \left(E(s) + \frac{1}{T_i s} E(s) - T_d s Y(s) \right) \frac{1}{T_f s + 1} \quad (5.60)$$

where $U(s)$, $E(s)$ and $Y(s)$ are clearly the Laplace transform of the control variable, control error and process output respectively. The required output transition is from $y_0 = 0$ to $y_1 = 1$.

FOPDT Process

The following FOPDT process is considered:

$$P_1(s) = \frac{1}{10s + 1} e^{-5s}. \quad (5.61)$$

The PID controller has then been tuned by fixing $K_p = 2.61$, $T_i = 10.05$, $T_d = 2.51$, and $T_f = 0.01$, and the sampling time has been fixed to 0.5 s. In order to obtain the closed-loop system model, $N = 161$ samples of the step response have been evaluated and matrix G has been constructed accordingly. By fixing $\varepsilon = 0.01$, it results $k = 11$, i.e., the first 11 rows and the last 11 columns of G have been removed, therefore obtaining matrix $\hat{G}$ of dimension 150×150. The desired output array has been constructed by selecting $\tau = 5$ s. Then, the command input that substitutes the step signal has been determined by inverting the system. The response of the control system in the two cases is plotted in Figure 5.38, while the determined command input is shown in Figure 5.39. A significant improvement of the set-point following performance appears, although the control effort does not increase by applying the inversion-based command input.

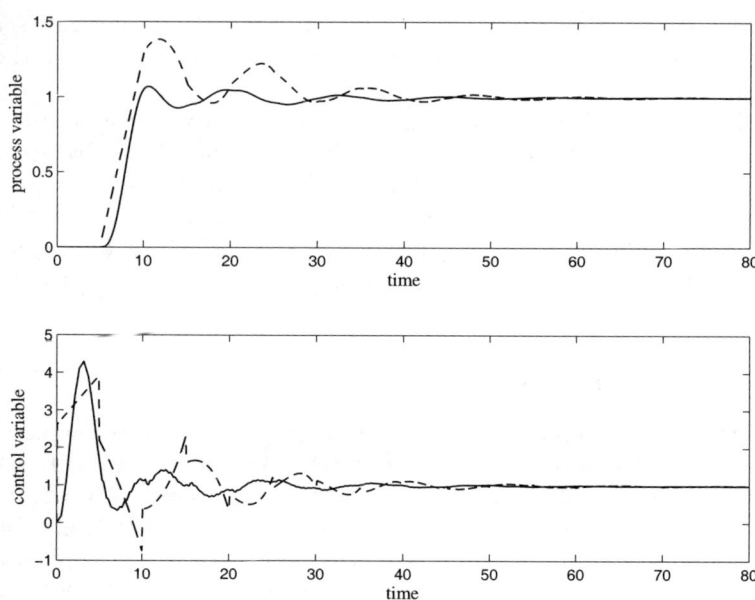

Fig. 5.38. Response before and after the application of the noncausal feedforward approach for process $P_1(s)$. Solid line: step response; dashed line: noncausal feedforward command input response.

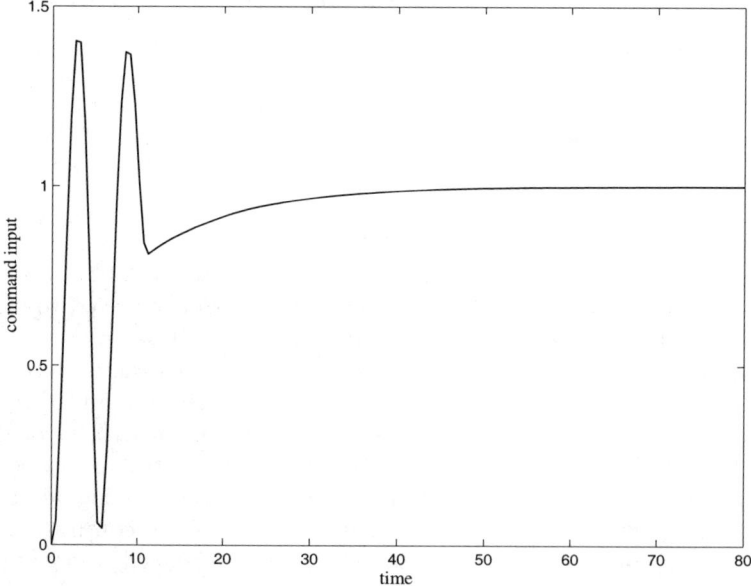

Fig. 5.39. Inversion-based command input for process $P_1(s)$

High-order Process

As a second example, the high-order process

$$P_2(s) = \frac{1}{(s+1)^8} e^{-5s} \quad (5.62)$$

is considered. The selected PID parameters are $K_p = 0.37$, $T_i = 19.93$, $T_d = 4.98$ and $T_f = 0.01$. The sampling time has been fixed to 2 s. A number $N = 101$ of output samples has been used to model the closed-loop system. By fixing again $\varepsilon = 0.01$ it results $k = 5$ and therefore the obtained $\hat{G}$ is of dimension 96×96. The (initial) set-point step response and the one obtained by applying the noncausal feedforward technique after having selected a transition time $\tau = 20$ are plotted in Figure 5.40. The determined command input is reported in Figure 5.41

As in the previous example, the dynamic-inversion-based systems outperforms the one with the step signal, by providing a much less value of the rise time.

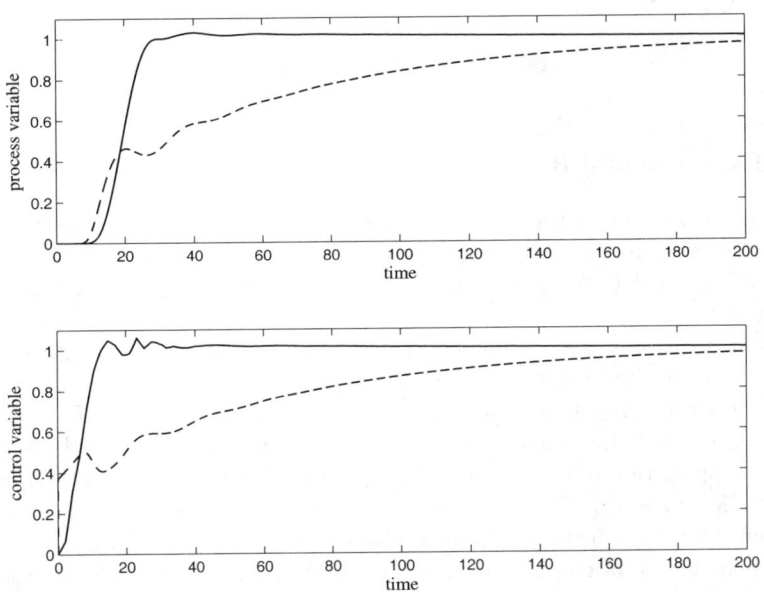

Fig. 5.40. Response before and after the application of the noncausal feedforward approach for process $P_2(s)$. Solid line: step response; dashed line: noncausal feedforward command input response.

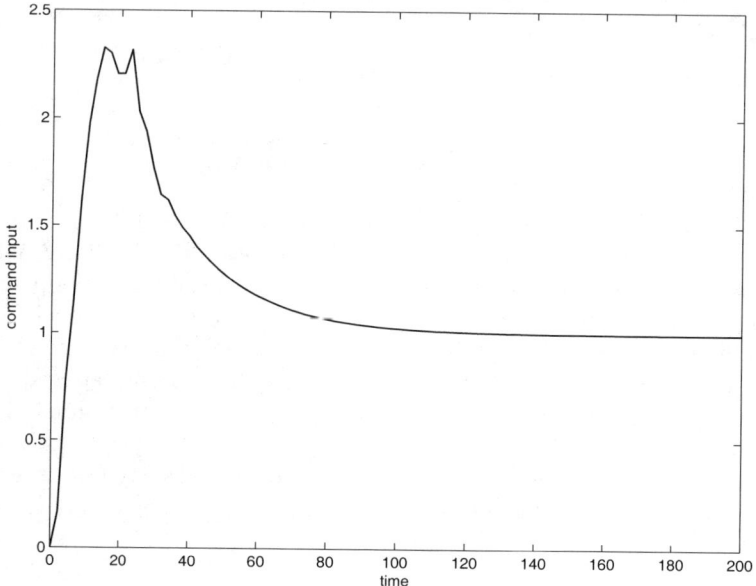

Fig. 5.41. Inversion-based command input for process $P_2(s)$

5.5.3 Experimental Results

As for the continuous-time case, also for the discrete-time case the effectiveness of the noncausal feedforward approach has been verified by applying it to a level control task by means of the laboratory equipment described in Section A.1.

The same tank has been considered and a transition from the initial level $y_0 = 2$ V to the final level $y_1 = 3$ V has been required.

The PI controller has been tuned again by fixing $K_p = 2.5$ and $T_i = 8.3$ and the sampling time has been chosen as 0.01 s. The process variable resulting from the application of a step set-point signal has been filtered before applying the inversion procedure. In particular, a median filter has been first employed and then data have been interpolated by means of a polynomial.

The noncausal feedforward action has been determined by selecting a transition time of $\tau = 10$ s and $\tau = 20$ s and it has been applied to the closed-loop system. The process variable and controller output obtained in the two experiments, together with the one that results from the application of the step set-point signal, are plotted in Figures 5.42 and 5.43 respectively.

Note that the actual process input saturates at 5 V. The resulting command input $r(t)$ for both $\tau = 10$ s and $\tau = 20$ s is reported in Figure 5.44. As expected, the rise time obtained for a desired transition time of $\tau = 10$ s is less than that obtained for $\tau = 20$ s and it is very similar to the one obtained with a step signal. However, the overshoot is significantly decreased (and, consequently, also the settling time) when the feedforward action is used. It can be also evaluated how parameter τ handles the trade-off between aggressiveness and control effort (actually, for $\tau = 10$ s, the manipulated variable saturates for a small time interval).

Again as for the continuous-time case, the methodology has also been applied by adding (via software) an additional dead time of 10 s to the process input (the PI parameters have been selected as $K_p = 1.24$, $T_i = 31$). Again, the two values of $\tau = 10$ s and $\tau = 20$ s for the transition time have been selected. Results are reported in Figures 5.45–5.47 (note that the same sampling time and the same filtering method as before has been used).

It can be seen that for $\tau = 10$ s the saturation of the control variable (and possibly the nonlinear dynamics of the system) prevents the achievement of a monotonic process variable. In any case with both $\tau = 10$ s and $\tau = 20$ s the control system response presents a much smaller overshoot and a similar rise time.

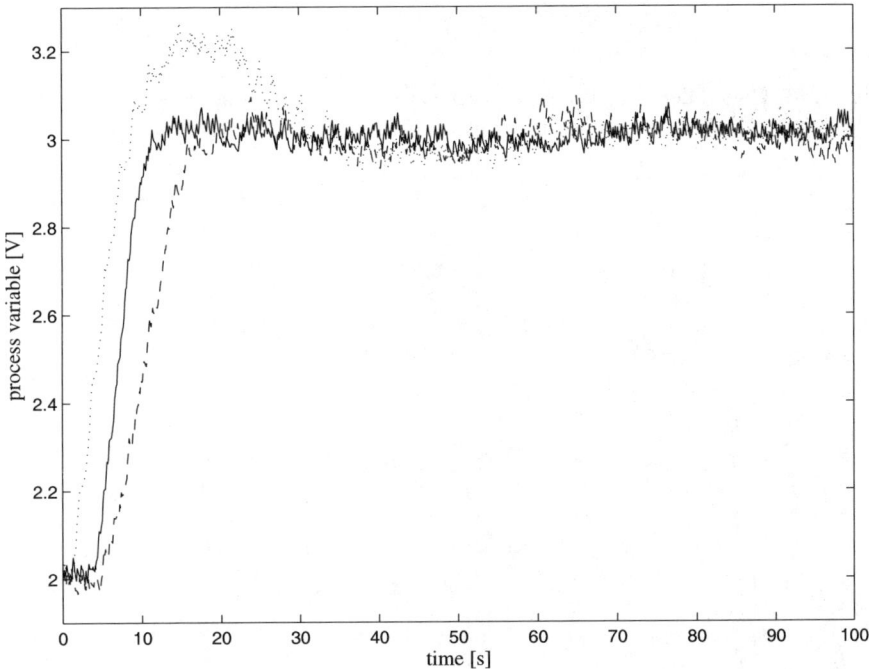

Fig. 5.42. Process variable for the level control task. Solid line: noncausal approach with $\tau = 10$; dashed line: noncausal approach with $\tau = 20$; dotted line: step response.

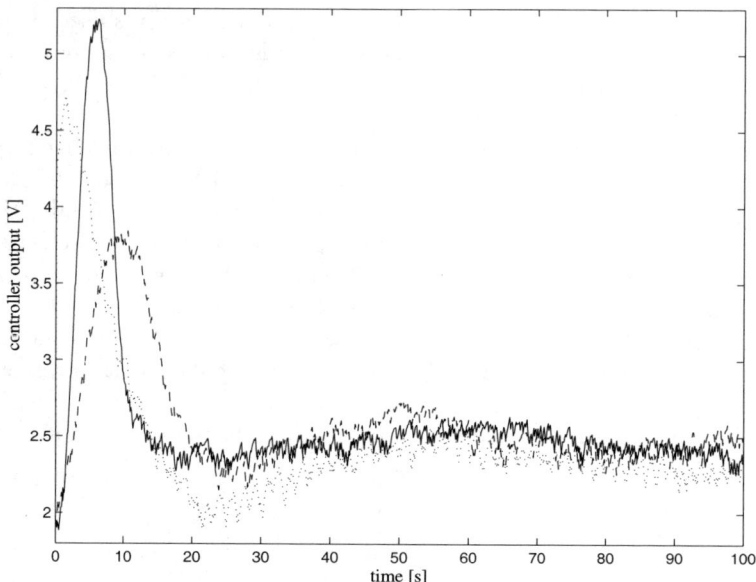

Fig. 5.43. Controller output for the level control task. Solid line: noncausal approach with $\tau = 10$; dashed line: noncausal approach with $\tau = 20$; dotted line: step response.

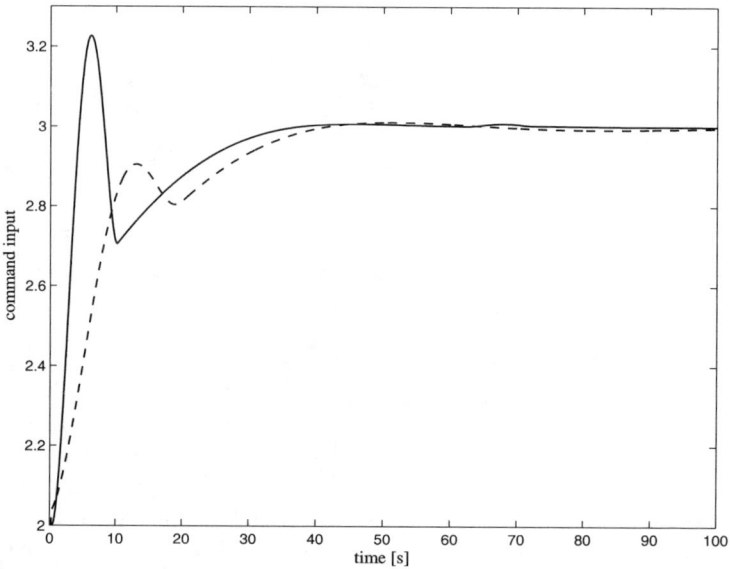

Fig. 5.44. Inversion-based command input for the level control task. Solid line: $\tau = 10$; dashed line: $\tau = 20$.

5.5 Noncausal Feedforward Action: Discrete-time Case   139

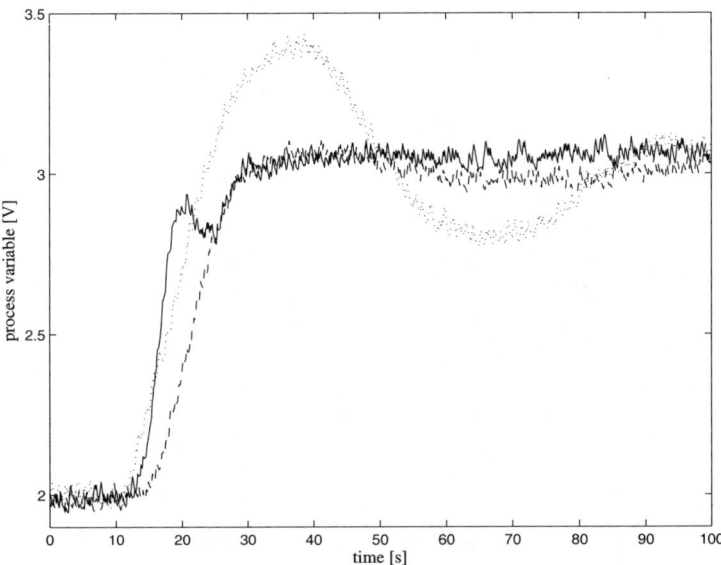

Fig. 5.45. Process variable for the level control task with additional dead time. Solid line: noncausal approach with $\tau = 10$; dashed line: noncausal approach with $\tau = 20$; dotted line: step response.

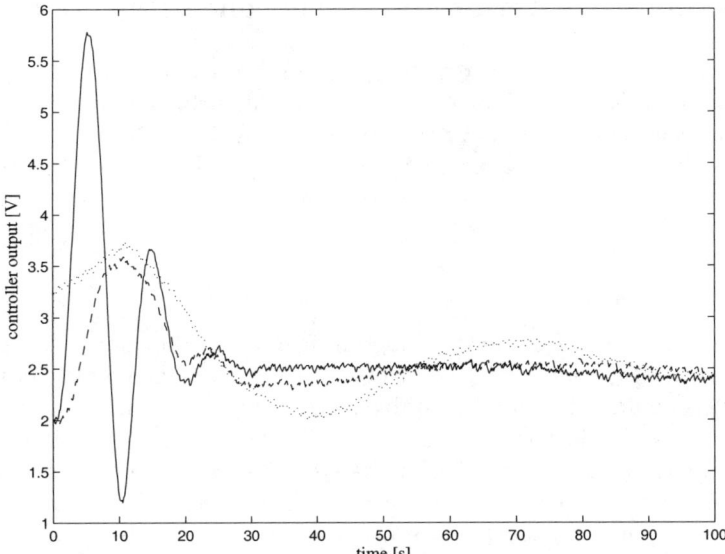

Fig. 5.46. Controller output for the level control task with additional dead time. Solid line: noncausal approach with $\tau = 10$; dashed line: noncausal approach with $\tau = 20$; dotted line: step response.

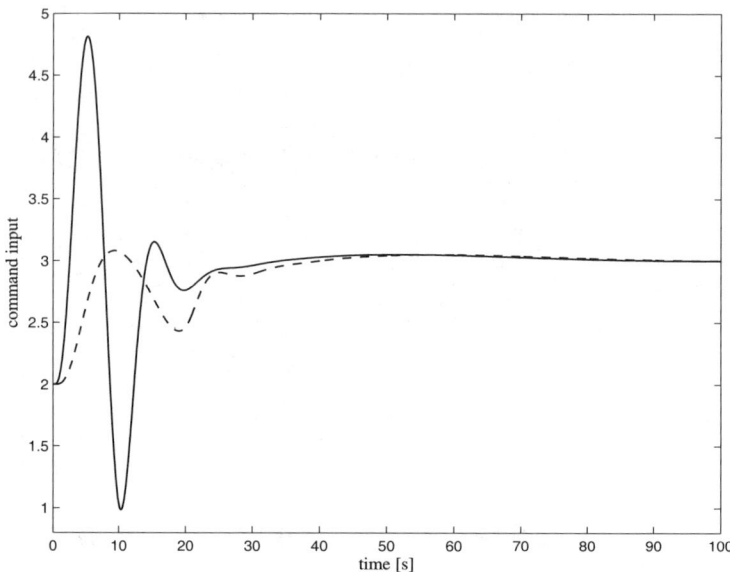

Fig. 5.47. Inversion-based command input for the level control task with additional dead time. Solid line: $\tau = 10$; dashed line: $\tau = 20$.

5.6 Feedforward Action for Disturbance Rejection

A feedforward action can be very beneficial also for compensating load disturbances, if the model of the process and of the disturbance are known with a sufficient accuracy (Lewin and Scali, 1988). In this case, the standard control scheme is that shown in Figure 5.48, where $H(s)$ is the transfer function that expresses the influence of the disturbance on the process and

$$G(s) = \frac{H(s)}{P(s)}, \qquad (5.63)$$

so that, in principle, the transfer function from d to y results to be zero.

It has to be noted that, since the inverse of $P(s)$ has to be employed, the process model should not contain unstable zeros, otherwise an internally unstable system results. Further, transfer function $G(s)$ has to be physically realisable, namely, the dead time term of $H(s)$ has to be greater than or equal to that of $P(s)$, otherwise a negative time delay results. Finally, the determined $G(s)$ has to be proper, $i.e.$, the relative order of $H(s)$ has to be greater than or equal to that of $P(s)$.

Often, in practical situations, both $H(s)$ and $P(s)$ are selected as FOPDT transfer functions, $i.e.$,

$$P(s) = \frac{K}{Ts+1}e^{-Ls}$$

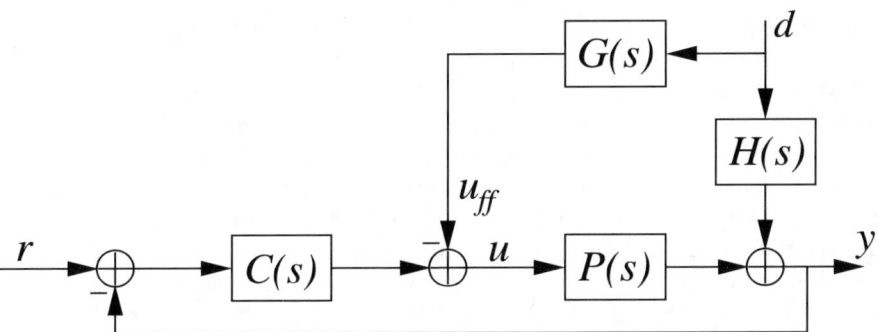

Fig. 5.48. Block diagram for the standard implementation of feedforward action for load disturbance rejection task

and
$$H(s) = \frac{K_H}{T_H s + 1} e^{-L_H s}.$$
If the dead time terms of the two transfer functions is the same, i.e., $L = L_H$, then the resulting feedforward controller results to be a lead-lag unit:
$$G(s) = \frac{K_H}{K} \frac{Ts + 1}{T_H s + 1}.$$

A lead-lag element is used also in other situations when a physically unrealisable controller results and therefore an approximate solution has to be found (Seborg *et al.*, 2004). In fact, if $L_G := L_H - L$ is greater than zero, then the term e^{+L_G} can be approximated by increasing the lead time constant from T to $T + L_G$, i.e.,
$$G(s) = \frac{K_H}{K} \frac{(T + L_G)s + 1}{T_H s + 1}.$$
Further, if $L_H = 0$ a SOPDT transfer function is estimated for the process, namely,
$$P(s) = \frac{K}{(T_1 s + 1)(T_2 s + 1)},$$
then the resulting improper transfer function of the feedforward controller can be approximated by
$$G(s) = \frac{K_H}{K} \frac{(T_1 + T_2)s + 1}{T_H s + 1}.$$

It is worth noting that transfer function $H(s)$ expresses where the disturbance enters in the process, i.e., which part of the process dynamics is excited by the disturbance. Actually, it is known that a disturbance is most detrimental when it affects all the process dynamics, that is, when it enters at the input of the process (thus, $H(s) = P(s)$). In this case it is worth employing a feedforward action (provided that the model of the process and the model of the disturbances are known with a sufficient accuracy), while if the disturbance enters late in the process, a properly designed feedback controller is sufficient to compensate it effectively. In this context, a simple method to assess if a feedforward action is worth to being used has been proposed in (Petersson et al., 2001; Petersson et al., 2003). It consists in comparing the control variable that results when the actual (step) disturbance occurs with the control variable signals that results when a (step) disturbance occurs at the input and at the output of the process. In particular, consider the two signals u_b and u_a that results by applying a step disturbance signal (without feedforward action) at the process input and at the process output respectively (i.e., by setting $H(s) = P(s)$ and $H(s) = 1$ respectively). These signals can be obtained in practical cases by applying the disturbances to the actual process or via simulation if an accurate process model is available. Denote the (reference) area between the two signals as A_r. The time interval to be considered for this computation is the one from the application of the disturbance to the average residence time of the process, which is defined as (Åström and Hägglund, 1995)

$$T_{ar} = \frac{A_0}{K} \qquad (5.64)$$

where K is the process gain and

$$A_0 = \int_0^\infty (s(\infty) - s(t))dt \qquad (5.65)$$

where $s(t)$ is the (open-loop) step response (note that T_{ar} is a rough estimate of the time taken for the process input to have a significant influence on the output and for FOPDT processes this is equal to the sum of the dead time and the time constant). Then, consider the control variable u that results when the true load disturbance occurs in the process and calculate the (disturbance) area A_d between u and u_b. The index that determines the suitability of an additional feedforward control is determined by dividing the disturbance area by the reference area, namely, by calculating A_d/A_r. This fact is depicted in Figure 5.49. If the index is close to or greater than one, then this means that the disturbance signal enters before or early in the process and therefore an additional feedforward control action is likely to improve the regulation performance. Conversely, if the index is close to zero, then the disturbance enters late in the process and therefore the feedback action is sufficient. Note that, for a correct application of the technique, the signals adopted have to be scaled properly.

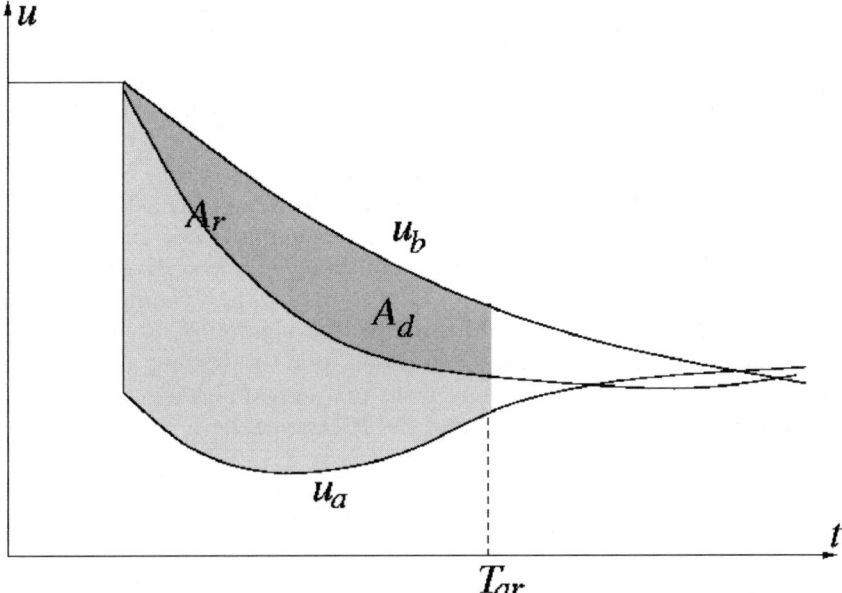

Fig. 5.49. Illustration of technique for the evaluation of the suitability of a feedforward action for disturbance rejection. Solid line: the disturbance enter at the process input; dashed line: the disturbance enters at the process output; dotted line: measured disturbance.

5.7 Conclusions

A feedforward controller can be very beneficial in solving the problem of achieving a satisfactory performance both in the set-point following and in the load disturbance rejection task. Different methodologies for the design of a feedforward controller have been described in this chapter. In particular, the set-point following performance have been addressed. Features of the standard approach have been discussed. It has been shown that when the control task does not involve the tracking of a reference signal but only the transition of the process variable from a set-point value to another one is of concern, different alternative methods can be considered. The use of a nonlinear feedforward action allows to improve considerably the control system performance by taking into account explicitly the actuator constraints. Its implementation requires indeed a modest extra design effort. The great advantage of the noncausal approach is that a predefined performance can be actually obtained almost "independently" of the tuning of the (PID) controller and of the actual process dynamics. In fact, by looking at the results, it appears that very similar responses are obtained with very different values of the PID parameters (namely, with PID parameters that provide very different set-point step responses) and with processes of different dynamics. This advantage is paid by an increased implementation complexity. It can be remarked that each of the considered methodologies has in any case a tuning parameter that allows to handle the trade-off between aggressiveness and robustness.

Finally, it is worth highlighting that, if only the set-point following task is of concern, in many practical cases a fine tuning of the controller could allow to obtain a high performance and the improvement provided by the use of a feedforward control system is not significant. However, the selection of the correct parameters can be a difficult and time consuming task. In this context the feedforward control action can be used to achieve (in a relatively easy way) a satisfactory performance despite a not very appropriate tuning of the PID parameters and therefore to reduce the overall design effort. It is important to note that, in contrast with the constant set-point weighting approach, the use of a feedforward allows to recover the set-point following performances even in the case of a sluggish tuning of the PID controller.

6
Plug&Control

6.1 Introduction

One of the main reasons for the success of PID controllers in the industrial context is their relative ease of use. Indeed, the fast commissioning of the controller is essential in many applications, where a tight performance is not required, in order to reduce the implementation costs. In this context, the availability of the so-called Plug&Control function (*i.e.*, to automatically make the controller work properly after simply connecting it in the control architecture, without further intervention from the operator) is highly desirable.
With respect to the classic automatic tuning procedures, this function has the advantage that a dedicated identification (possibly time-consuming) experiment is not required, since the estimation of the process parameters is performed during the normal start-up of the process. This might allow a significant saving of time, energy and material.
Methodologies related to this topic are described in this chapter. Although this can be considered a relatively recent subject of research, results are very encouraging and industrial implementations are already available. The aim of the following analysis is also to provide a characterisation of the considered techniques in order to verify its applicability in different contexts.

6.2 Self-tuning Temperature Control

The algorithm described hereafter has been presented in (Pfeiffer, 1999; Pfeiffer, 2000), and is implemented as a functional block in Programmable Logic Controllers (PLCs) made by Siemens.
It is particularly suitable for temperature processes (although it can be employed also in other contexts), where the dead time is small and a pole of the system transfer function is close to the origin of the complex plane and therefore causes an integrator-like dynamics in the relevant working range. Further, there is often the requirement of avoiding the overshoot because a

long time might be necessary to recover it if an active cooling is not present. The rejection of disturbances is also of concern. In this framework a tuning of the PID controller based on pole-zero cancellation is not recommended, as it would provide a very sluggish load disturbance response. The problem can be solved partially by setting a low value of the integral time constant and by applying the derivative and the proportional action to the process output (*i.e.*, by setting the set-point weight to zero).

The Plug&Control algorithm consists of initially applying a constant heating energy u_{max} when a set-point change from y_0 (initial process value) to y_1 is required (at time t_0). The constant value of the manipulated variable is maintained until the inflection point of the process output is detected. This occurs when the ascent ratio of the step response decreases in two successive cycles or when the process value attains 70% percent of the step amplitude. It has to be noted that an adaptive low-pass filtering has to be applied in order to effectively suppress the measurement noise. Then, an integrating plus first-order lag model of the process is initially estimated as

$$P(s) = \frac{K}{s(Ts+1)} \tag{6.1}$$

with

$$K = \frac{\frac{dy_w}{dt}}{u_{max} - u_0} \tag{6.2}$$

and

$$T = t_w - t_0 - \frac{y_w - y_0}{\frac{dy_w}{dt}} \tag{6.3}$$

where y_w is the value of the process output at the inflection point, t_w is the corresponding time instant and u_0 is the initial steady-state value of the control variable.

Based on this process model, a PI controller is initially tuned according to the symmetric optimum principle (Åström and Hägglund, 1995), namely

$$K_p = \frac{1}{2TK} \qquad T_i = 4T. \tag{6.4}$$

It is then slightly detuned in order to avoid overshoots.

The designed PI controller is immediately applied to the process and this allows to attain the desired set-point value y_1 (see Figure 6.1) for an illustration of the technique). At this point, the process gain can be correctly estimated as

$$K = \frac{y_1 - y_0}{u_1 - u_0}, \tag{6.5}$$

where u_1 is the final steady-state value of the control variable. With this new value, a more accurate process model can be estimated. In particular, a

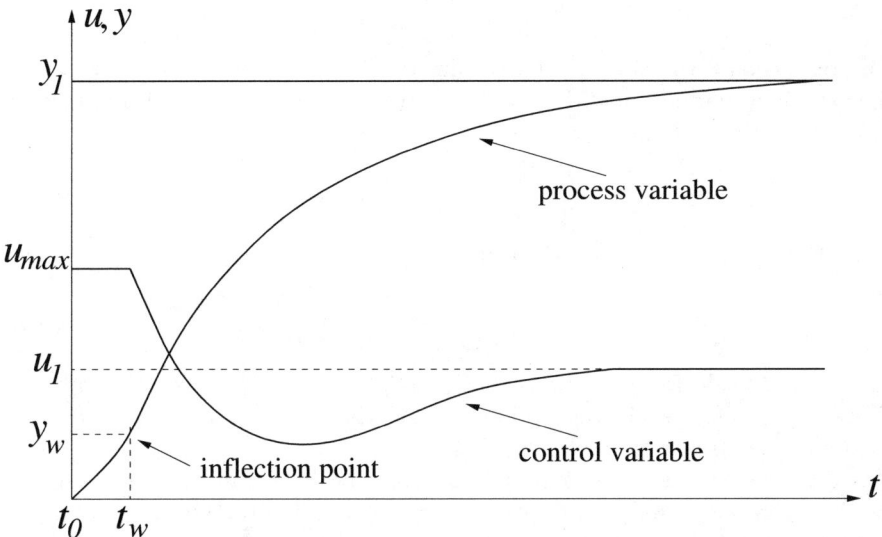

Fig. 6.1. An illustrative example of the application of the self-tuning temperature control methodology

second-order transfer function is initially selected, namely,

$$P(s) = \frac{K}{(T_1 s + 1)(T_2 s + 1)}. \tag{6.6}$$

In order to estimate the two time constants T_1 and T_2, the so-called recovery time is first determined:

$$t_r = \frac{K(u_{max} - u_0)}{\dfrac{dy_w}{dt}}. \tag{6.7}$$

The inflection point in the step response of $P(s)$ (6.6) can be determined analytically as

$$t_w = f \frac{\ln f}{f - 1} T_2. \tag{6.8}$$

where f is the ratio between T_1 and T_2. Then, considering the intersection of the inflection tangent with the lines $y = y_0$ and $y = y_1$ allows to write the following equations:

$$\frac{t_r}{T} = \frac{1}{f^{-\frac{f}{f-1}}\left(1 + f + \dfrac{f \ln f}{f - 1}\right) - 1}, \tag{6.9}$$

$$T_2 = t_r f^{-\frac{f}{f-1}}. \tag{6.10}$$

An approximate solution of this nonlinear system can be found by replacing the exponential function with a linear approximation in the relevant parameter space $f \in (2, 20)$. Thus, Equation (6.9) can be rewritten as

$$\frac{t_r}{T} = 1.1919f + 8.0633, \tag{6.11}$$

while Equation (6.10) can be rewritten as

$$T_2 = t_r \frac{1}{1.0722f + 2.0982}. \tag{6.12}$$

Then, by considering Equations (6.3), (6.7), (6.11) and (6.12), the values of f and T_2 can be easily determined, while the value of T_1 can be trivially calculated by applying the relation $T_1 = fT_2$.

Based on the estimated second-order model, the PID controller in ideal form (with the proportional and derivative actions applied to the process output) is tuned definitely according to a tuning rule derived from the minimisation of a quadratic performance index:

$$K_p = 1.5 \frac{21.4}{K}, \tag{6.13}$$

$$T_d = T_2 \left(0.985 - \frac{8.417}{f + 10.66} \right), \tag{6.14}$$

$$T_i = (0.1236f + 3.322)T_d. \tag{6.15}$$

A first-order filter, whose time constant is fixed to $T_d/5$, is then used for the derivative action. It is worth noting at this point that if $t_r/T > 9.64$, the second-order transfer function (6.6) is considered to be no more suitable for an accurate modelling of the process. Thus, in this case the transfer function

$$P(s) = \frac{K}{(T_1 s + 1)^n} \tag{6.16}$$

is employed. By following a reasoning similar to the previous one, the two model parameters n and T_1 can be determined as

$$n = \frac{7.9826}{\frac{t_r}{T} - 0.3954} + 1.1099. \tag{6.17}$$

and

$$T_1 = \frac{T}{0.0165n^2 + 0.5078n + 0.8387}. \tag{6.18}$$

Then, a different tuning rule for the PID controller is employed.

Simulation and experimental results for the devised self-tuning temperature control methodology have been presented in (Pfeiffer, 1999; Pfeiffer, 2000)

6.3 Time-optimal Plug&Control

6.3.1 Methodology

A time-optimal Plug&Control strategy for FOPDT and IPDT processes has been first proposed in (Visioli, 2003b) and then developed in (Visioli, 2005b). It is based on the combined use of three-state and PID control to perform a transition from one set-point value to another, as required by the process start-up operation, and it is applicable without *a priori* knowledge of the process model parameters, with the exception of the sign of the process gain, which will be assumed to be positive from now on, without loss of generality. Basically, the methodology consists of initially setting the controller output at its upper limit, when the step on the set-point signal is applied. Afterwards, when the process output leaves its previous value, the dead time L of the process is estimated. Then, from this instant the process parameters are estimated through a least squares procedure. Once the process model is estimated, a time-optimal control strategy, based on the saturation limits of the actuator, can be computed and applied. At the same time, the PID controller (which is not adopted in this phase) can be properly tuned according to a selected tuning rule. If the process parameters are perfectly estimated, then at the end of the time-optimal control, the process output would be exactly at its desired steady-state value. However, estimation inaccuracies are not avoidable in practical cases, mainly due to the presence of measurement noise and numerical approximations. Therefore, at the end of the time-optimal strategy, when the process output is actually close to its desired value, the controller is set to the PID mode. In this way, the desired output value is actually attained and possible subsequent load disturbances can be compensated.

6.3.2 Algorithm

FOPDT processes

Consider a process described by a first-order plus dead-time model:

$$P(s) = \frac{K}{Ts+1} e^{-Ls} \quad K > 0, \ T > 0 \tag{6.19}$$

and denote u as the controller output and y as the process output. Suppose now that an output transition from y_0 to $y_{sp} = y_0 + y_1$ is then required to be performed, starting from time t_0 (assume that the process is at an equilibrium point with $u_0 := u(t_0)$ and $y_0 := y(t_0)$). For the sake of simplicity and without loss of generality, hereafter it will be assumed $y_1 > 0$.

Then, the following algorithm can be applied. For the sake of clarity it refers to the ideal noise-free case. Modifications to be carried out in order to cope with measurement noise are discussed in Section 6.3.3. The sampling time is denoted by Δt.

TOPC algorithm for FOPDT processes

1. Set u_{max} and u_{min} as the maximum and minimum values respectively of the control variable u during the three-state control and calculate $u^+ = u_{max} - u_0$ and $u^- = u_{min} - u_0$.
2. Set flag=1.
3. At time $t = t_0$ set $u = u_{max}$.
4. When $y > y_0$ set $t_1 = t$ and $\hat{L} = t_1 - t_0$ (estimated dead time of the process).
5. At time $t = t_1$ start the recursive least squares algorithm (Åström and Wittenmark, 1995, page 51).
6. When $|\hat{K}(t) - \hat{K}(t - \Delta t)| < \varepsilon$ and $|\hat{T}(t) - \hat{T}(t - \Delta t)| < \varepsilon$ ($\hat{K}$ and $\hat{T}$ are the estimated gain and time constant of the process):
   a) Set $t_2 = t$.
   b) Set $\hat{K} = \hat{K}(t_2)$ and $\hat{T} = \hat{T}(t_2)$.
   c) Apply a PI(D) tuning rule based on the model identified.
   d) Calculate
   $$t_{s1} = t_0 - \hat{T} \ln \left(\frac{u^+ - \frac{y_1}{\hat{K}}}{u^+} \right). \qquad (6.20)$$
   e) If $t_{s1} < t_2$ then set $t_{s1} = t_2$, flag=0 and calculate
   $$t_{s2} = t_{s1} - \hat{T} \ln \left(\frac{\frac{y_1}{\hat{K}} - (u^-)}{-u^+ \exp\left(-\frac{t_{s1} - t_0}{\hat{T}}\right) + u^+ - u^-} \right). \qquad (6.21)$$
7. If flag=1 then set $u = u_{max}$ when $t \le t_{s1}$ and $u = u_0 + y_1/\hat{K}$ when $t > t_{s1}$, else set $u = u_{min}$ when $t \le t_{s2}$ and $u = u_0 + y_1/\hat{K}$ when $t > t_{s2}$.
8. When $t > \hat{L} + t_{s1}$ (if flag=1) or when $t > \hat{L} + t_{s2}$ (if flag=0) apply the PI(D) controller.

It can be seen that the algorithm requires that when a set-point change is required at time $t = t_0$ the control variable is set to its maximum level $u = u_{max}$. Then, when the process output leaves its initial value y_0 at time $t = t_1$, the dead time L of the process is detected. A standard recursive least-squares algorithm is then applied. When the estimation of the parameters converges (i.e., when the difference of two successive estimation is less than a predefined threshold ε) at time $t = t_2$, a model of the process is available. This allows to tune the PID controller and to determine a time-optimal strategy to attain the set-point value. In particular, the time interval for which the control variable has to be kept at its maximum value u_{max} in order for the process output to attain the set-point value in the minimum time, namely, the time instant

t_{s1} when the value of the control variable has to be switched from u_{max} to the final steady-state value $u_0 + y_1/\hat{K}$ is determined (Equation (6.20) can be trivially derived in the context of the optimal control theory (Lewis, 1996)). It may happen that, because of the large dead time of the process or because of the large time interval required for the identification procedure to converge, it follows that $t_{s1} < t_2$, i.e., that the control variable has been set to its maximum value for a larger time than requested by the time-optimal control (this condition determines the setting flag=0). This means that the output, even in the perfect match case (i.e., even if the process parameters are perfectly estimated) presents an overshoot. Hence, the control variable must be set immediately at its minimum level and kept at this value for a determined time interval $t_{s2} - t_{s1}$ (the switching time t_{s2} is also derived in the context of optimal theory). Then, the control variable is set at its (new) steady-state value $u = u_0 + y_1/\hat{K}$. At the end of the (three-state) time-optimal control strategy, if the process model is perfectly estimated the process output attains its setpoint value.
Since in practical cases this does not occur because of the unavoidable estimation inaccuracies, the PID controller is applied to cancel the remaining steady-state error and to cope with subsequent possible load disturbances.

IPDT Processes

Consider a integrator plus dead-time process:

$$P(s) = \frac{K}{s}e^{-Ls} \quad K > 0, \quad T > 0. \tag{6.22}$$

The algorithm is very similar to the FOPDT case. The only differences are the formulae for the determination of the optimal switching times t_{s1} and t_{s2} and the final steady-state value of the control variable. Also here the noise-free case is described for the sake of clarity, leaving the discussion of practical cases to Section 6.3.3.

TOPC algorithm for IPDT processes

1. Set u_{max} and u_{min} as the maximum and minimum values respectively of the control variable u during the three-state control and calculate $u^+ = u_{max} - u_0$ and $u^- = u_{min} - u_0$.
2. Set flag=1.
3. At time $t = t_0$ set $u = u_{max}$.
4. When $y > y_0$ set $t_1 = t$ and $\hat{L} = t_1 - t_0$ (estimated dead time of the process).
5. At time $t = t_1$ start the recursive least squares algorithm (Åström and Wittenmark, 1995, page 51).
6. When $|\hat{K}(t) - \hat{K}(t - \Delta t)| < \varepsilon$ ($\hat{K}$ is the estimated gain of the process):
    a) Set $t_2 = t$.

b) Set $\hat{K} = \hat{K}(t_2)$.
c) Apply a PI(D) tuning rule based on the model identified.
d) Calculate

$$t_{s1} = t_0 + \frac{y_1}{\hat{K}u^+}. \tag{6.23}$$

e) If $t_{s1} < t_2$ then set $t_{s1} = t_2$, flag=0 and calculate

$$t_{s2} = -\frac{u^+(t_{s1} - t_0) - u^-(t_{s1} - t_0) - \frac{y_1}{\hat{K}}}{u^-}. \tag{6.24}$$

7. If flag=1 then set $u = u_{max}$ when $t \leq t_{s1}$ and $u = 0$ when $t > t_{s1}$, else set $u = u_{min}$ when $t \leq t_{s2}$ and $u = 0$ when $t > t_{s2}$.
8. When $t > \hat{L} + t_{s1}$ (if flag=1) or when $t > \hat{L} + t_{s2}$ (if flag=0) apply the PI(D) controller.

The same considerations for the FOPDT case can be applied also in this case.

6.3.3 Practical Considerations

A few technical problems have to be solved in order to effectively apply the TOPC algorithm in practical cases. First, since real measurements are always corrupted with noise, the condition $y > y_0$ at step 4 (both for IPDT and FOPDT processes) has be substituted with $y > y_0 + NB$ where NB is the estimated noise band (Åström et al., 1993) (as already mentioned in Section 5.5.1, this estimation can be performed in a time interval before the application of the TOPC technique).

It has also to be noted that it is not strictly necessary for the control constraints u_{min} and u_{max} to correspond to the actual physical limits of the actuator. Actually, more conservative bounds can be selected for various operating reasons or to preserve the linearity of the model.

Then, the recursive least squares algorithm in step 5 has to be initialised. This can be easily done by selecting a very rough estimate of the process gain and of the process time constant, denoted respectively as $\hat{K}_0$ and $\hat{T}_0$ (see Section 6.3.5). The value of ε has to be fixed as well. Actually, by fixing it at a low value the user is confident that the identification phase is ended with a satisfactory accuracy.

Then, a bumpless transfer (Åström and Hägglund, 1995) has to be applied at step 8 at the time of switching from the three-state to the PID controller.

Finally, it has to be highlighted that the proposed method could be applied in the context of feedforward control of set-point steps with "full power" plus PID control (Pfeiffer, 2000). In other words, for sufficiently large set-point steps, a control zone has to be defined: inside a narrow band around the setpoint a closed-loop PID control is employed, while outside the control zone the controller output is set at its maximum value. Note that with such a strategy, the integral time constant of the PID controller can be increased without the occurrence of overshoots and yielding at the same time a faster load disturbance rejection. The identification procedure can be skipped if the process parameters have already been estimated or it can be applied if variations of the process parameters are detected.

6.3.4 Simulation Results

In order to understand better the TOPC algorithm and to verify its effectiveness, a few simulation results are presented hereafter.
First, consider the FOPDT process

$$P_1(s) = \frac{1}{1.4s+1}e^{-0.4s}. \qquad (6.25)$$

The initial conditions at $t_0 = 0$ are $y_0 = 0$, $u_0 = 0$. Then, it is set $y_{sp} = y_1 = 1$ and $u_{max} = 1.5$ and $u_{min} = -1.5$. Since no measurement noise is considered (see Section 6.3.5 for the case where measurement noise is present), the noise band NB is not employed. Then, it is fixed $\varepsilon = 10^{-3}$. The recursive least squares algorithm is initialised with $\hat{K}_0 = 0.5$ and $\hat{T}_0 = 1$. Finally, the following tuning rule has been used for the PI controller (the derivative action has not been employed) (Rivera et al., 1986):

$$K_p = \frac{\hat{T}}{\hat{K}\lambda}, \quad \lambda = \max\{0.1\hat{T}, 1.7\hat{L}\},$$
$$T_i = \hat{T}. \qquad (6.26)$$

The result of the application of the Plug&Control strategy is shown in Figure 6.2. It can be seen that the condition $y > 0$ is verified at time $t_1 = \hat{L} = 0.4$ and therefore $\hat{L} = 0.4$ is fixed (the dead time is correctly estimated). Then, the recursive least squares algorithm starts. It converges at time $t_2 = 1.2$ with $\hat{K} = 0.98$ and $\hat{T} = 1.36$. The optimal switching strategy is consequently determined by calculating $t_{s1} = 1.554$ (see Equation (6.20)). Since $t_{s1} > t_2$, it is flag=1 and therefore the control variable is kept at the maximum level $u = u_{max} = 1.5$ for $t < t_{s1}$. Then, for $t_{s1} < t < t_{s1} + L$ it is set $u = y_1/\hat{K} = 1.02$. In the meantime, the PI controller parameters are fixed to $K_p = 2.04$ and $T_i = 1.36$. Finally, at time $t = t_{s1} + L = 1.954$ the PI controller is applied to the control system and the effects of the slight model mismatch are compensated. In order to evaluate the performance of the designed PI controller, a load unitary step disturbance has been applied to the process at time $t = 10$.
Overall, a satisfactory performance emerges since the overshoot is negligible and the PI controller appears to be well tuned (note that other tuning rules could have been applied).
Now, consider the process

$$P_2(s) = \frac{1}{s+1}e^{-1.4s}, \qquad (6.27)$$

6 Plug&Control

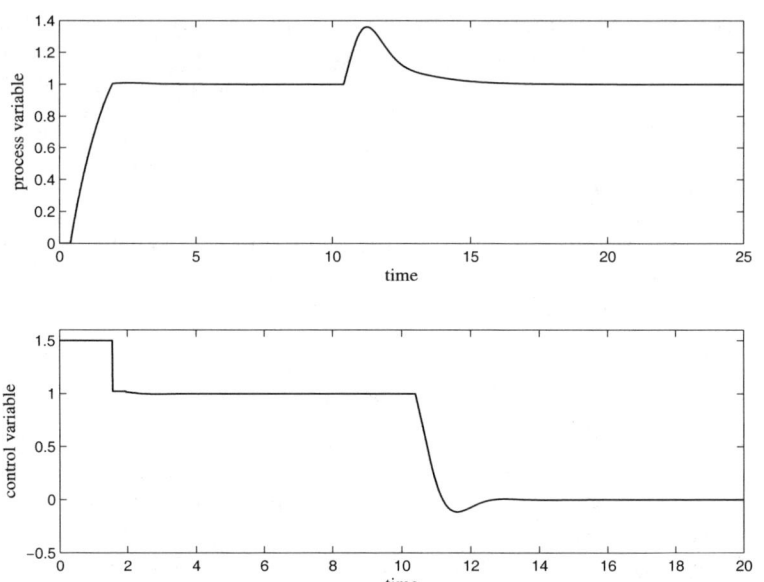

Fig. 6.2. Result of the application of the TOPC strategy with the FOPDT process $P_1(s)$

where the dead time is greater than the time constant (a fast response without overshoot is therefore difficult to obtain with a PID controller). The same initial conditions and the same design parameters as before are fixed. The result obtained in this case is shown in Figure 6.3. The dead time $\hat{L}$ is correctly estimated at time $t = t_1 = 1.4$. Then, the recursive least squares procedure starts and it ends at time $t = t_2 = 1.98$ with $\hat{K} = 1.008$ and $\hat{T} = 1.015$. Based on the estimated parameters, the optimal switching time $t_{s1} = 1.1$ is determined. Since it is $t_{s1} < t_2$, it is fixed $t_{s1} = t_2 = 1.98$ and flag=0. Then, the control variable is set at its minimum value $u_{min} = -1.5$ and the optimal switching time t_{s2} is calculated by means of Expression (6.21). It results $t_{s2} = 2.093$. Thus, at time $t = t_{s2} = 2.093$ the control variable is set to $u = y_1/\hat{K} = 0.99$ and at time $t = t_{s2} + \hat{L} = 3.493$ the PI controller (which has been tuned by setting $K_p = 0.42$ and $T_i = 1.0015$) is applied. A unitary step load disturbance has then been applied at time $t = 10$ in order to verify the effectiveness of the PI controller designed by means of the TOPC strategy.

It can be seen that, actually, the TOPC strategy produces an overshoot in the first transient response. This is due to the fact that the process has a significant time delay and therefore the control variable has been kept at the maximum level for a too long time interval when the identification procedure converges. This implies that the control variable has to be set to the minimum level for a given interval, but this does not suffices to avoid the overshoot (which is in

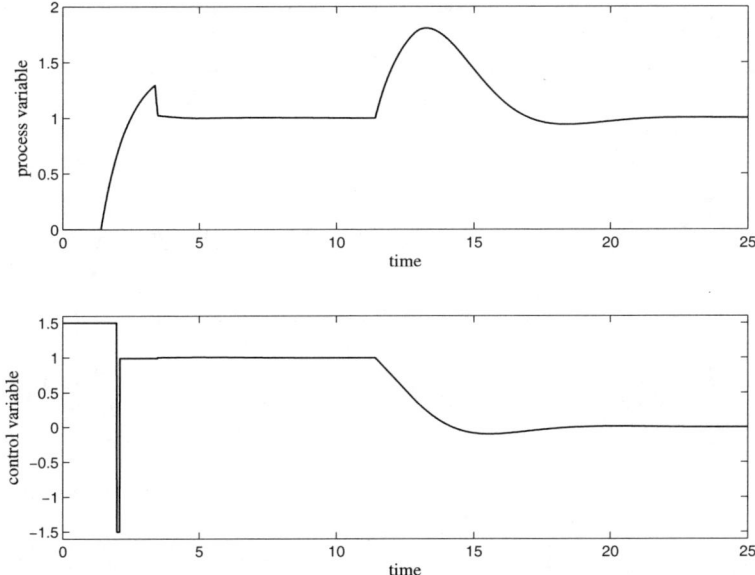

Fig. 6.3. Result of the application of the TOPC strategy with the FOPDT process $P_2(s)$

any case kept at a reasonable level).
In order to verify the effectiveness of the TOPC strategy also for integral processes, the following process is considered:

$$P_3(s) = \frac{0.1}{s}e^{-s}. \tag{6.28}$$

As for the FOPDT processes, the initial conditions are fixed to $t_0 = 0$, $y_0 = 0$ and $u_0 = 0$. Then, it is set $y_{sp} = y_1 = 1$ and $u_{max} = 1.5$ and $u_{min} = -1.5$. Here, a noise band of $NB = 0.01$ is set. The recursive least squares procedure parameters are $\hat{K}_0 = 0.5$ and $\varepsilon = 10^{-3}$. The tuning formula employed for the PI controller is (Shinskey, 1994)

$$K_p = \frac{0.9259}{\hat{K}\hat{L}}, \tag{6.29}$$
$$T_i = 4\hat{L}.$$

The result obtained by applying the TOPC strategy is shown in Figure 6.4. It is $t_1 = \hat{L} = 1.06$ and $t_2 = 2.33$ (the estimated process parameter is $\hat{K} = 0.1$). By means of Expression (6.23) the optimal switching time is determined as $t_{s1} = 6.65$, where the control variable is set to zero. Then, at time $t = t_{s1} + \hat{L} = 7.71$ the PI controller is applied ($K_p = 8.71$ and $T_i = 4.24$) and its performance

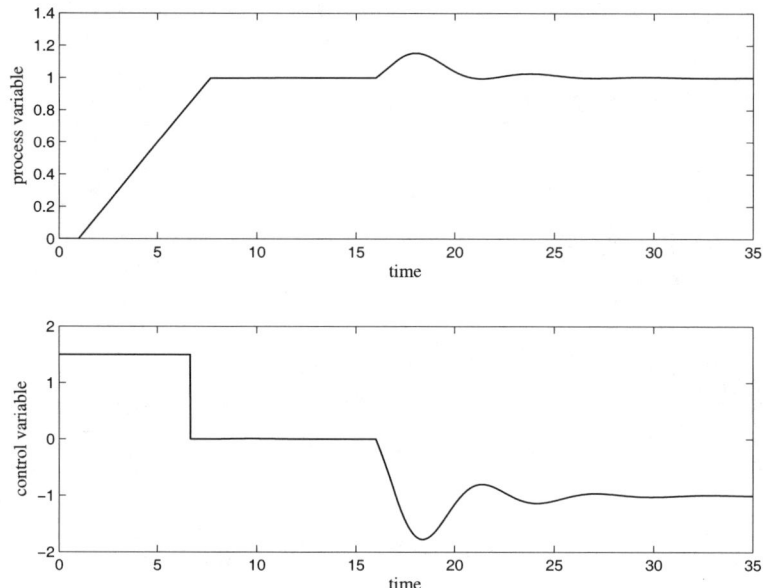

Fig. 6.4. Result of the application of the TOPC strategy with the IPDT process $P_3(s)$

in the load disturbance rejection task is evaluated at time $t = 15$.
It appears that a time optimal transition is achieved and, at the same time, the PI controller is tuned satisfactorily.
As a last illustrative example, the process

$$P_4(s) = \frac{1}{s}e^{-s} \qquad (6.30)$$

is considered (note that the value of the dead time is significant with respect to the time constant). The same design parameters of the previous case (process $P_3(s)$) are employed. The result obtained is shown in Figure 6.5. It is $t_1 = \hat{L} = 1.0$ and $t_2 = 1.13$ (the estimated process parameter is $\hat{K} = 1.012$). The optimal switching time is calculated as $t_{s1} = 0.66$. Since it is $t_{s1} < t_2$, t_{s1} is fixed equal to t_2 and then the optimal switching time t_{s2} is calculated by means of Expression (6.24). It results $t_{s2} = 1.601$. Thus, the control variable is set to zero for $t_{s2} < t < t_{s2} + \hat{L}$ before applying the PI controller, whose parameters have been set to $K_p = 0.915$ and $T_i = 4$.
It is worth underlying that the resulting overshoot is due to the high value of the dead time of the process, for which the control variable is set to its maximum value u_{max} for a too long time interval.

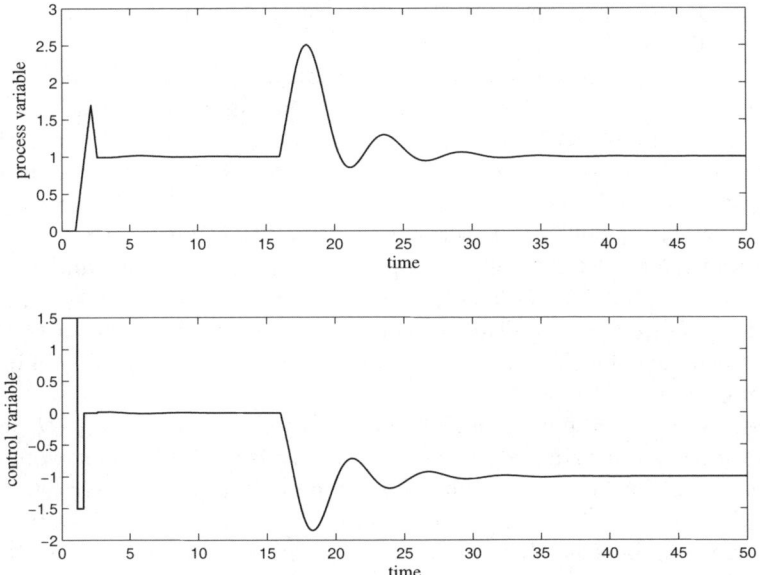

Fig. 6.5. Result of the application of the TOPC strategy with the IPDT process $P_4(s)$

6.3.5 Experimental Results

Level Control

A few experiments related to the application of the TOPC strategy to a level control problem are presented hereafter. Results have been obtained with the laboratory setup described in Section A.1. In all cases it has been set $\varepsilon = 10^{-3}$ and the following tuning rule (Skogestad, 2003) has been selected:

$$K_p = \frac{0.3\hat{T}}{\hat{K}\hat{L}} \qquad T_i = \min\{\hat{T}, 8\hat{L}\}. \tag{6.31}$$

Note that, with respect to the original tuning rule in (Skogestad, 2003), here the proportional gain K_p has been conveniently detuned in order to take into account system nonlinearities and the unavoidable unmodelled dynamics.
In the first experiment, the set-point value has been fixed at $y_{sp} = 3$ V and it has been fixed $u_{max} = 4.5$ V and $u_{min} = 0$ V. For the first two seconds of the experiment the control variable has been set to zero in order to estimate the noise band and to measure the value $y_0 = 0.68$ V (note that $u_0 = 0$ V and therefore $u^+ = u_{max} = 4.5$ V and $u^- = u_{min} = 0$ V). Thus, the value of y_1 is easily determined as $y_{sp} - y_0 = 2.32$ V. Then, at time $t = t_0 = 2$ s the control variable is set to $u = u_{max}$ (see step 3 of the TOPC algorithm

for FOPDT processes) and, subsequently, the dead time is estimated at time $t = t_1 = 3.68$ s as $\hat{L} = 1.68$ s (see step 4).
The recursive least squares algorithm (initialised with $\hat{K}_0 = 1$ and $\hat{T}_0 = 10$) converges at time $t_2 = 6.19$ s with the gain and the time constant estimated as $\hat{K} = 0.94$ and $\hat{T} = 8.76$ (see steps 6(a) and 6(b) of the TOPC algorithm). The resulting value of t_{s1} is 6.96 s (see Expression (6.20)). Since t_{s1} is greater than t_2, then there is no need to set $u = u_{min}$ for a given time interval. Hence, at time $t = t_{s1} = 8.96$ s it is set $u = y_1/\hat{K} = 2.47$ V for a time interval of 1.68 s and then the PI controller is applied (it results $K_p = 1.66$ and $T_i = 8.76$, see Expression (6.31)). The resulting process output and control variable are plotted in Figure 6.6. Note that at time $t = 100$ s and $t = 200$ s a (software) step disturbance of -0.5 V has been applied to the control variable in order to test the designed controller.
A satisfactory overall performance appears. Note that at time $t_{s1} + \hat{L}$ the process output has not attained the desired steady-state value (as it would be in the ideal case) because of the unavoidable modelling inaccuracies. Indeed, it is the PI controller that immediately compensates the residual system error. Actually, it can be seen that no overshoot is practically present and the load disturbances are rejected with a low settling time, demonstrating that the PI controller is well tuned.
In order to verify what happens with a lower set-point value, the case with $y_{sp} = 2$ V (and $u_{max} = 4.5$ V) has been then considered (note that $u_0 = 0$ V and $t_0 = 2$ s as before). In this case, it results $y_0 = 0.60$ V, $y_1 = 1.4$ V, $t_1 = \hat{L} = 1.44$ s and $t_2 = 8.85$ s (with the estimated parameters $\hat{K} = 0.52$ and $\hat{T} = 4.64$). Since the first switching time $t_{s1} = 6.23$ s results to be lower than t_2, the control variable is immediately set to $u_{min} = 0$ until the time $t = t_{s2} = 10.03$ s (see step 6(e) of the TOPC algorithm) and then set to 2.69 V for a time interval equal to $\hat{L}$ before the PI controller is applied ($K_p = 1.84$ and $T_i = 4.64$). Results are shown in Figure 6.7. Note that an overshoot is present as expected, and the tuning of the PI controller is again satisfactory, as demonstrated by the load disturbance rejection performance.
In any case, it appears that if a limitation of the overshoot is required it is sensible to select lower values of u_{max} for lower values of y_{sp}. For this reason, the case where $y_{sp} = 2$ V and $u_{max} = 3$ V ($u_{min} = 0$ V) has been also considered. The result is shown in Figure 6.8, where $t_0 = 2$ s, $y_0 = 0.64$ V, $y_1 = 1.36$ V, $t_1 = \hat{L} = 1.76$ s, $t_2 = 9.64$ s, $\hat{K} = 0.88$, $\hat{T} = 9.11$, $t_{s1} = 6.59$ s, $t_{s2} = 8.52$ s, $K_p = 1.76$ and $T_i = 9.11$.
As expected, it appears that the reduction of the overshoot is obviously paid with a higher rise time. Once again, however, the (automatically) designed PI controller achieves a satisfactory performance.

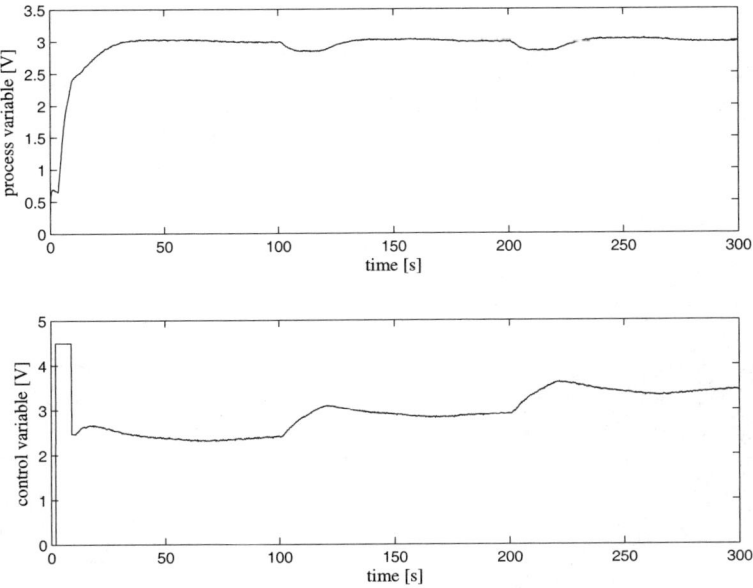

Fig. 6.6. Results of the TOPC algorithm for level control task with $y_{sp} = 3$ V and $u_{max} = 4.5$ V

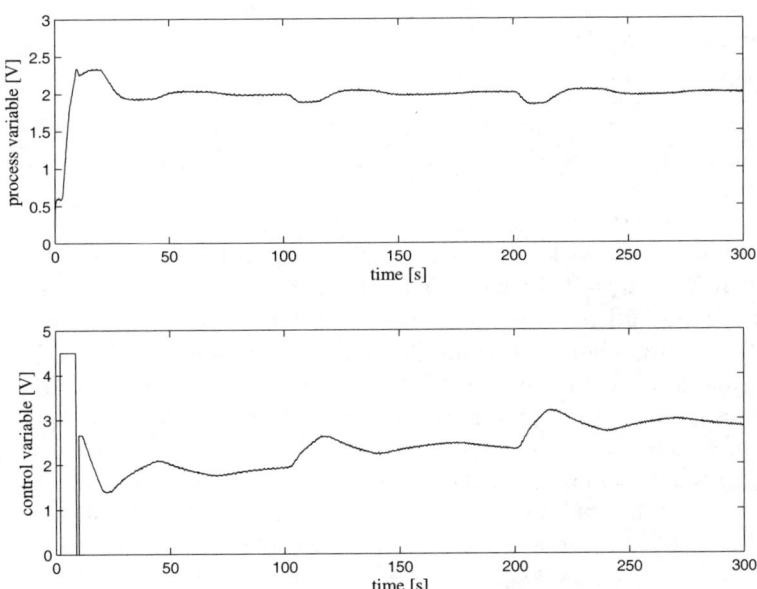

Fig. 6.7. Results of the TOPC algorithm for level control task with $y_{sp} = 2$ V and $u_{max} = 4.5$ V

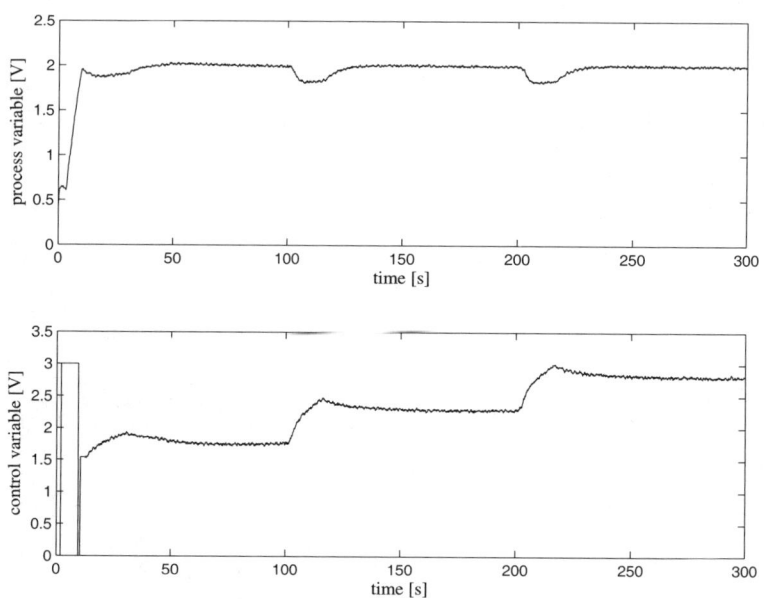

Fig. 6.8. Results of the TOPC algorithm for level control task with $y_{sp} = 2$ V and $u_{max} = 3$ V

Temperature Control

The TOPC strategy has been also applied to a temperature control task by means of the experimental setup described in Section A.2. As in the case of level control, it has been set $\varepsilon = 10^{-3}$ and the PI controller is tuned according to the rule (6.31).

For the first experiment the values $y_{sp} = 3.5$ V and $u_{max} = 4$ V, $u_{min} = 0$ V have been selected. As for the level control task, the first two seconds of the experiment (*i.e.*, $t_0 = 2$ s) have been adopted (by setting $u_0 = 0$ V) to measure the noise band and the value of $y_0 = 0.58$ V (thus, $y_1 = y_{sp} - y_0 = 2.92$ V). Then, the TOPC algorithm is applied by initialising $\hat{K}_0 = 1$ and $\hat{T}_0 = 100$. The estimated dead time is $\hat{L} = 26.20$ s and the convergence of the recursive least squares algorithm occurs at time $t_2 = 1351.7$ s with $\hat{K} = 1.16$ and $\hat{T} = 1308.4$. The switching time t_{s1} results to be 1300.4 s and therefore the second switching time is calculated as $t_{s2} = 1381.0$ s. The control variable is therefore set to zero for a period of $t_{s2} - t_2 = 29.3$ s, before being set to $y_1/\hat{K} = 2.52$ V for a period of $\hat{L} = 26.20$ s. Then, the PI controller ($K_p = 12.91$ and $T_i = 209$) is applied and in order to test it, a (software) load step disturbance of -0.5 V has been applied at time $t = 2000$ s. The resulting process output and control variable are plotted in Figure 6.9. Note that the process is clearly lag-dominant, but the adopted tuning rule, which avoids the pole-zero cancellations, guarantees a somewhat fast load rejection.

Note also that a very low overshoot occurs.

For a second experiment, the values $y_{sp} = 3$ V and $u_{max} = 4.5$ V have been selected. Being $y_0 = 0.58$ V it results $y_1 = 2.42$ V. Then, the dead time is estimated as $\hat{L} = 25.35$ s and the gain and time constant estimation ends at time $t_2 = 1250.1$ s with $\hat{K} = 1.13$ and $\hat{T} = 1301.5$. It is therefore $t_{s1} = 840.9$ s (which is less than t_2) and $t_{s2} = 1585.4$ s. The PI controller is then tuned with $K_p = 13.65$ and $T_i = 202.8$. Results are shown in Figure 6.10, where the expected overshoot appears as well as the satisfactory tuning of the controller, as shown by the load disturbance rejection performance. Note that, although a significant overshoot occurs, the process variable is still far from its saturation, as expected since a low value of the set-point (with respect to the value of u_{max}) has been selected.

To clarify better this fact, in the third experiment the value of y_{sp} has been raised to 4 V, while the value of u_{max} has been kept to 4.5 V. The result is reported in Figure 6.11, where it is $y_0 = 0.49$ V, $y_1 = 3.51$ V, $t_1 - t_0 = \hat{L} = 23.56$ s, $t_2 = 1163.6$ s, $\hat{K} = 1.16$, $\hat{T} = 1331.5$, $t_{s1} = 1485.9$ s, $K_p = 14.60$ and $T_i = 188.4$. In this case there is no need of setting the control variable to zero for a determined period and the process output attains its set-point value monotonically. As in the previous cases, the load disturbance rejection is effective, demonstrating a satisfactory PI controller design.

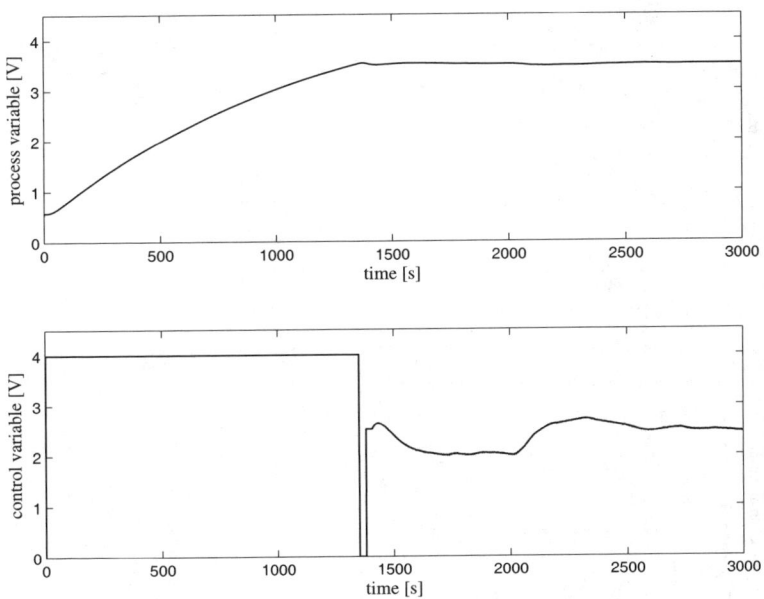

Fig. 6.9. Results of the TOPC algorithm for temperature control task with $y_{sp} = 3.5$ V and $u_{max} = 4$ V

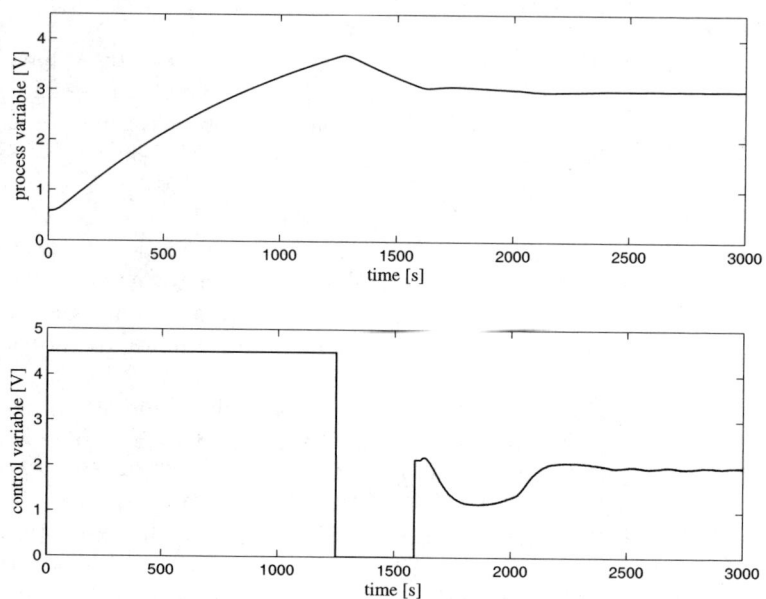

Fig. 6.10. Results of the TOPC algorithm for temperature control task with $y_{sp} = 3$ V and $u_{max} = 4.5$ V

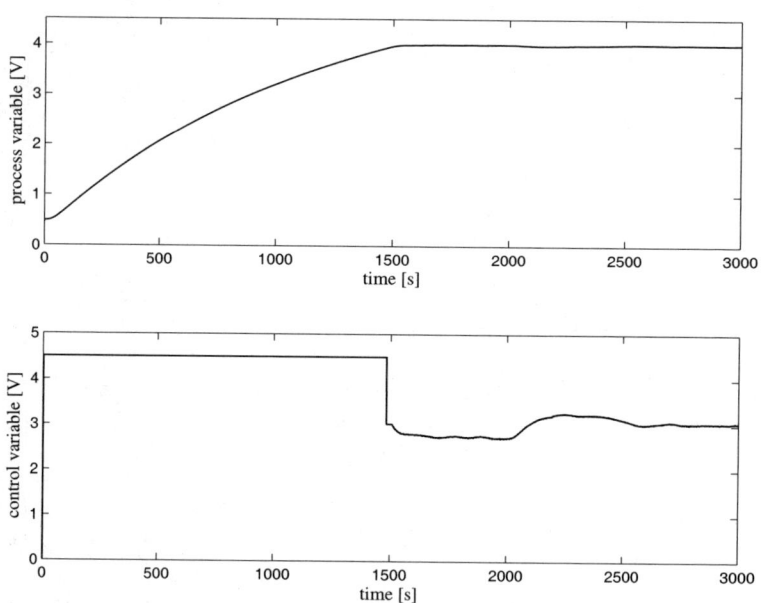

Fig. 6.11. Results of the TOPC algorithm for temperature control task with $y_{sp} = 4$ V and $u_{max} = 4.5$ V

6.3.6 Discussion

From the results presented it appears that the time-optimal Plug&Control strategy is effective in providing a fast commissioning of the control loop when a tight performance is not required. Indeed, the technique is suitable for those processes where the dominant dynamics is not of high order and where possibly somewhat large overshoots are allowed (at least in the start-up phase of the process). It is worth stressing, however, that by a suitable choice of the design parameters (namely, the maximum and minimum level of the control variable during the three-state control phase) the overshoot can be significantly reduced (at the expense of the rise time). In fact, it has been shown that the design parameters have a clear physical meaning and technical problems can be solved in a practical context by exploiting a reasonable knowledge of the plant.

Finally, it has to be noted that, instead of the recursive least squares algorithm, a batch least squares algorithm (Sung *et al.*, 1998) can be applied for the identification purpose (Visioli, 2003*b*). Although in this case the methodology is more capable of coping with a high-order dynamics, it has the disadvantage that the user has to select the part of the transient for which data are collected for the estimation of the parameters. Although this choice somehow allows the handling of the trade-off between estimation accuracy and the resulting overshoot, it might not be intuitive to the user.

6.4 Conclusions

In this chapter different Plug&Control strategies have been presented. It has been shown how this approach is very promising since, provided that the methodology is applied in a suitable context, it is capable of providing a fast and effective design of the control loop. Obviously, this feature is more relevant in large plants, when there are many (simple) loops to tune.

It is believed that in the future new techniques in this framework can be devised since there is much room for the improvement of the performance, especially for high-order processes.

# 7

Identification and Model Reduction Techniques

7.1 Introduction

In this chapter, the issue of the system identification, in the context of PID tuning and control, is addressed. Rather than present an exhaustive review of the existing methodologies for the estimation of a (parametric or non parametric) model of the process, which would be a very huge task, the aim of the following sections is to point out possible issues that might arise when selecting the identification procedure. For this purpose, some techniques are presented and their main features are highlighted. In particular, techniques based on the evaluation either of an open-loop step response or of a relay feedback test are considered, in order to estimate the parameters of a FOPDT or a SOPDT transfer function. This choice is motivated by the fact that methods of this kind are the most adopted in practical cases because of their simplicity. The analysis focuses on self-regulating processes which do not exhibit an oscillatory dynamics.

In addition, the issue of designing a PID controller when a high-order model of the process is available is addressed. In particular, two approaches in the Internal Model Control (IMC) framework (Morari and Zafiriou, 1989) are analysed and discussed. In the first, the (high-order) controller that results from considering the high-order process model is reduced through a Maclaurin series expansion in order to obtain a PID controller. In the second, the process model is first reduced (different techniques are considered for this purpose) in order to naturally obtain a PID controller.

7.2 FOPDT Systems

The great majority of PID tuning rules actually assume that a FOPDT model of the process is available, namely the process is described by the following transfer function:

$$P(s) = \frac{K}{Ts+1}e^{-Ls} \qquad T>0, \quad L>0 \qquad (7.1)$$

where K is the estimated gain, T is the estimated time constant and L is the estimated (apparent) dead-time. This is motivated by the fact that many processes can be described effectively by this dynamics and, most of all, that this suits well with the simple structure of a PID controller.

Different methods have been therefore proposed in the literature to estimate the three parameters by performing a simple experiment on the plant. They are typically based either on an open-loop step response or on a closed-loop relay feedback experiment.

7.2.1 Open-loop Identification Techniques

The identification techniques based on an open-loop experiment generally derive the FOPDT transfer function parameters based on the evaluation of the process step response (often denoted as the process reaction curve). This can be done in many ways. Some techniques proposed in the literature are explained hereafter with the aim of highlighting their main features.

The Tangent Method

The tangent method consists of drawing the tangent of the process response at the inflection point. Then, the process gain can be determined simply by dividing the steady-state change in the process output y by the amplitude of the input step A. Then, the apparent dead time L is determined as the time interval between the application of the step input and the intersection of the tangent line with the time axis. Finally, the value of $T+L$ is determined as the time interval between the application of the step input and the intersection of the tangent line with the straight line $y = y_\infty$ where y_∞ is the final steady-state value of the process output. Alternatively, the value of $T+L$ can be determined as the time interval between the application of the step input and the time when the process output attains the 63.2% of its final value y_∞. From this value the time constant T can be trivially calculated by subtracting the previously estimated value of the time delay L. The method is sketched in Figure 7.1.

It is worth stressing that the method is based on the fact that it gives exact results for a true FOPDT process. The main drawback of this technique is that it relies on a single point of the reaction curve (*i.e.*, the inflection point) and that it is very sensible to the measurement noise. In fact, the measurement noise might cause large errors in the estimation of the point of inflection and of the first time derivative of the process output.

The Area Method

A technique that is more robust to the measurement noise is the so-called area method. By taking into account that the process gain K can be determined

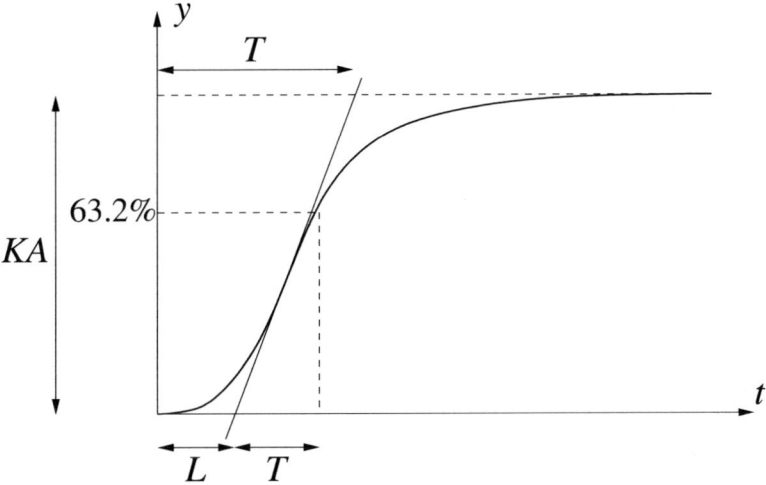

Fig. 7.1. Application of the tangent method for the estimation of a FOPDT transfer function

as for the tangent method, it consists of first calculating the area between the process output and the straight line $y = y_\infty$, namely:

$$A_1 := \int_{t_0}^{\infty} (y_\infty - y(t))dt \qquad (7.2)$$

where t_0 is the time instant of the input step change. Then, the value of $T + L$ can be determined by the following expression:

$$L + T = \frac{A_1}{K}. \qquad (7.3)$$

Subsequently, the area A_2 between the process output and the time axis in the time interval from t_0 to $T + L$ is evaluated, namely,

$$A_2 := \int_{t_0}^{T+L} (y(t) - y_0)dt \qquad (7.4)$$

where y_0 is the initial process output steady-state value. Finally, the values of T and L are determined by means of the following expressions:

$$T = \frac{eA_2}{K} \qquad L = \frac{A_1 - KT}{K} \qquad (7.5)$$

The procedure is depicted in Figure 7.2. It is worth noting that the previous expressions are derived by considering the response of a FOPDT system. In other words, as for the tangent method, a perfect parameter estimation occurs

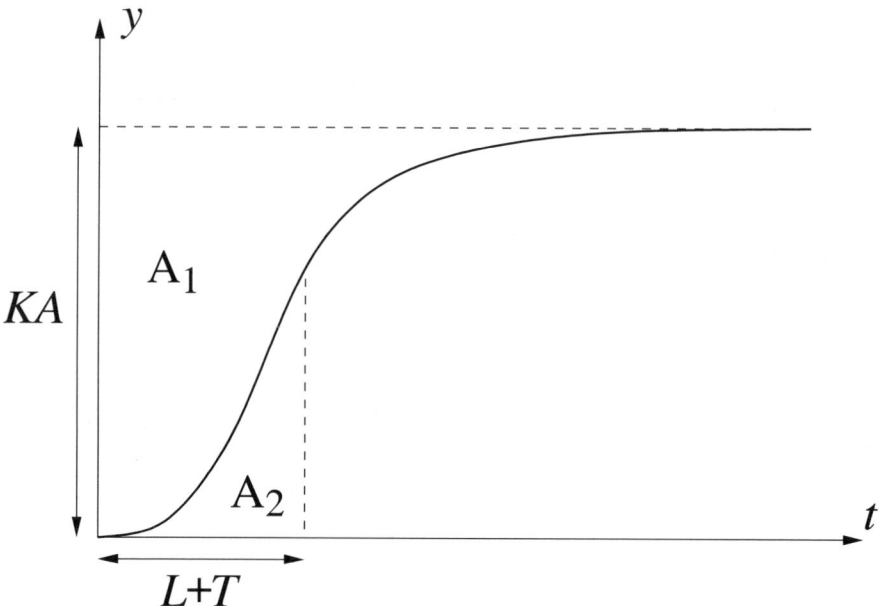

Fig. 7.2. Application of the area method for the estimation of a FOPDT transfer function

if the process has exactly a FOPDT dynamics. In any case it is also possible to apply it for (moderately) undershooting, overshooting or oscillatory responses, provided that the part of $y(t)$ that is less than y_0 be truncated to y_0 and the part of $y(t)$ that is greater than y_∞ be "mirrored" with respect to y_∞ (Leva et al., 2001).

Being based on the calculus of integrals, this approach is more relevant from the computational point of view (the final result is difficult to derive by hand) but has the remarkable feature of being much more robust to the measurement noise than the tangent method. However, it has a drawback in the possible determination of a negative value of the time delay L when the process exhibits a nonlinear lag-dominant dynamics.

Consider for example the nonlinear process described by the following differential equation (note that this can be a model of a tank system where the process variable y is the fluid level, the manipulated variable u is the inflow and $Q_o = 1.2\sqrt{y}$ is the outflow):

$$\dot{y}(t) = \frac{1}{16}(u(t-1) - 1.2\sqrt{y}). \qquad (7.6)$$

The unitary step response is plotted in Figure 7.3 (note that there is no measurement noise). The straightforward application of the area method gives the following results: $K = 0.69$, $T = 18.8$ and $L = -0.15$. Obviously, if

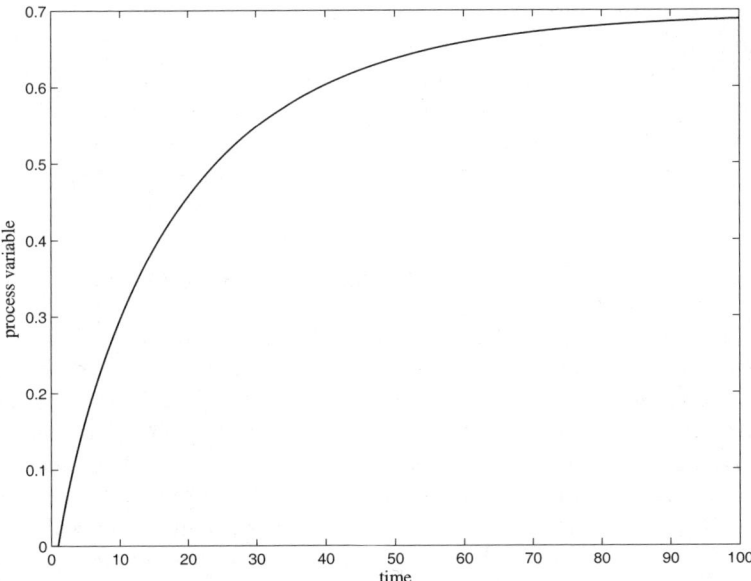

Fig. 7.3. Example of a step response for which the application of the area method gives a negative dead time

the process is not lag-dominant, a significant underestimation of the dead time results in any case. This might be a problem from the point of view of the tuning of the PID parameters because a more aggressive controller than expected might result.

Two-points-based Method

A method that is based on the estimation of two time instants of the reaction curve has been proposed in (Sundaresan and Krishnaswamy, 1978) (it is also reported in (Seborg *et al.*, 2004)). It consists in determining the time instants t_1 and t_2 when the process output attains 35.3% and 85.3% of its final steady-state respectively. Then, the dead time and the time constant are calculated by means of the following formulae:

$$T = 0.67(t_2 - t_1) \qquad L = 1.3t_1 - 0.29t_2. \tag{7.7}$$

The gain of the process is determined as in the previous methods. The previous formulae have been found, by considering many data sets, in order to minimise the difference between the experimental process response and the model response. It is worth noting that the method is very simple (indeed, it can be applied by hand easily).

This technique, in addition to the problem of being sensible to the measurement noise in the estimation of the two times t_1 and t_2, suffers from the same problem as the area method. Indeed, if it is applied to the same transient response obtained by Process (7.6) (see Figure 7.3), it results in $K = 0.69$, $T = 19.69$ and $L = -0.64$. Thus, the same considerations for the area method apply also in this case. As a consequence, care should be devoted in choosing the appropriate context for applying these techniques.

Least-squares-based (with Model Reduction) Methods

A possible way to obtain a FOPDT model is to obtain first a high-order model and then to reduce it to FOPDT form. A method that can be exploited in this context has been proposed in (Sung et al., 1998). The first step is to estimate an arbitrarily high-order transfer function (denoted by $G(s)$) by means of a least-squares approach. A remarkable robustness with respect to measurement noise is achieved by considering the integrals of the input and output signals instead of their derivatives. Then, a low-order model can be derived by applying a model-reduction algorithm. A salient feature of the methodology is that it does not require any special input to the process, but it can be applied in different operating conditions.

Here the case where a step input is applied to the process and a FOPDT model is determined starting from the obtained high-order model is considered. While the value of the gain K can be found as usual by dividing the steady-state change in the process output y by the amplitude of the input step, again a least-squares-based approach is employed in order to find the time constant of the process model (7.1) that minimise the difference between the magnitude of the frequency response of the high-order model and of the FOPDT one. Formally, the value of T that satisfies the following equation is determined as:

$$|G(j\omega_i)| = \frac{K}{\sqrt{T^2\omega_i^2 + 1}}, \qquad 0 < \ldots < \omega_i < \ldots < \omega_u \qquad (7.8)$$

where ω_u is the ultimate frequency of $G(s)$. Finally, the apparent dead time of the process is determined as the value that gives the same phase angle (i.e., $-\pi$) of the high-order model at the ultimate frequency ω_u:

$$L = \frac{\pi - \arctan(T\omega_u)}{\omega_u}. \qquad (7.9)$$

It appears that this method requires much more computational effort than the previous ones (note that ω_u has to be calculated since it is not available). Critical choices in this context are the selection of the order (and of the relative order) of the rational transfer function $G(s)$ and the portion of the step response to be considered (it is obviously meaningless to use data after the steady-state has been attained). For the first issue, by taking into account that eventually a FOPDT model is determined, a sensible choice is to select a fourth-order transfer function with a relative order equal to one (this means that the dead time term is approximated by three zeros and three poles). For the second issue, it is sufficient to consider the settling time at 2% of the steady-state value.

It is worth noting at this point that other model-reduction methodologies (that result in a rational transfer function) will be presented in Section 7.5.

Method Based on Laguerre Functions

Laguerre functions are a set of complete orthonormal functions defined as:

$$l_1(t) = \sqrt{2p}e^{-pt}$$

$$l_2(t) = \sqrt{2p}(-2pt + 1)e^{-pt}$$

$$\vdots$$

$$l_i(t) = \sqrt{2p}\left[(-1)^{i-1}\frac{(2p)^{i-1}}{(i-1)!}t^{i-1} + (-1)^i\frac{(i-1)(2p)^{i-2}}{(i-2)!}t^{i-2} + (-1)^{i-1}\frac{(i-1)(i-2)(2p)^{i-3}}{2!(i-3)!}t^{i-3} + \cdots + 1\right]e^{-pt} \quad (7.10)$$

where $p > 0$ is called the time scaling factor. In the context of system identification, the property that an arbitrary function $g(t)$ can be expanded with respect to a set of functions that is orthonormal and complete over the interval $(0, \infty)$ can be exploited. In particular, if $g(t)$ is the unit impulse response of a process, it can be written as

$$g(t) = c_1 l_1(t) + c_2 l_2(t) + \cdots + c_i l_i(t) + \cdots \quad (7.11)$$

where the c_i's are the coefficients of the expansion. By applying the Laplace transform we obtain

$$G(s) = c_1 L_1(s) + c_2 L_2(s) + \cdots + c_i L_i(s) + \cdots \quad (7.12)$$

where

$$L_i(s) = \frac{\sqrt{2p}(s-p)^{i-1}}{(s+p)^i} \quad i = 0, 1, \ldots \quad (7.13)$$

are often referred as the Laguerre filters. In theory, the expansion expresses in Equation (7.11) requires an infinite number of terms to converge to the true impulse response. However, an arbitrarily good approximation can be obtained by truncating the series after N terms. In any case, from Expression (7.13) it can be easily deduced that the estimated transfer function has coincident poles.

Starting from the measured step response $y(t)$, the coefficients of the expansion can be calculated as:

$$c_1(t) = p \int_0^{T_s} y(t)l_1(t)dt + \bar{y}l_1(T_s)$$

$$c_2(t) = 2p \int_0^{T_s} y(t)l_1(t)dt + p \int_0^{T_s} y(t)l_2(t)dt + \bar{y}l_2(T_s)$$

$$\vdots$$

$$c_i(t) = 2p \int_0^{T_s} y(t)l_1(t)dt + 2p \int_0^{T_s} y(t)l_2(t)dt + \ldots + p \int_0^{T_s} y(t)l_i(t)dt + \bar{y}l_i(T_s) \tag{7.14}$$

where T_s is the time at which the process attains the steady-state and $\bar{y}$ is the steady-state value. The choice of the time scaling factor p (*i.e.*, of the location of the approximating system poles) affects the accuracy of the approximation, in the sense that a poor choice of p requires more terms in order to provide a desired model accuracy. For this reason, methodologies for a sound selection of p have been investigated (Wang and Cluett, 1994). In particular, it is proposed to search for the optimal value of p (in the sense that it gives the best approximation for a given value of N) in the interval $[p_{min}, p_{max}]$, where $p_{min} = 4/T_s$ and $p_{max} = 5p_{min}$ if $N \leq 4$ and $p_{max} = 10p_{min}$ if $N > 4$. This interval is then discretised and for any value of p the Laguerre coefficients c_N and c_{N+1} are determined by means of Formulae (7.14). The values of p for which $c_N c_{N+1} = 0$ are selected as possible candidates. Among them, the one that produces the maximum value of $\sum_{i=1}^N c_i^2$ is selected as the best one.

A detailed analysis of the use of Laguerre functions in this context can be found in (Wang and Cluett, 2000). It is shown that this modelling technique based on the step response has nice statistical properties: it is very robust to the measurement noise and a simple strategy for the pretreating of the data can be implemented in order to cope with disturbances.

In any case, it has to be stressed that the technique requires a somewhat computational effort and the high-order model that results has to be subsequently reduced to a FOPDT model. The method presented in the previous subsection (or others presented in Section 7.5) can be used for this purpose. By applying a similar reasoning, the unique user-chosen parameter N can be chosen as equal to four.

Finally, it is worth noting that a closed-loop approach based on the use of Laguerre functions and a least-squares technique has been proposed in (Park et al., 1997).

Optimisation-based Method

Another technique that is worth to being considered is to estimate the three transfer function parameters K, T and L by solving the following optimisation problem:

$$\min_{K,T,L} \int_0^\infty |y(t) - y_m(t)|dt \tag{7.15}$$

where $y(t)$ denotes the experimental step response and $y_m(t)$ denotes the model step response. In other words, the model parameters are searched in order to minimise the difference between the experimental step response and the model step response.

In order to solve the posed optimisation problem, genetic algorithms (Mitchell, 1998) can be employed. In any case, obviously, the computational effort is significant and this is the major drawback of this method.

7.2.2 Closed-loop Identification Techniques

The closed-loop identification techniques employed in industrial settings typically rely on a relay-feedback experiment. The initial idea of the use of the relay-feedback controller (Åström and Hägglund, 1984) is to evaluate the obtained process output oscillation (see Section 1.3) in order to obtain a nonparametric model of the process, namely its ultimate gain K_u and the ultimate frequency ω_u, in analogy with the original idea of the ultimate sensitivity experiment of Ziegler–Nichols (Ziegler and Nichols, 1942), where the control system is led to the stability limit.

However, recently, different techniques for the determination of a FOPDT parametric model based on a relay-feedback experiment have been also devised. A few of them are presented hereafter, again with the aim of highlight possible issues that might arise when they are applied in a practical context.

Standard Relay-feedback Method

The original relay-feedback experiment proposed in (Åström and Hägglund, 1984) involves the use of a standard symmetrical relay in order to generate a persistent oscillatory response of the process output. Denoting by h the amplitude of the relay and by A the amplitude of the output oscillations, the value of the ultimate gain can be derived, by applying the describing function theory, as:

$$K_u = \frac{4h}{\pi A}. \qquad (7.16)$$

The ultimate period T_u is simply the period of the obtained output oscillation. Based on these two values, many PID tuning rules can be applied (O'Dwyer, 2006). Only the amplitude h of the relay has to be selected by the user. This should be done in order to provide an output oscillation of sufficient amplitude to be well distinguished from the measurement noise, but at the same time it has not to be too high so that the process is perturbed as less as possible (and the normal production is not interrupted). Indeed, it is worth stressing that the estimation of the output oscillation is sensible to the measurement noise and therefore some filtering technique has to be applied (Wang et al., 1999c) (this is a drawback with respect to the open-loop least-squares-based methods considered in the previous section). In addition to having just one parameter to be selected by the user and to be performed in closed-loop, so that the process is kept close to the set-point value, the main advantage of this identification technique is that a short time is necessary to run the test (with respect to the use of a pseudo-random binary sequence (PBRS) (Ljung, 1996)). Further, possible load disturbances that might occur during the experiment can be easily detected by the change to asymmetric pulses in the control variable.

In any case, the obtained values of the ultimate gain and ultimate period are approximated, because of the adoption of the describing function theory and the estimation may not be accurate enough for some applications, for example when the process exhibits a long dead time (Li et al., 1991). In order to improve the estimation of the actual values of K_u and T_u, different methods have been proposed in the literature (see, for example, (Majhi and Atherton, 2000; Atherton, 2000)). In any case, if it is desired to implement a model-based controller, the knowledge of a transfer function is required. A FOPDT transfer function can be derived by employing the following two relations, which can be derived by calculating the ultimate gain and period for Process (7.1) (Luyben, 1987):

$$T = \frac{\tan(\pi - L\omega_u)}{\omega_u}, \tag{7.17}$$

$$T = \frac{\sqrt{(KK_u)^2 - 1}}{\omega_u}. \tag{7.18}$$

It can be noted that there are two equations for three parameters. Thus, the gain of the process has to be estimated in an other way. Then, Equation (7.18) can be employed to estimate the value of T and subsequently the value of L can be determined by means of Equation (7.17). Alternatively, the dead time of the process can be estimated in an other way (for example at the beginning of the experiment, with considerations analogous to those made for the open-loop experiments) and then the time constant T and the process gain K are subsequently calculated. However, in this case the resulting time constant and process gain might incorrectly result to be negative (Vivek and Chidambaram, 2005a) and therefore this approach should be avoided.

In order to cope with the inaccuracies due to the presence of the describing function approximation, in (Yu, 1999) it is proposed to substitute Equation (7.17) with the following one:

$$T = \frac{\pi}{\omega_u \ln(2e^{\frac{L}{T}} - 1)}. \tag{7.19}$$

Alternative Calculation of the FOPDT Parameters

An alternative way of identifying the FOPDT transfer function by means of a symmetrical relay-feedback experiment has been proposed in (Vivek and Chidambaram, 2005a). It consists of first evaluating the integral

$$y(s_1) = \int_0^\infty y(t)e^{-st}dt \qquad (7.20)$$

for $s_1 = 8/t_s$, where t_s is the time at which three repeated cycles of oscillations appear in the process output after the initial transient has ended. Analogously, the integral

$$u(s_1) = \int_0^\infty u(t)e^{-st}dt \qquad (7.21)$$

is also evaluated for $s_1 = 8/t_s$. With the resulting values, the following equation can be posed:

$$\frac{K}{Ts_1+1}e^{-Ls_1} = \frac{y(s_1)}{u(s_1)}. \qquad (7.22)$$

Then, the frequency response of the process transfer function can be written as

$$P(j\omega_u) = \frac{y(j\omega_u)}{u(j\omega_u)} = \frac{c_1 - jd_1}{c_2 - jd_2} \qquad (7.23)$$

where

$$\begin{aligned} c_1 &= \int_0^{T_u} y(t)\cos(\omega_u t)dt \\ d_1 &= \int_0^{T_u} y(t)\sin(\omega_u t)dt \\ c_2 &= \int_0^{T_u} u(t)\cos(\omega_u t)dt \\ d_2 &= \int_0^{T_u} u(t)\sin(\omega_u t)dt \end{aligned} \qquad (7.24)$$

where ω_u is the frequency of the oscillation obtained in the process output and $T_u = 2\pi/\omega_u$. The values of c_1, d_1, c_2 and d_2 can be evaluated numerically based on the process input and output data $u(t)$ and $y(t)$ obtained from the relay test. Thus, Equation (7.23) can be rewritten as

$$P(j\omega_u) = p + jq \qquad (7.25)$$

where

$$p = \frac{c_1 c_2 + d_1 d_2}{c_2^2 + d_2^2} \qquad q = \frac{d_2 c_1 - d_1 c_2}{c_2^2 + d_2^2}. \qquad (7.26)$$

By taking into account that

$$P(j\omega_u) = \frac{K}{Tj\omega_u + 1}e^{-Lj\omega_u} \qquad (7.27)$$

it can be easily deduced that

$$p + jq = \frac{K(\cos(L\omega_u) - j\sin(L\omega_u))}{Tj\omega_u + 1}. \qquad (7.28)$$

Finally, by equating the real and imaginary parts, it can be written

$$p - qw_uT - K\cos(Lw_u) = 0, \tag{7.29}$$

$$q + pw_uT + K\sin(Lw_u) = 0. \tag{7.30}$$

The three process parameters can be obtained by means of Equations (7.22), (7.29) and (7.30). It is worth stressing that a numerical solution has to be derived. In any case the merit of the methodology is that all the three parameters can be found with a single relay test.

Use of an Asymmetrical Relay

If a biased relay is adopted for the experiment, the process gain K can be determined by using the process input and output data $u(t)$ and $y(t)$ according to the expression (Shen et al., 1996):

$$K = \frac{\int_0^{2\pi} e(t)d(\omega_u t)}{\int_0^{2\pi} u(t)d(\omega_u t)}. \tag{7.31}$$

Then, the other process parameters T and L can be calculated by means of Equations (7.29) and (7.30), for which an analytical solution exists (Srinivasan and Chidambaram, 2003). Obviously, in this case both the up-amplitude and the down-amplitude of the relay have to be selected. Further, the use of an asymmetrical relay represents a sort of disturbance to the process since it cause the operating point to drift.

Use of a Relay with Hysteresis

As already mentioned, the relay-feedback test is sensitive to the measurement noise. The easiest way to reduce the influence of the noise is to employ a relay with a hysteresis, whose width is usually chosen as twice the noise band. Denoting again by A and T_u the amplitude and the period of the resulting oscillation, and assuming that the process gain K is known, the process time constant can be determined as (Wang et al., 1997):

$$T = \frac{1}{2}T_u \left(\ln \frac{hK + A}{hK - A} \right)^{-1} \tag{7.32}$$

where h is the amplitude of the (symmetrical) relay. Then, the dead time is estimated as

$$L = \frac{1}{2}T_u \left(\ln \frac{hK - \varepsilon}{hK - A} \right) \left(\ln \frac{hK + A}{hK - A} \right)^{-1} \tag{7.33}$$

where ε is the width of the hysteresis. It is worth stressing that the method requires a previous estimation of the process gain.

Use of a Biased Relay with Hysteresis

The identification of a FOPDT process can be performed also by means of a biased relay (Wang *et al.*, 1997) (see Figure 7.4). By denoting the periods and the amplitudes of oscillations as shown in Figure 7.5, the process parameters can be determined as follows. First, the process gain is calculated again as:

$$K = \frac{\int_0^{T_{u1}+T_{u2}} y(t)dt}{\int_0^{T_{u1}+T_{u2}} u(t)dt}. \quad (7.34)$$

Then, the normalised dead time $\Theta = L/T$ is obtained as:

$$\Theta = \ln \frac{(h+h_0)K - \varepsilon}{(h+h_0)K - A_u} \quad (7.35)$$

or

$$\Theta = \ln \frac{(h-h_0)K - \varepsilon}{(h+h_0)K + A_d}. \quad (7.36)$$

Then, the process time constant can be calculated as

$$T = T_{u1} \left(\ln \frac{2hKe^{\Theta} + h_0 K - hK + \varepsilon}{h_0 K + hK - \varepsilon} \right)^{-1} \quad (7.37)$$

or

$$T = T_{u1} \left(\ln \frac{2hKe^{\Theta} - h_0 K - hK + \varepsilon}{h_0 K - hK - \varepsilon} \right)^{-1}. \quad (7.38)$$

The dead time is finally determined by simply calculating $L = \Theta T$. As already mentioned for the simple asymmetrical relay, the technique suffers from the drawback of drifting the process away from the operating point.

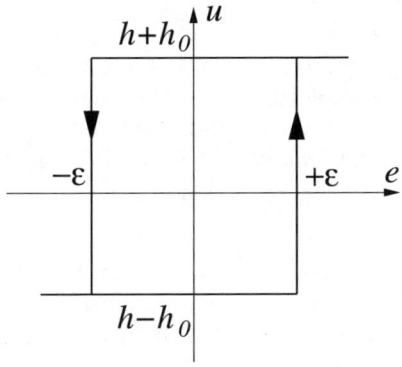

Fig. 7.4. The biased relay

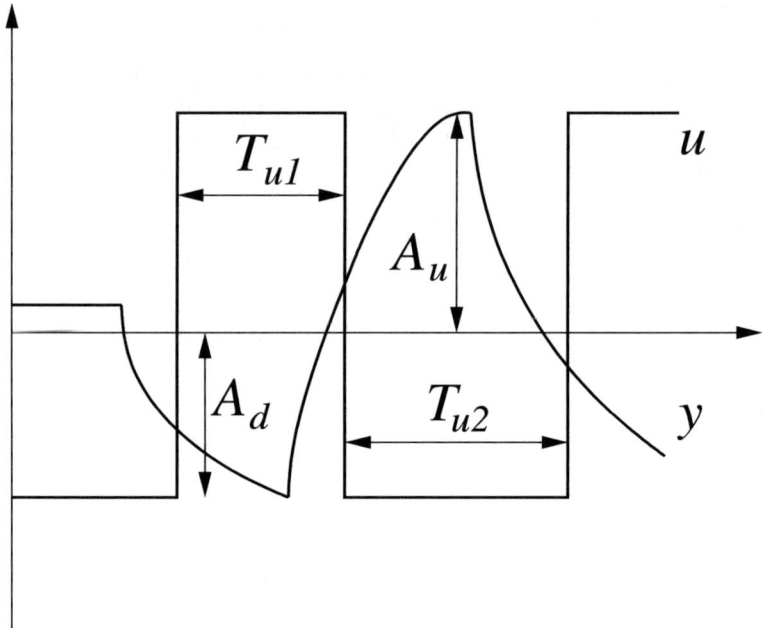

Fig. 7.5. Illustrative example of the use of a biased relay

Method Based on the Output Curve Shape

In the previous sections, it has been shown how to estimate a FOPDT transfer function starting from the data collected after a unique relay-feedback test. However, it has been recognised by many author that for some processes, in particular those with a large dead time, the knowledge of just the ultimate gain and of the ultimate frequency is actually insufficient for the effective design of the (PID) controller.

Indeed, it has been shown that the shape of the output oscillation depends on the process dynamics and analytical expressions are derived in (Panda and Yu, 2003). This fact has been exploited in (Luyben, 2001b), where the curve shape obtained by a standard (not biased and without hysteresis) relay is analysed in order to derive a FOPDT model. In particular, the following algorithm is proposed (see Figure 7.6), where A denotes, as usual, the amplitude of the oscillations.

1. Determine $K_u = 4h/(\pi A)$, where h is the relay amplitude, and evaluate the ultimate period T_u (equivalently, the ultimate frequency ω_u).
2. Draw a vertical line passing through the peak in the curve and denote the corresponding time as t_2.
3. Draw a horizontal line at $A/2$.

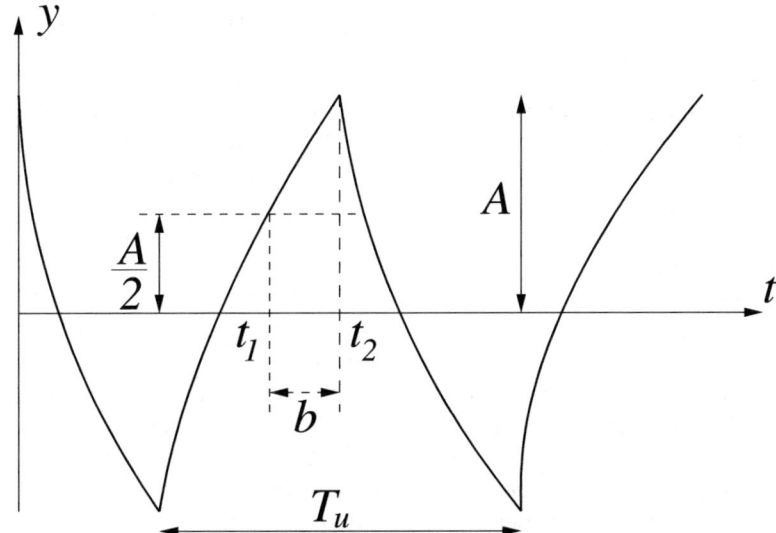

Fig. 7.6. Illustration of the method based on the output curve shape

4. Draw a vertical line passing through the intersection of the curve with the line drawn at step 3 and denote the corresponding time as t_1.
5. Set $b = t_2 - t_1$.
6. Calculate a curvature factor F as $F = 4b/T_u$. This actually indicates if the curve has a shape more similar to a triangle (this results when the dead time is small with respect to the time constant) or more similar to a rectangle (when the dead time is big).
7. Calculate $R := L/T$ by means of the following expression (determined by interpolating results for different processes):

$$\ln\left(\frac{L}{T}\right) = -5.2783 + 12.7147F - 9.8974F^2 + 2.6788F^3. \qquad (7.39)$$

8. Substitute $L = RT$ in the following equation

$$-\omega_u L - \arctan(\omega_u T) = -\pi. \qquad (7.40)$$

and solve iteratively for T. Then, determine $L = RT$.
9. Determine K by means of the equation

$$\frac{K}{\sqrt{1 + (\omega_u T)^2}} = \frac{1}{K_u}. \qquad (7.41)$$

The technique has the great merit of exploiting, in a simple way, the shape of process variable oscillation. Possible drawbacks of the method are its sensitivity to the noise, the somewhat significant computational effort and a possibly inaccurate estimation of the process gain (for example, for the noise-free step response of the process $P(s) = 1/(s+1)^2 e^{-s}$ the results are $K = 2.97$, $T = 5.71$ and $L = 1.29$). Thus, an appropriate tuning procedure should be applied in this context (Scali et al., 1999).

It is worth stressing that the idea of exploiting the shape factor has been developed in (Thyagarajan and Yu, 2003; Panda and Yu, 2005) where the model structure (FOPDT or SOPDT) is also conveniently selected based on the shape of the obtained oscillation.

7.3 SOPDT Systems

Even if the majority of the existing tuning rules are based on FOPDT transfer functions of the process, there are also many rules that relies on the estimation of SOPDT transfer functions (Panda et al., 2004), as they include overdamped, critically damped and underdamped systems and the presence of the two poles can be handled by the two zeros of the controller.

Usually, such a transfer function can be expressed in two ways, namely:

$$P(s) = \frac{K}{(T_1 s + 1)(T_2 s + 1)} e^{-Ls} \qquad T_1 > T_2 > 0, \quad L > 0 \qquad (7.42)$$

or, alternatively,

$$P(s) = \frac{K}{T^2 s^2 + 2\xi T s + 1} e^{-Ls} \qquad T > 0, \quad \xi > 0, \quad L > 0. \qquad (7.43)$$

It has to be noted that Expression (7.43) is more general than Expression (7.42) since it includes the cases of both real and complex conjugate poles (when $\xi \geq 1$ and $\xi < 1$ respectively), while in (7.42) the poles are assumed to be real. However, this latter case is highlighted since it is significant in the context of PID control, as many tuning rules assume that the PID controller is in a series form and the derivative time constant is selected in order to cancel the pole associated with the smallest time constant (see for example (Skogestad, 2003)). Further, processes that present an oscillatory dynamics are rarely found in industrial settings, although the case is of concern when a closed-loop dynamics is considered (for example, the secondary loop of a cascade control system, see Section 9.2.1).

Note also that, for the sake of simplicity, it has been assumed that the process has no zeros. This fact is in any case briefly addressed hereafter. As for the estimation of a FOPDT model, both (step-based) open-loop identification techniques and (relay-feedback-based) closed-loop identification techniques are addressed.

7.3.1 Open-loop Identification Techniques

As for FOPDT models, different techniques have been devised for the estimation of a SOPDT transfer function by evaluating open-loop process step responses (not necessarily for the purpose of tuning a PID controller). Some of them are reviewed hereafter, always with the aim of highlighting their practical issues in the context of PID control.

Two-points-based Method

The method described in (Åström and Hägglund, 1995) is based on the numerical solution of two equations that, in case of an overdamped (monotonic) step response, imposes that the experimental step responses and that provided by Model (7.42) matches exactly when the process output attains 33% and 67% of its final value. The (apparent) dead-time L is previously determined by applying the tangent method (*i.e.*, by considering the intersection between the baseline and the tangent line of the response in its inflection point), while the process gain K is previously calculated as usual by dividing the steady-state change in the process output by the amplitude of the input step.

Being based on the selection of single points in the step response, the method is sensitive to the measurement noise and some filtering technique might be required. Possible problems with this technique arise when a process with two coincident poles and a time delay is considered. For example, if the (noise-free) step response of the process

$$P(s) = \frac{1}{(2s+1)^2} e^{-s} \qquad (7.44)$$

is considered, the estimated parameters of Model (7.42) are $K = 1$, $T_1 = 3.14$, $T_2 = 0.52$ and $L = 1.56$, which are quite different from the actual ones (indeed, it seems that the dominant dynamics is of first order). Some problems occur also when the dominant dynamics of the process is of first order. For example, consider the step response of the process

$$P(s) = \frac{1}{(0.1s+1)(0.1^2 s+1)(0.1^3 s+1)(0.1^4 s+1)}. \qquad (7.45)$$

In this case the result of the application of the method is $K = 1$, $T_1 = 0.05$, $T_2 = 0.05$, $L = 0.01$. It appears that the estimated process has a dominant dynamics of second order. Obviously, the effectiveness of the identification methodology has to be evaluated in conjunction with the employed tuning procedure. However, from the above considerations it might be useful to employ this technique with another one devoted to the estimation of FOPDT processes and to evaluate which of the two estimated transfer functions fits better the experimental data. In case an oscillatory response is detected, in order to estimate T and ξ, two solutions can be adopted. In the first one, the parameters of Model (7.43) are determined by imposing that the step response of the estimated model attains the same peak amplitude y_M at the same time t_M of the experimental response. The time constant T and the damping ratio ξ are therefore determined as:

$$\xi = \frac{\eta}{\sqrt{1+\eta^2}} \tag{7.46}$$

and

$$T = \frac{1}{\pi}\sqrt{\frac{t_M}{1+\xi^2}} \tag{7.47}$$

where

$$\eta = \left|\frac{\ln(y_M - y_\infty)}{\pi}\right| \tag{7.48}$$

where y_∞ is the final steady-state value of the step response.

In the second case, the values of the first minimum y_m (attained at time t_m) and of the second maximum y_{M2} (attained at time t_{M2}) are also considered. Once the decay ratio is determined as

$$d = \frac{y_{M2} - y_M}{y_m - y_M}, \tag{7.49}$$

the two model parameters are calculated by means of the following equations:

$$\varphi = \left|\frac{\log(1-d)}{\pi}\right|, \tag{7.50}$$

$$\xi = \frac{\varphi}{\sqrt{1+\varphi^2}}, \tag{7.51}$$

$$T = \frac{(t_{M2} - t_M)\sqrt{1-\xi^2}}{2\pi}. \tag{7.52}$$

Note that in both cases the dead time is determined as for overdamped responses.

Harriot's Method

The method proposed in (Harriot, 1964), and described also in (Johnson and Moradi (eds.), 2005), is based on the fact that almost all the step responses of processes described by transfer function (7.42) reach 73% of their steady-state values approximately at a time of $1.3(T_1 + T_2)$ and separate from each other most widely at time $0.5(T_1 + T_2)$. Oscillatory responses are not addressed in this case. Thus, the technique consists of first determining the value of A_1 according to Expression (7.2). Then, the value of the dead time can be estimated from the following equation

$$L = A_1 - \frac{t_{73}}{1.3} \tag{7.53}$$

where t_{73} is the time at which the process output attains the 73% of its final value. Then, the sum of the two time constants $T_1 + T_2$ can be derived as

$$T_1 + T_2 = A_1 - L. \tag{7.54}$$

At this point it is possible to evaluate the value y^* of the step response at time $t = 0.5(T_1 + T_2)$. From the plot of Figure 7.7, the value of the ratio $r = T_1/(T_1+T_2)$ can be derived (note that the plot can be easily reconstructed by considering different systems with different values of r). Finally, the values of T_1 and T_2 are determined as

$$T_1 = r \frac{t_{73} - t_0}{1.3} \tag{7.55}$$

and

$$T_2 = (1-r) \frac{t_{73} - t_0}{1.3}. \tag{7.56}$$

Although Harriot's method is somewhat robust to the measurement noise, its main drawback is that it might result in an estimation of a small dead time value, with respect to other methods. This might imply that the resulting controller is more aggressive than expected and this fact is actually detrimental in practical cases.

Indeed, in some cases it might occur that a negative value of the time delay

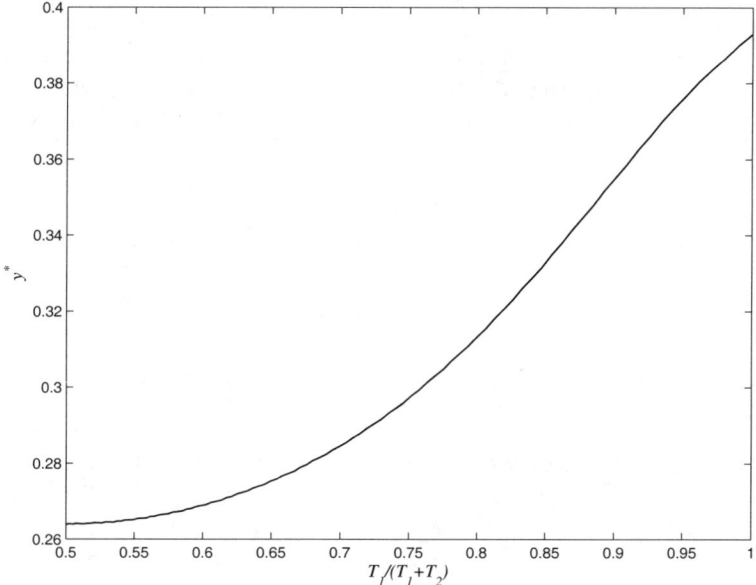

Fig. 7.7. Relation between y^* and $T_1/(T_1 + T_2)$

results (as in the case of the area method for FOPDT processes). For example, if the (noise-free) step response of the process

$$P(s) = \frac{1}{(10s+1)^2} e^{-s} \tag{7.57}$$

is considered, the resulting estimated parameters of the Model (7.42) are $K = 1$, $T_1 = 10.17$, $T_2 = 10.17$ and $L = -0.23$.

Area–Tangent Method

In the technique described in (Sundaresan et al., 1978) both monotonic and oscillatory step responses are considered and one of the two models (7.42)–(7.43) is automatically selected (as already mentioned, this implies that only Model (7.43) can be adopted and if $\xi \geq 1$ then Model (7.42) can be easily derived). After having calculated the process gain K as usual by looking at the input and output steady-state values, the estimation procedure consists of first determining the area between the process output and the straight line $y = y_\infty$, namely:

$$A_1 := \int_{t_0}^{\infty} (y_\infty - y(t)) dt \tag{7.58}$$

where t_0 is the time instant of the input step change. Then, the tangent of the process response is drawn at the inflection point and its slope is denoted by M_i and its intersection with the straight line $y = y_\infty$ is denoted as t_m. An auxiliary variable $\lambda = (t_m - A_1) M_i$ can be easily calculated for the purpose of selecting Model (7.42) or Model (7.43). In particular, if $\lambda < e^{-1}$, then Model (7.42) is considered and its parameters are determined by means of the following formulae (which are derived starting from the analytical expression of the step response):

$$T_1 = \frac{\eta^{\frac{\eta}{1-\eta}}}{M_i} \tag{7.59}$$

$$T_2 = \frac{\eta^{\frac{1}{1-\eta}}}{M_i} \tag{7.60}$$

$$L = A_1 - \frac{\eta^{\frac{1}{1-\eta}}}{M_i} \frac{\eta+1}{\eta} \tag{7.61}$$

where the auxiliary variable η is determined as the solution of the equation

$$\lambda = \ln\left(\frac{\eta}{\eta-1}\right) \exp\left(-\frac{\eta}{\eta-1}\right). \tag{7.62}$$

The case $\lambda = e^{-1}$ corresponds to a critically damped system, for which the previous expressions reduce to

$$T_1 = T_2 = \frac{1}{M_i e} \qquad (7.63)$$

and

$$L = A_1 - \frac{2}{M_i e}. \qquad (7.64)$$

Finally, if $\lambda > e^{-1}$, an oscillatory dynamics results and Model (7.43) is selected. Its parameters are selected by solving the following equations:

$$\lambda = \frac{\cos^{-1} \xi}{\sqrt{1-\xi^2}} \exp\left(\frac{-\xi}{\sqrt{1-\xi^2}} \cos^{-1} \xi \right), \qquad (7.65)$$

$$T = \frac{\sqrt{1-\xi^2}}{\cos^{-1} \xi}(t_m - A_1), \qquad (7.66)$$

$$L = A_1 - 2\xi T. \qquad (7.67)$$

It appears that the technique, being based also on the drawing of the tangent line in the inflection point, has a somewhat high noise sensitivity. It requires also a somewhat significant computational effort (a few equations have to be solved numerically). Further, it has to be stressed that Model (7.43) may result even if a monotonic step response occurs. For example, if the (noise-free) step response of the process

$$P(s) = \frac{1}{(s+1)^3} \qquad (7.68)$$

is considered, the resulting estimated parameters are $K = 1$, $T = 0.69$, $\xi = 0.77$ and $L = 1.94$. As already mentioned, this fact is relevant especially if it is intended to employ a PID controller in series (interacting) form, since in this case the design is often based on pole-zero cancellation.

Four-points-based Method

The methodology proposed in (Huang and Huang, 1993) provides a SOPDT model expressed in the form (7.43) by evaluating four points of the process step response. The algorithm can be summarised as follows (equations are derived by applying a least-squares method).

1. Determine the process gain K by dividing the steady-state change in the process output by the amplitude of the step input.
2. Calculate

$$\alpha = \frac{t_9 - t_6}{t_3 - t_1} \qquad (7.69)$$

where t_1, t_3, t_6, t_9 are the time at which the step response attains 10%, 30%, 60%, 90% of its final value.

3. Calculate ξ as

$$\xi = 7.40898 \cdot 10^{-40} e^{16.3329\alpha} + \frac{100\alpha}{4.55048\alpha + 1.57083} + 1.79015 \cdot 10^{-2}\alpha^3$$
$$+2.25401 \cdot 10^{-2}\alpha^2 - 1.14789\alpha - 16.007 \qquad (7.70)$$

which has a usable range $2.005 \leq \alpha \leq 5.508$ ($0.707 \leq \xi \leq 3.0$).

4. Calculate T as
$$T = \frac{4\sum t_i f_i(\xi) - \sum f_i(\xi) \sum t_i}{4\sum f_i^2(\xi) - (\sum f_i(\xi))^2} \qquad (7.71)$$

and L as
$$L = \frac{\sum t_i \sum f_i^2(\xi) - \sum f_i(\xi) \sum t_i f_i(\xi)}{4\sum f_i^2(\xi) - (\sum f_i(\xi))^2} \qquad (7.72)$$

where
$$\sum t_i = t_1 + t_3 + t_6 + t_9, \qquad (7.73)$$

$$\sum f_i(\xi) = f_1(\xi) + f_3(\xi) + f_6(\xi) + f_9(\xi), \qquad (7.74)$$

$$\sum f_i^2(\xi) = f_1^2(\xi) + f_3^2(\xi) + f_6^2(\xi) + f_9^2(\xi), \qquad (7.75)$$

$$\sum t_i f_i(\xi) = t_1 f_1(\xi) + t_3 f_3(\xi) + t_6 f_6(\xi) + t_9 f_9(\xi), \qquad (7.76)$$

and
$$f_1(\xi) = 0.45465 + 0.06033\xi + 0.01674\xi^2, \qquad (7.77)$$

$$f_3(\xi) = 0.848967 + 0.071809\xi + 0.19753\xi^2 - 0.021823\xi^3, \qquad (7.78)$$

$$f_6(\xi) = 1.08111 + 0.40977\xi + 0.634313\xi^2 - 0.093324\xi^3, \qquad (7.79)$$

$$f_9(\xi) = 0.581618 + 0.875726\xi + 3.64626\xi^2 - 1.35143\xi^3 + 0.173916\xi^4. \qquad (7.80)$$

It is worth noting that, as for the area–tangent method, Model (7.43) (with $\xi \geq 0.707$) may result even if the method deals only with monotonic step responses. Thus, it is not possible to apply a tuning rule for a series PID controller where the derivative action is employed to cancel a pole of the process. Indeed, this is in accordance to the fact that the four-points-based method proposed in (Huang and Huang, 1993) aims at estimating a SOPDT transfer function without any relationship with the tuning of a PID controller and it is recognised that the range $0.707 \leq \xi < 1$ can be applied also to nonoscillatory processes.

Three-points-based Method

Similar to the four-points-based method, a three-points-based method has been developed (on a more theoretical basis) in (Rangaiah and Krishnaswamy, 1994). It consists of finding the parameters of the Model (7.43) by applying the following algorithm:

1. Determine the process gain K by dividing the steady-state change in the process output by the amplitude of the step input.
2. Calculate
$$\alpha = \frac{t_3 - t_2}{t_2 - t_1} \tag{7.81}$$
where t_1, t_2 and t_3 are the time at which the step response attains 14%, 55%, and 91% of its final value.
3. Calculate β and ξ as
$$\beta = \ln\left(\frac{\alpha}{2.485 - \alpha}\right) \tag{7.82}$$

$$\xi = 0.50906 + 0.51743\beta - 0.076284\beta^2 + 0.041363\beta^3$$
$$-0.0049224\beta^4 + 0.00021234\beta^5 \tag{7.83}$$

which has a usable range $1.2323 < \alpha < 2.4850$ that corresponds to $0.707 < \xi < 3.0$.
4. Calculate T and L from the following equations

$$\frac{t_2 - t_1}{T} = 0.85818 - 0.62907\xi + 1.2897\xi^2 - 0.36859\xi^3 + 0.038891\xi^4, \tag{7.84}$$

$$\frac{t_2 - L}{T} = 1.3920 - 0.52536\xi + 1.2991\xi^2 - 0.36859\xi^3 + 0.037605\xi^4. \tag{7.85}$$

The method appears to be simpler than the four-points-based methods but similar considerations can be applied, since also in this case it is assumed that that a model with two complex-conjugate poles (with a damping factor greater than 0.707) can accurately model a process with an overdamped step response.

Method with Model Structure Identification

In the method proposed in (Huang et al., 2001), the model structure is selected according to the shape of the step response. In particular, two model structures are considered, namely,

$$P(s) = \frac{K(as + 1)}{(Ts + 1)(\eta Ts + 1)} e^{-Ls} \qquad 0 < \eta \leq 1, \tag{7.86}$$

and

$$P(s) = \frac{K(as+1)}{T^2 s^2 + 2\xi T s + 1} e^{-Ls} \qquad 0 < \xi < 1. \tag{7.87}$$

It can be remarked that processes with a positive zero (*i.e.*, with inverse response) and with a negative zero can be addressed by this method. Here, for the sake of simplicity, the analysis is restricted to the case of processes with a monotonic step response, for which it is set $a = 0$. Note that in this case it is trivial to derive Model (7.42) from Model (7.86) by simply setting $T_2 = \eta T_1$. Then, the following algorithm is applied.

1. Determine the process gain K by dividing the steady-state change in the process output by the amplitude of the step input.
2. Calculate

$$R_{0.5} = \frac{A_1 - t_{0.3}}{t_{0.5} - t_{0.3}} \tag{7.88}$$

$$R_{0.9} = \frac{A_1 - t_{0.7}}{t_{0.9} - t_{0.7}} \tag{7.89}$$

where A_1 is determined as in Equation (7.58) and t_x is the time when $y(t_x)/y_\infty = x$ (*i.e.*, the time when the process output attains the $x\%$ of its steady-state value).

3. If $1.5573 < R_{0.5} < 1.9108$ and $-0.303 < R_{0.9} < -0.0736$ then select Model (7.86) and determine $\eta_{0.5}$ and $\eta_{0.9}$ by solving the following equations:

$$R_{0.5} = 1.9108 + 0.2275\eta_{0.5} - 5.5504\eta_{0.5}^2 + 12.8123\eta_{0.5}^3$$
$$- 11.8164\eta_{0.5}^4 + 3.9735\eta_{0.5}^5 \tag{7.90}$$

$$R_{0.9} = -0.1871 + 0.0736\eta_{0.9} - 1.2329\eta_{0.9}^2 + 2.1814\eta_{0.9}^3$$
$$- 1.5317\eta_{0.9}^4 + 0.3937\eta_{0.9}^5 \tag{7.91}$$

else, select Model (7.87) and determine $\xi_{0.5}$ and $\xi_{0.9}$ by solving the following equations:

$$R_{0.5} = -3.1623 + 9.3343\xi_{0.5} - 5.7804\xi_{0.5}^2 + 1.1588\xi_{0.5}^3 \tag{7.92}$$

$$R_{0.9} = -6.1991 + 14.6087\xi_{0.5} - 12.1250\xi_{0.5}^2 + 3.4080\xi_{0.5}^3 \tag{7.93}$$

4. Calculate

$$\eta = \frac{\eta_{0.5} + \eta_{0.9}}{2} \tag{7.94}$$

or

$$\xi = \frac{\xi_{0.5} + \xi_{0.9}}{2}. \tag{7.95}$$

5. If Model (7.86) is selected, then determine

$$\bar{t}_{0.3} = 0.3548 + 1.1211\eta - 0.5914\eta^2 + 0.2145\eta^3 \tag{7.96}$$

$$\bar{t}_{0.5} = 0.6862 + 1.1682\eta - 0.1704\eta^2 + 0.0079\eta^3 \qquad (7.97)$$

$$\bar{t}_{0.7} = 1.1988 + 1.0818\eta - 0.4043\eta^2 - 0.2501\eta^3 \qquad (7.98)$$

$$\bar{t}_{0.9} = 2.3063 + 0.9017\eta + 1.0214\eta^2 + 0.3401\eta^3 \qquad (7.99)$$

else, (if Model (7.87) is selected) determine

$$\bar{t}_{0.3} = 0.7954 + 0.2204\xi + 0.0631\xi^2 + 0.0184\xi^3 \qquad (7.100)$$

$$\bar{t}_{0.5} = 1.0472 + 0.3952\xi + 0.1577\xi^2 + 0.0784\xi^3 \qquad (7.101)$$

$$\bar{t}_{0.7} = 1.2662 + 0.6045\xi + 0.2834\xi^2 + 0.2868\xi^3 \qquad (7.102)$$

$$\bar{t}_{0.9} = 1.4655 + 0.9862\xi - 0.1236\xi^2 + 1.5732\xi^3 \qquad (7.103)$$

6. Calculate

$$T = \frac{1}{3}\left(\frac{t_{0.9} - t_{0.7}}{\bar{t}_{0.9} - \bar{t}_{0.7}} + \frac{t_{0.7} - t_{0.5}}{\bar{t}_{0.7} - \bar{t}_{0.5}} + \frac{t_{0.5} - t_{0.3}}{\bar{t}_{0.5} - \bar{t}_{0.3}}\right). \qquad (7.104)$$

7. Calculate

$$L = \frac{t_{0.9} + t_{0.7} + t_{0.5} + t_{0.3}}{4} - \frac{\bar{t}_{0.9} + \bar{t}_{0.7} + \bar{t}_{0.5} + \bar{t}_{0.3}}{4} \qquad (7.105)$$

In the application of this method it has to be taken into account that, although the model structure is automatically selected, a model with two complex conjugate poles might result even for a monotonic step response. Further, the resulting time delay can be negative. For example, if the process described by the transfer function (7.57) is considered again, the resulting values of the estimated SOPDT transfer function are $K = 1$, $T = 12.24$, $\xi = 0.85$ and $L = -0.97$ (note that a noise-free step response has been evaluated). Similarly to Harriot's method, it can be deduced that, in general, the estimated dead time is quite small. This might imply that the resulting controller is more aggressive than expected (especially if a tuning rule based on the normalised dead time is selected).

Least-squares-based Method

A least-squares method that provides directly a SOPDT transfer function from a step response without any iteration has been presented in (Wang et al., 2001; Wang and Zhang, 2001). Indeed, with respect to the method presented in (Sung et al., 1998) (see Section 7.2.1), it does not require a model reduction phase, since the process parameters are determined directly from the least-squares equations. It can be remarked that the approach is very robust to the measurement noise, being based on the use of process output integrals in the regression equations.

The method assumes that the process is described by the following SOPDT transfer function

$$P(s) = \frac{b_1 s + b_2}{s^2 + a_1 s + a_2} e^{-Ls} \qquad (7.106)$$

and therefore processes with a stable and an unstable zero are considered. The following definitions are required:

$$\gamma(t) = y(t), \qquad (7.107)$$

$$\phi^T(t) = \left[-\int_0^t y(\tau)d\tau, -\int_0^t \int_0^\tau y(\tau_1)d\tau_1 d\tau, A, tA, \frac{1}{2}t^2 A \right], \qquad (7.108)$$

$$\theta^T = \left[a_1, a_2, -b_1 L + \frac{1}{2} b_2 L^2, b_1 - b_2 L, b_2 \right]. \qquad (7.109)$$

After choosing t_i, $i = 1, 2, \ldots, N$ such that $L \leq t_1 < t_2 < \cdots < t_N$, let

$$\Gamma = [\gamma(t_1), \gamma(t_2), \ldots, \gamma(t_N)], \qquad (7.110)$$

and

$$\Phi = [\phi(t_1), \phi(t_2), \ldots, \phi(t_N)]^T. \qquad (7.111)$$

Then, the equation

$$\Gamma = \Phi \theta \qquad (7.112)$$

can be solved by applying the ordinary least-squares approach. From the resulting vector θ, the process parameters can be found from Equation (7.109):

$$\begin{bmatrix} a_1 \\ a_2 \\ b_1 \\ b_2 \\ L \end{bmatrix} = \begin{bmatrix} \theta_1 \\ \theta_2 \\ \beta \\ \theta_5 \\ \frac{-\theta_4 + \beta}{\theta_5} \end{bmatrix} \qquad (7.113)$$

where

$$\beta = \begin{cases} -\sqrt{\theta_4^2 - 2\theta_5 \theta_3} & \text{if an inverse response is detected} \\ \sqrt{\theta_4^2 - 2\theta_5 \theta_3} & \text{otherwise} \end{cases} \qquad (7.114)$$

Practical issues, such as how to cope effectively with the measurement noise and how to choose t_1, t_N and N are given in (Wang et al., 2001; Wang and Zhang, 2001). Further, the knowledge of the process gain can be easily exploited by slightly modifying the technique. In any case, the methodology can be easily applied also by assuming a process model of second order without any zero (see (7.42)). However, it is worth stressing that, as for other techniques, it is not guaranteed that a process with two real poles results when a monotonic step response is considered.

Optimisation-based Method

Analogously to what has been explained for FOPDT transfer functions (see Section 7.2.1), the transfer function parameters K, T_1, T_2 and L (or, alternatively, K, T, ξ and L) can be obtained by solving the following optimisation problem:

$$\min_{K,T_1,T_2,L} \int_0^\infty |y(t) - y_m(t)| dt \qquad (7.115)$$

where $y(t)$ denotes the experimental step response and $y_m(t)$ denotes the model step response. Actually, the model parameters are searched in order to minimise the difference between the experimental step response and the model step response.

A practical way to solve the posed optimisation problem is to use genetic algorithms (Mitchell, 1998). In this way, it can be easily imposed that the resulting process poles be real, but, in any case, as already mentioned the computational effort is significant and this is the major drawback of this method.

7.3.2 Closed-loop Identification Techniques

Identification techniques based on closed-loop experiments can be employed also for the estimation of SOPDT transfer functions. For example, the use of an asymmetrical relay-feedback experiment is suggested in (Ramakrishnan and Chidambaram, 2003). The approach is similar to that described in (Srinivasan and Chidambaram, 2003) for FOPDT system (see Section 7.2.2). It consists of assuming that the process is described by Model (7.42) (the extension to Model (7.43) is trivial). Then, the obtained (ultimate) period of oscillations is denoted by T_u. The process gain can be determined by calculating

$$K = \frac{\int_{t_0}^{t_0+T_u} y(t) dt}{\int_{t_0}^{t_0+T_u} u(t) dt}. \qquad (7.116)$$

At this point, the following equation can be written:

$$\frac{K}{(T_1 s_1 + 1)(T_2 s_1 + 1)} e^{-Ls_1} = \frac{y(s_1)}{u(s_1)} \qquad (7.117)$$

where $u(s_1)$ and $y(s_1)$ are determined as in (7.20) and (7.21). Then, other two equations can be considered, namely,

$$p(1 - T_1 T_2 \omega_u^2) - q\omega_u(T_1 + T_2) - K \cos(L\omega_u) = 0 \qquad (7.118)$$

and

$$q(1 - T_1 T_2 \omega_u^2) + p\omega_u(T_1 + T_2) + K \sin(L\omega_u) = 0 \qquad (7.119)$$

where p and q are determined as in (7.26). The solution of the (nonlinear) system given by Equations (7.116)–(7.119) provides the four process parameters.

It is worth stressing that a numerical procedure is necessary to find the parameters and therefore the computational complexity is somewhat relevant. Further, two complex conjugate poles might result even when the actual process dynamics is not oscillatory.

7.4 Discussion

A review of some (step-based) open-loop and (relay-feedback) closed-loop identification techniques for FOPDT and SOPDT transfer functions has been presented in the previous sections with the aim of highlighting the importance of the choice of the identification method in the context of PID design. A short description of each of the considered methods has been given in order to understand the rationale of the approach and its complexity. Details have been omitted (they can be found in the references).

In general, it has been shown that different options are available both for the open-loop and the closed-loop approach, each of them showing interesting features and possible drawbacks in a given application. In particular, it is worth highlighting that when an integral is employed in the context of the techniques based on step responses, the robustness to measurement noise increases significantly, but it might be possible that the final result is incorrect, because a negative (or, in any case, a too small) value of the time delay might result. This fact has been actually often overlooked in the literature.

The techniques based on a relay-feedback experiment are in general less robust to the measurement noise with respect to the step-based ones (but performing a closed-loop experiment might be significantly advantageous since the operating point of the process does not change). In general, it has been shown that there is a variety of strategies in the context of relay-feedback-based methodologies and it is actually difficult to choose the most appropriate one in a given application. At this point it is worth noting that the technique presented in the previous sections generally provide the process transfer function from the estimated values of the (approximate) ultimate gain and frequency of the process. An alternative procedure has been presented in (Friman and Waller, 1997) where a two-channel relay is employed. In particular, the adoption of two relays operating in parallel on the process output and on the integral of the process output allows to estimate a user-chosen point in the third quadrant of the complex plane. This feature can be exploited for an effective tuning of the PID controller. In any case, the model of the process is still basically determined starting from the knowledge of a single point of the frequency response of the process.

Obviously, the identification of multiple points would improve the accuracy of the obtained model. For this purpose, different techniques have been proposed in the literature. For example, the use of a delay term in addition to the relay element provides the identification of a point in the frequency response that is different from the ultimate one (Li et al., 1991; Leva, 1993). This fact has been exploited in (Scali et al., 1999) for the identification of a completely unknown process (a suitable model order is selected automatically in the devised procedure). The obvious drawback of this method is that a multiple experiment has to be run. This can be avoided if a technique based on the Discrete Fourier Transform is applied (Wang et al., 1999a; Wang et al., 1999b), but a significant additional computational effort is required.

It has also to be noted again that in the previous sections the case of processes that can be well described by FOPDT and SOPDT models (7.1) and (7.42) have been particularly addressed. However, it has to be taken into account that there exists also a variety of methodologies capable to deal with processes with an oscillatory dynamics (see, for example, (Huang and Chou, 1994; Rangaiah and Krishnaswamy, 1996; Panda, 2006) in addition to the works referenced in Section 7.3), with processes with a stable and an unstable zero, with integral processes (see, for example, (Kwak et al., 1997)) and with unstable processes (see, for example, (Vivek and Chidambaram, 2005b)).

Summarising, from the above (very simple) analysis it turns out that the choice of the identification strategy is indeed a crucial issue in the context of PID controllers if the tuning of the parameters is based on an estimated model of the plant. Different aspects have been pointed out and they have to be considered in order to provide the most satisfactory performance from a cost/benefit point of view.

7.5 PID Control of High-order Systems

In the previous sections it has been mentioned that the great majority of PID tuning rules assumes that the process model is described as a FOPDT or a SOPDT transfer function. In this context different techniques that aim at obtaining such a models directly from simple experiments on the process have been analysed.

From another point of view, it is recognised that many identification techniques are available nowadays in order to obtain accurate high-order models when the process exhibits a somewhat complex dynamics. Actually, it has to be taken into account that, in many cases, an apparent time delay is indeed due to the presence of a high-order dynamics (Leva, 2005). This has motivated a significant research interest in the last years for the design of PID controllers for high-order processes. It is realised that, because of the relative low-order of the controller, a model reduction has necessarily to be performed. In this context, two approaches can be actually followed:

1. design a model-based high-order controller by considering the (full) high-order dynamics of the process and then reduce the controller to a PID form;

2. reduce first the process model to an appropriate low-order form so that a model-based controller results directly to be in PID form.

In this section this two approaches are analysed and compared in the Internal Model Control (IMC) framework (Morari and Zafiriou, 1989), which has been extensively adopted for the purpose of PID controller tuning, in order to assess their advantages and disadvantages from the point of view of the achievable performance and of the ease of use.

7.5.1 Internal Model Control Design

The IMC methodology has been widely adopted for the purpose of PID controller tuning (though, being based on a pole-zero cancellation approach it is not suitable for lag-dominant processes subject to load disturbances (Scali and Semino, 1991; Shinskey, 1996)). Indeed, it provides the user with a desirable feature as a tuning parameter that handles the trade-off between robustness and aggressiveness of the controller.

In a general form, the IMC design can be applied to a standard unity-feedback control system (see Figure 7.8). The (stable) process to be controlled can be described by the model:

$$P(s) = p_m(s) p_a(s) \qquad (7.120)$$

where $p_a(s)$ is the all-pass portion of the transfer function containing all the nonminimmum phase dynamics (note that it has to be $p_a(0) = 1$ in order to add the integral action to the resulting controller). The controller transfer function is then chosen as

$$C(s) = \frac{f(s) p_m^{-1}(s)}{1 - f(s) p_a(s)} \qquad (7.121)$$

in which

$$f(s) = \frac{1}{(\lambda s + 1)^n} \qquad (7.122)$$

is the IMC filter where λ is the adjustable time constant and n is an appropriate order so that the controller is realisable. It has to be noted that

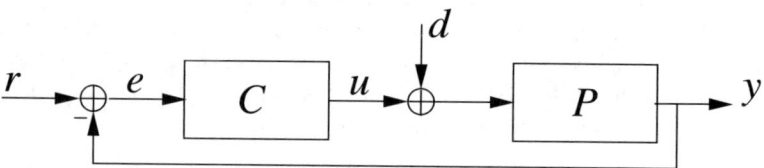

Fig. 7.8. Standard unity-feedback control scheme

the nominal closed-loop transfer function, *i.e.*, the transfer function from the set-point signal r and the process output y, results to be

$$T(s) = \frac{p_a(s)}{(\lambda s + 1)^n} \qquad (7.123)$$

and this makes the role of the free design parameter λ clear in selecting the desired closed-loop dynamics (and therefore in handling the trade-off between robustness and aggressiveness, as unavoidable mismatches between the true process dynamics and its model have to be taken into account).
Obviously, in general, the resulting controller is not in PID form, *i.e.*:

$$C(s) = K_p \left(1 + \frac{1}{T_i s} + T_d s\right) \frac{1}{T_f s + 1}, \qquad (7.124)$$

if the output-filtered ideal form is implemented. This occurs if the process model has one positive zero and two poles (note that this results if a FOPDT transfer function is considered and a first-order Padè approximation is adopted for the delay term (Rivera *et al.*, 1986)), while a PI controller results if the plant has a simple first-order dynamics. Thus, if a high-order process model is considered, this must be reduced to this suitable form before applying the IMC design or, alternatively, the resulting high-order controller has to be subsequently reduced to a PID form.

7.5.2 Process Model Reduction

Skogestad's Half Rule

The method proposed by Skogestad in (Skogestad, 2003) considers a process model reduction based on the so-called "half rule", which states that the largest neglected (denominator) time constant is distributed evenly to the effective dead time and the smallest retained time constant. In practice, given a high-order transfer function, each numerator term $(\tau_0 s + 1)$ with $\tau_0 > 0$ is first simplified with a denominator term $(T_0 s + 1)$, $T_0 > 0$ using the following rules:

$$\frac{\tau_0 s + 1}{T_0 s + 1} \approx \begin{cases} \tau_0/T_0 & \text{for } \tau_0 \geq T_0 \geq L \\ \tau_0/L & \text{for } \tau_0 \geq L \geq T_0 \\ 1 & \text{for } L \geq \tau_0 \geq T_0 \\ \tau_0/T_0 & \text{for } T_0 \geq \tau_0 \geq 5L \\ \frac{(\tilde{T}_0/T_0)}{(\tilde{T}_0 - \tau_0)s + 1} & \text{for } \tilde{T}_0 \stackrel{\text{def}}{=} \min(T_0, 5L) \geq \tau_0 \end{cases} \qquad (7.125)$$

where L is the final effective delay (to be determined subsequently). It has to be noted that T_0 is normally chosen as the closest larger denominator time constant $(T_0 > \tau_0)$, except when a larger denominator time constant does not exist or there is a smaller denominator time constant closer to τ_0; this is true if the ratio between τ_0 and the smaller denominator time constant is less than the ratio between the larger denominator time constant and τ_0 and (both conditions must be satisfied) less than 1.6.

Once this procedure has been terminated for all the positive numerator time constants, the process transfer function is in the following form:

$$\tilde{P}(s) = \frac{\prod_j (-\tau'_{j0} s + 1)}{\prod_i (T_{i0} + 1)} e^{-L_0 s} \tag{7.126}$$

where $\tau'_{j0} > 0$ and the time constants are ordered according to their magnitude. Then, a SOPDT transfer function

$$P(s) = \frac{k}{(T_1 s + 1)(T_2 s + 1)} e^{-Ls} \tag{7.127}$$

is obtained by applying the half rule, *i.e.*, by setting

$$T_1 = T_{10}, \quad T_2 = T_{20} + \frac{T_{30}}{2}, \tag{7.128}$$

$$L = L_0 + \frac{T_{30}}{2} + \sum_{i \geq 4} T_{i0} + \sum_j \tau'_{j0}. \tag{7.129}$$

It appears that, being the rules (7.125) based on the final apparent time delay L, in the first part of the algorithm there is the need to guess this final value and to iterate in case at the end the result is incorrect.

Once the SOPDT process model is obtained, the parameters of a series PID controller expressed by the transfer function

$$C(s) = K_p \left(\frac{T_i s + 1}{T_i s} \right) \frac{T_d s + 1}{T_f s + 1} \tag{7.130}$$

are determined by applying the IMC design procedure (and by approximating the delay term as $e^{-Ls} = 1 - Ls$) and by possibly modifying the value of T_i in order to address the case of lag-dominant processes (Skogestad, 2003) (note that this fact in not of concern in the examples presented in Section 7.5.4). It results that the PID parameters in (7.130) are selected as

$$K_p = \frac{T_1}{k(\lambda + L)}, \quad T_i = T_1, \quad T_d = T_2, \quad T_f = 0.01 T_d. \tag{7.131}$$

Note that the conversion of the tuning rule (7.131) for the PID controller in the ideal form (7.124) is straightforward and a recommended choice for the desired closed-loop time constant is $\lambda = L$ (Skogestad, 2003).

Summarising, the method is based on simple, easy to remember, tuning rules (indeed, this is one of the main features of the method). However, the possible iterations in the model reduction algorithm makes the overall procedure somewhat difficult to automate.

Isaksson and Graebe's Method

The technique proposed by Isaksson and Graebe in (Isaksson and Graebe, 1999) is also based on a suitable process model reduction before applying the IMC design. The model reduction is performed as follows. Let the initial (high-order) process model be described by the transfer function

$$\tilde{P}(s) = \frac{B(s)}{A(s)} \quad (7.132)$$

Then, the numerator and denominator polynomials are considered separately and the polynomials $B_1(s)$ and $A_1(s)$ that retain only the slowest roots are determined. Subsequently, the polynomials $B_2(s)$ and $A_2(s)$ that retain the low-order coefficients are calculated. Finally, the reduced-order model is obtained as

$$P(s) = \frac{\frac{1}{2}(B_1(s) + B_2(s))}{\frac{1}{2}(A_1(s) + A_2(s))} \quad (7.133)$$

By choosing $B_1(s)$ and $B_2(s)$ of first order and $A_1(s)$ and $A_2(s)$ of second order and by subsequently applying the IMC design (with a first-order filter (7.122)), a PID controller (7.124) naturally arises. If there are no zeros, two solutions can be applied:

1. a second-order denominator is calculated in the reduction procedure and a second-order filter (7.122) is applied in the IMC design, resulting in a PID controller;
2. a first-order denominator is calculated in the reduction procedure and a first-order filter (7.122) is applied in the IMC design, resulting in a PI controller.

It has to be noted that, differently from the Skogestad's half rule, the case of complex conjugate roots is also addressed in (Isaksson and Graebe, 1999), but it will not be considered hereafter (see Section 7.5.4).
Summarising, the Isaksson and Graebe's method can be easily automated, although it is not explicitly based on tuning formulae.

Model Approximation with Step Response Data

The least-squares method presented in (Wang et al., 2001; Wang and Zhang, 2001) and explained in Section 7.3.1 can be employed also for the purpose of model reduction. In particular, given a high-order model of the process, a SOPDT transfer function (with one zero) can be obtained directly from the open-loop step response. Note that there is no need of time consuming and costly experimental results, since a simulation can be performed (Huang, 2003). In this way the overall procedure can be easily automated, although a

relatively significant computational effort is actually necessary.
Starting from the identified model, the tuning rule (7.131) has been adopted. However, for this purpose, the obtained SOPDT model must have real poles and no zeros. Thus, if a zero is determined the half rule is then adopted, while if complex conjugate poles occur, a FOPDT model (obtained with the same identification method) is actually employed. In this latter case a PI controller results.

7.5.3 Controller Reduction

The methods described in Section 7.5.2 are based on the reduction of the process model before applying the IMC design. Conversely, it is possible to apply the IMC procedure described in Section 7.5.1 by considering the full process dynamics and then reduce the obtained high-order controller to a PID controller form. For this purpose, a Maclaurin series expansion can be employed. The expression of the resulting controller can be always written as (Lee et al., 1998b; Lee et al., 1998a):

$$C(s) = \frac{k(s)}{s} \tag{7.134}$$

and expanding $C(s)$ in a Maclaurin series in s it results:

$$C(s) = \frac{1}{s}\left[k(0) + k'(0)s + \frac{k''(0)}{2}s^2 + \cdots \right] \tag{7.135}$$

It turns out that the first part of the series expansion contains a proportional term, an integral term and a derivative term and therefore, if the high-order terms are neglected, a PID controller (7.124) results (a first-order filter can be easily added in order to make the controller proper and its time constant can be selected sufficiently small so that its dynamics is not significant). Indeed, the following relations hold:

$$K_p = k'(0)$$
$$T_i = \frac{k'(0)}{k(0)} \tag{7.136}$$
$$T_d = \frac{k''(0)}{2k'(0)}$$

Hence, the overall procedure can be easily automated, although it is not based on tuning formulae and its computational burden is somewhat considerable. However, it has to be stressed that a wrong choice of the design parameter λ can result in the overall control system being unstable (see Section 7.5.5). Although this can be easily checked before applying the controller, it can be considered as a major drawback of the method.

7.5.4 Simulation Results

In order to analyse and compare the different methodologies, the following processes with high-order dynamics have been considered:

$$P_1(s) = \frac{(15s+1)^2(4s+1)(2s+1)}{(20s+1)^3(10s+1)^3(5s+1)^3(0.5s+1)^3}, \qquad (7.137)$$

$$P_2(s) = \frac{(-0.3s+1)(0.08s+1)}{(2s+1)(s+1)(0.4s+1)(0.2s+1)(0.05s+1)^3}, \qquad (7.138)$$

$$P_3(s) = \frac{(-45s+1)(4s+1)(2s+1)}{(20s+1)^3(18s+1)^3(5s+1)^3(10s+1)^2(16s+1)(14s+1)(12s+1)}, \qquad (7.139)$$

$$P_4(s) = \frac{1}{(s+1)^4}, \qquad (7.140)$$

$$P_5(s) = \frac{1}{(s+1)^8}, \qquad (7.141)$$

$$P_6(s) = \frac{1}{(s+1)^{20}}. \qquad (7.142)$$

The main characteristics of the processes are summarised in Table 7.1. It has to be noted that transfer functions $P_1(s)$ and $P_3(s)$ have been taken from (Wang and Cluett, 2000) (actually, $P_3(s)$ has been modified in order to obtain an inverse response), $P_2(s)$ from (Skogestad, 2003) and $P_4(s) - P_6(s)$ are representative of typical industrial processes (Åström and Hägglund, 2000a; Shinskey, 2000).

The reduced-order model that have been adopted for the PI(D) tuning are shown in Table 7.2. Note that two transfer functions might occur for the Isaksson and Graebe's technique, whereas the process dynamics has no zeros, as explained in Section 7.5.2. Indeed, the first one results in a PID controller (with a second-order IMC filter), while the second one results in a PI controller (with a first-order IMC filter). Besides, whereas a FOPDT transfer function is reported for the step response based method, this means that the resulting SOPDT model has complex conjugate poles and has not therefore been employed (see Section 7.5.2).

In order to make a fair comparison, for each method and for each process the value of λ that minimises the integrated absolute error (4.23) for both the set-point and the load disturbance step responses have been selected (*i.e.*, a unit step has been applied on signals r and d separately, see Figure 7.8). For those methods that do not provide the value of the filter time constant T_f explicitly, this has been selected in such a way its dynamics is negligible. The resulting values of the integrated absolute error and the corresponding optimal values of λ are shown in Tables 7.3–7.8. Note again that for the Isaksson and Graebe's method two cases (PID and PI control) might emerge, depending on the fact that a second-order or first-order IMC filter respectively has been adopted (since there are no zeros in the process to be controlled). Analogously, a PID or a PI controller results from the technique based on the step response, depending on the use of a SOPDT model or a FOPDT model (the latter in case the identified SOPDT model has complex conjugate poles). Finally, the resulting (set-point and load) unit step responses are plotted in Figures 7.9–7.14. The process responses obtained with a PI controller resulting from the Isaksson and Graebe's method are not reported.

Table 7.1. Main characteristics of the considered processes

$P_1(s)$	Minimum phase dynamics
$P_2(s)$	Presence of a nondominant positive zero
$P_3(s)$	Presence of a dominant positive zero
$P_4(s)$	Minimum phase dynamics with a small number of coincident poles
$P_5(s)$	Minimum phase dynamics with a medium number of coincident poles
$P_6(s)$	Minimum phase dynamics with a high number of coincident poles

Table 7.2. Resulting reduced models for the different methods

Process	Skogestad	Isaksson and Graebe		step response
$P_1(s)$	$\dfrac{e^{-35.5s}}{(20s+1)(15s+1)}$	$\dfrac{25.5s+1}{2642s^2+73.25s+1}$		$\dfrac{e^{-27.38s}}{44.46s+1}$
$P_2(s)$	$\dfrac{e^{-0.77s}}{(2s+1)(1.2s+1)}$	$\dfrac{-0.26s+1}{3.21s^2+3.38s+1}$		$\dfrac{e^{-0.79s}}{2.48s^2+3.17s+1}$
$P_3(s)$	$\dfrac{e^{-180s}}{(30s+1)(20s+1)}$	$\dfrac{-42s+1}{8564s^2+115.7s+1}$		$\dfrac{e^{-127.7s}}{106.6s+1}$
$P_4(s)$	$\dfrac{e^{-1.5s}}{(1.5s+1)(s+1)}$	$\dfrac{1}{2.25s^2+3s+1}$	$\dfrac{1}{2.5s+1}$	$\dfrac{e^{-1.38s}}{2.71s+1}$
$P_5(s)$	$\dfrac{e^{-5.5s}}{(1.5s+1)(s+1)}$	$\dfrac{1}{14.52s^2+5.02s+1}$	$\dfrac{1}{4.51s+1}$	$\dfrac{e^{-3.88s}}{4.24s+1}$
$P_6(s)$	$\dfrac{e^{-17.5s}}{(1.5s+1)(s+1)}$	$\dfrac{1}{95.98s^2+11.40s+1}$	$\dfrac{1}{10.70s+1}$	$\dfrac{e^{-12.72s}}{7.76s+1}$

7.5 PID Control of High-order Systems

Table 7.3. Optimal IAE's (and corresponding values of λ) for process $P_1(s)$

Method		Task	
		set-point	load
Skogestad		71.38 (9.6)	54.45 (0.01)
Isaksson and Graebe	PID	63.08 (54.4)	48.13 (39.1)
	PI		
step response	PID		
	PI	73.49 (37.2)	60.16 (20.8)
Maclaurin		42.07 (4.4)	25.35 (2.8)

Table 7.4. Optimal IAE's (and corresponding values of λ) for process $P_2(s)$

Method		Task	
		set-point	load
Skogestad		1.852 (0.83)	0.869 (0.03)
Isaksson and Graebe	PID	1.963 (1.22)	0.985 (0.37)
	PI		
step response	PID	1.819 (0.78)	0.863 (0.01)
	PI		
Maclaurin		1.783 (0.18)	0.986 (0.07)

Table 7.5. Optimal IAE's (and corresponding values of λ) for process $P_3(s)$

Method		Task	
		set-point	load
Skogestad		376.5 (120.1)	384.3 (115.2)
Isaksson and Graebe	PID	251.7 (122.3)	253.6 (113.0)
	PI		
step response	PID		
	PI	301.1 (159.0)	307.7 (153.6)
Maclaurin		231.6 (11.5)	233.5 (10.9)

Table 7.6. Optimal IAE's (and corresponding values of λ) for process $P_4(s)$

Method		Task	
		set-point	load
Skogestad		3.133 (0.49)	1.952 (0.001)
Isaksson and Graebe PID		2.722 (0.87)	1.073 (0.39)
	PI	4.099 (3.44)	3.183 (2.33)
step response	PID		
	PI	3.998 (2.06)	3.063 (0.96)
Maclaurin		2.041 (0.41)	0.876 (0.19)

Table 7.7. Optimal IAE's (and corresponding values of λ) for process $P_5(s)$

Method		Task	
		set-point	load
Skogestad		11.58 (3.01)	10.95 (2.06)
Isaksson and Graebe PID		7.776 (2.78)	6.401 (2.34)
	PI	9.550 (8.83)	8.880 (7.88)
step response	PID		
	PI	9.662 (4.96)	9.030 (4.00)
Maclaurin		6.561 (0.74)	5.394 (0.61)

Table 7.8. Optimal IAE's (and corresponding values of λ) for process $P_6(s)$

Method		Task	
		set-point	load
Skogestad		36.86 (11.2)	36.49 (10.5)
Isaksson and Graebe PID		23.01 (8.83)	21.88 (8.42)
	PI	25.84 (24.6)	25.22 (23.8)
step response	PID		
	PI	27.36 (12.2)	26.95 (11.4)
Maclaurin		19.85 (0.90)	18.81 (0.86)

7.5 PID Control of High-order Systems

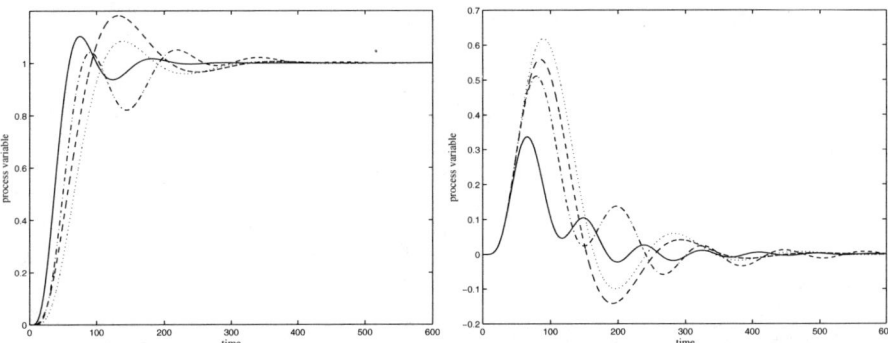

Fig. 7.9. Optimal set-point (left) and load disturbance (right) step responses for $P_1(s)$. Dashed line: Skogestad; dash-dot line: Isaksson and Graebe; dotted line: step response; solid line: Maclaurin.

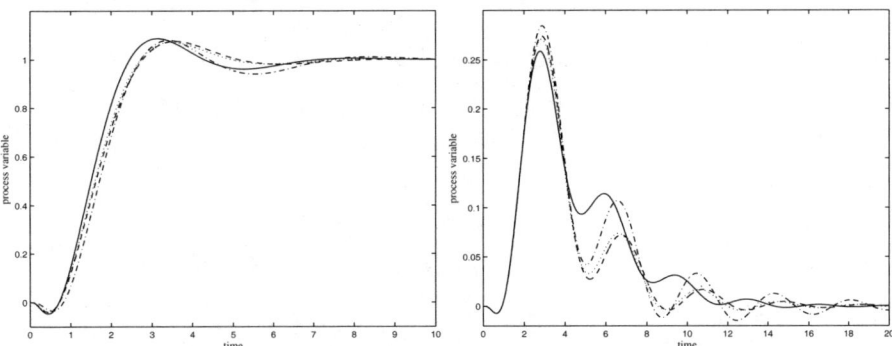

Fig. 7.10. Optimal set-point (left) and load disturbance (right) step responses for $P_2(s)$. Dashed line: Skogestad; dash-dot line: Isaksson and Graebe; dotted line: step response; solid line: Maclaurin.

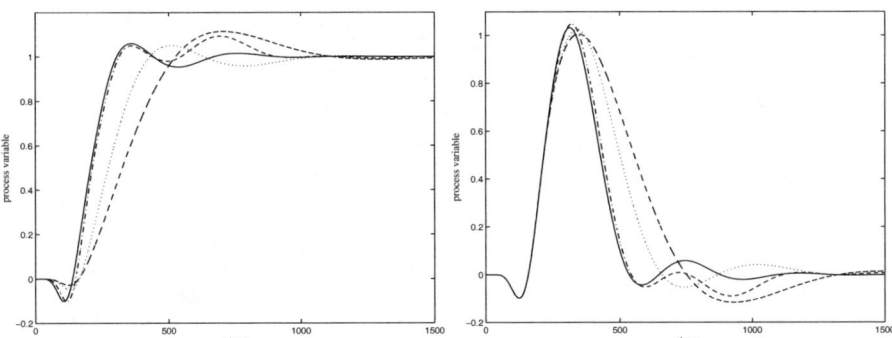

Fig. 7.11. Optimal set-point (left) and load disturbance (right) step responses for $P_3(s)$. Dashed line: Skogestad; dash-dot line: Isaksson and Graebe; dotted line: step response; solid line: Maclaurin.

204   7 Identification and Model Reduction Techniques

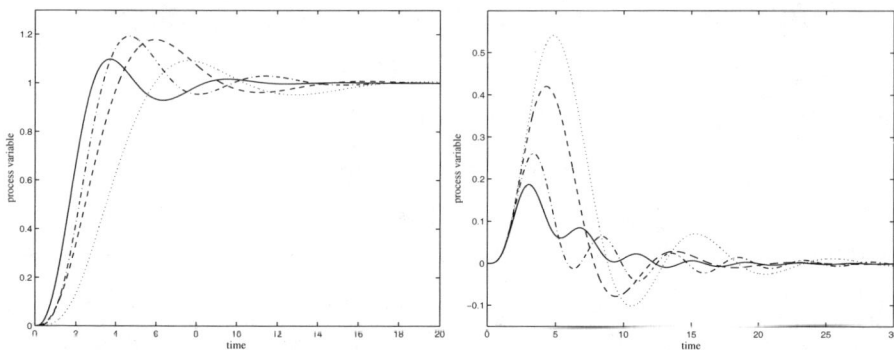

Fig. 7.12. Optimal set-point (left) and load disturbance (right) step responses for $P_4(s)$. Dashed line: Skogestad; dash-dot line: Isaksson and Graebe; dotted line: step response; solid line: Maclaurin.

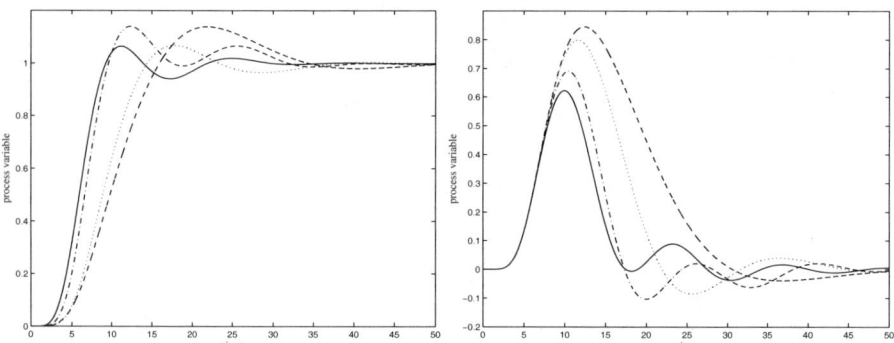

Fig. 7.13. Optimal set-point (left) and load disturbance (right) step responses for $P_5(s)$. Dashed line: Skogestad; dash-dot line: Isaksson and Graebe; dotted line: step response; solid line: Maclaurin.

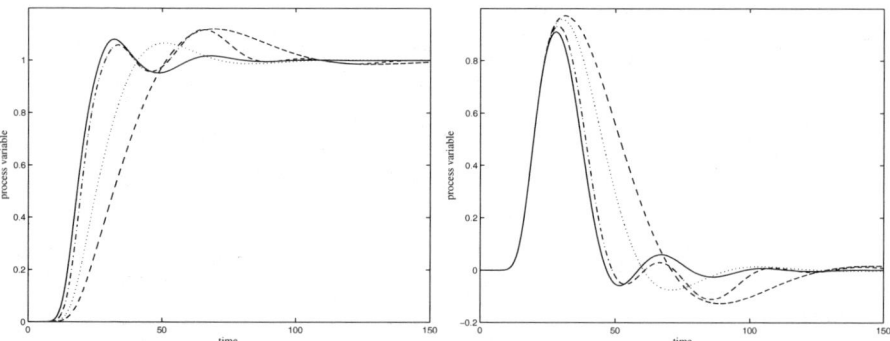

Fig. 7.14. Optimal set-point (left) and load disturbance (right) step responses for $P_6(s)$. Dashed line: Skogestad; dash-dot line: Isaksson and Graebe; dotted line: step response; solid line: Maclaurin.

7.5.5 Discussion

From the results obtained it appears that the approach based on the Maclaurin series expansion provides in general the best performance, both for the set-point following and the load disturbance rejection task. This is due to its capability of providing a higher open-loop crossover frequency without decreasing the phase margin with respect to the other methods (Visioli, 2005c). From another point of view, this means that in the set-point step responses a small rise time is achieved without impairing the overshoot and in the load disturbance step responses a small peak error results without the occurrence of significant oscillations.

It turns out that it is better to reduce the model of the controller than that of the plant, because the approximation introduced by adopting only the first three terms of the series expansion is not detrimental in the range of frequencies that is significant for the considered control system. However this is true only if an appropriate value of λ is selected. Indeed, a wrong choice of λ might imply negative parameters of the PID controller or, more remarkably, it might lead the system to instability (even if the PID parameters are positive). For example, for system $P_3(s)$, if $\lambda \leq 6$ or $\lambda \geq 162$ the resulting closed-loop system is unstable and, in any case, if $\lambda \geq 20$ at least one of the PID parameters results to be less than zero. In order to better clarify this fact, consider the value $\lambda = 5.0$. The Maclaurin series approximation of the IMC controller gives a PID controller with $K_p = 1.09$, $T_i = 158.4$ and $T_d = 68.32$ (all the parameters are positive) and the resulting closed-loop system is actually unstable. This fact can be understood by looking at Figures 7.15 and 7.16 where the Bode plots of the two controllers and of the open-loop system with the original IMC controller and with the PID one are reported respectively. It appears that the approximation of the IMC controller around the critical frequency is not sufficiently accurate (indeed, the series expansion is centered at the zero frequency) and therefore the crossover frequency increases too much to provide a positive phase margin.

In general, it might happen that a quite narrow range of values for λ is suitable for a given process. Despite the fact that an unappropriate value of λ can be easily recognised during the design phase, this can be considered as a major drawback of the method, which has been overlooked in the literature. Indeed, this makes the overall design more complicated and, most of all from a practical point of view, the physical meaning of the filter time constant, which should handle the trade-off between aggressiveness and robustness and control activity of the control system, is actually lost (increasing the value of λ does not necessarily correspond anymore to a more sluggish and stable control system).

It has also to be noted that the optimal values of λ differ significantly between the considered methodology, although it appears, as expected, that, in general, a higher order filter (*i.e.*, for the Maclaurin series based technique or when a PID controller is adopted instead of a PI controller in the Isaksson

206   7 Identification and Model Reduction Techniques

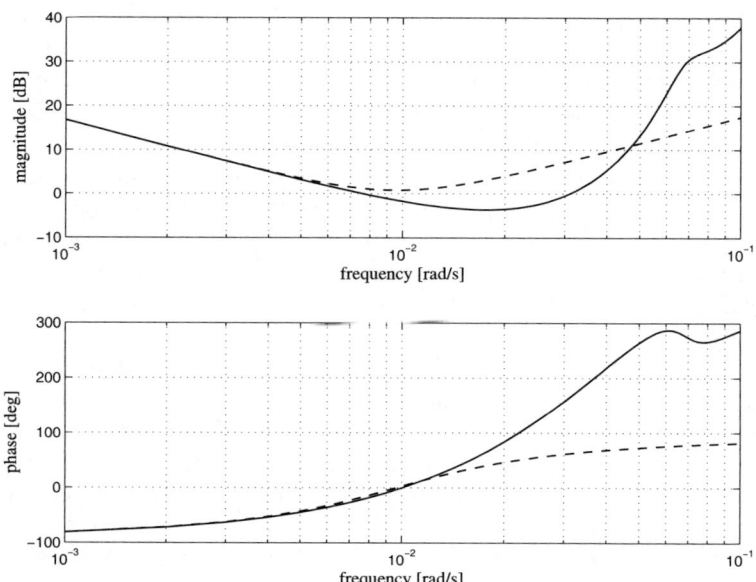

Fig. 7.15. Bode plot of the IMC controller (solid line) and of the approximating PID controller (dashed line)

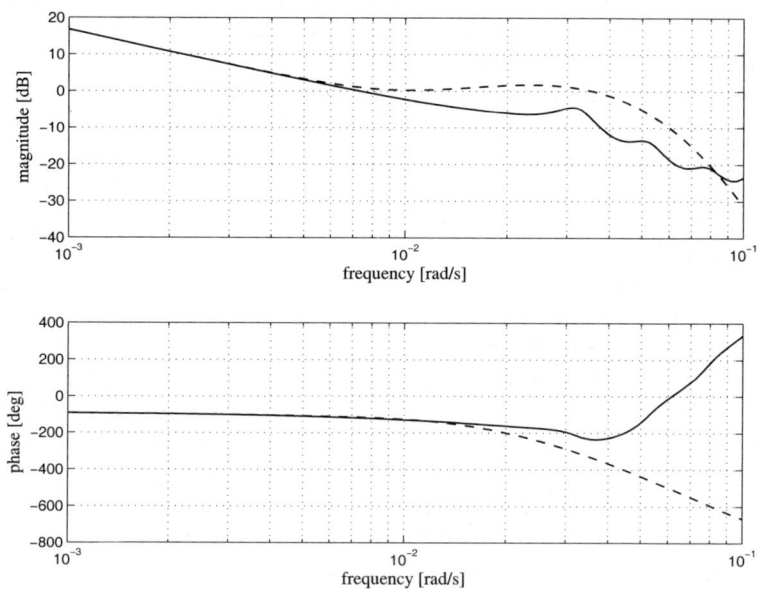

Fig. 7.16. Bode plot of the open-loop transfer function $C(s)P_3(s)$ with IMC controller (solid line) and with the approximating PID controller (dashed line)

and Graebe's method) implies a smaller value of λ (and the load disturbance rejection task requires a smaller value of λ than the set-point following task). From the results obtained, it also appears that the Isaksson and Graebe's method provides in general a better performance that the Skogestad's one and, as expected, the PID controller is better than the PI controller when the analytical PID design technique of Isaksson and Graebe is considered. Note, however, that the tuning rules (7.131) have been conceived with the aim of being applicable to a wide range of processes and of being easy to memorise. Regarding the method based on the step response data, it can be deduced that in general it provides worse performance than the Isaksson and Graebe's one, while no general conclusions can be drawn with respect to the Skogestad's method.

7.6 Conclusions and References

In this chapter, the issue of the identification of the process model and of the design of a PID controller when a high-order process model is available has been addressed. It has been shown that many methods for the estimation of a FOPDT and SOPDT transfer function have been devised, with different characteristics that have to be evaluated in a given application in order to provide a cost-effective PID design. Possible problems, often overlooked in the literature, have been indicated, mainly with the aim of highlighting that the identification method is indeed an integral part of the overall controller design and its choice is critical. It has also been shown that, in case a high-order process model is available, either the approach of reducing first the process model or of reducing the controller at the end can be applied. In the latter case a better performance can be achieved in general but at the expense of a more complicated and less intuitive design.

The review provided for the identification methods is surely not exhaustive. In the context of PID control, an up-to-date review of basic and advanced identification methods can be found in (Johnson and Moradi (eds.), 2005). The subject of relay-feedback is thoroughly addressed in (Wang *et al.*, 2003), while its use for system identification is analysed in (Yu, 1999), where advanced methodologies are also presented. A tutorial review of relay-feedback automatic tuning techniques for process controllers can be found in (Hang *et al.*, 2002).

8

Performance Assessment

8.1 Introduction

For plant safety and profitability it is essential to check continuously that the plant performance satisfies the required operating objectives. Thus, process monitoring plays a key role in running the plant effectively and economically (Seborg *et al.*, 2004). In this context, it is important to verify that the control system performs properly. Actually, because in large plants there are hundreds of control loops, it is almost impossible for operators to monitor each of them manually. Thus, it is important to have tools that are first able to determine automatically if an abnormal situation occurs and then to help the operator to understand the reason for it and possibly to suggest the way to solve the problem (for example, if a bad controller tuning is detected, then new appropriate values of controller parameters are determined).
In this chapter, techniques for control loop performance assessment are presented. Although they can be viewed under the same framework (Huang and Shah, 1999), they are conveniently divided in two categories (Qin, 1998): those related to the *stochastic performance monitoring*, in which the capability of the control system to cope with stochastic disturbances is of main concern, and those related to the *deterministic performance monitoring*, in which the performance related to more traditional design specifications such as set-point and load rejection disturbance step response parameters are taken into account (Eriksson and Isaksson, 1994). The main emphasis is devoted to this latter case, which is believed to be the most interesting in the PID control context. Further, it is worth stressing that many concepts can be applied in general, independently of the controller type. However, a special attention is paid to the case where a PID controller is employed.

8.2 Generalities

The assessment of the performance of a control loop is a complex task which should be generally performed by applying the following steps (Jelali, 2006).

1. A performance index is calculated based on the available data in order to measure the performance of the current control system.
2. A benchmark, which represents the desired performance, is selected, so that the current control performance can be evaluated against it.
3. The deviation of the current control performance from the selected benchmark is determined, so that it is evaluated if the control system performance can be considered as satisfactory or if it needs to be improved.
4. In case the current control performance is not satisfactory, the reasons for it are diagnosed. Indeed, this is the most difficult task, since there are different causes for a poor performance (inappropriate control structure, bad controller tuning, inappropriate equipment design, equipment malfunction).
5. Corrective actions are suggested for pursuing the selected benchmark performance.

It can be easily deduced that different approaches can be selected for each step, and they have to be properly integrated in order to design a consistent overall strategy.

8.3 Stochastic Performance Assessment

Many works in the field of performance monitoring are related to the assessment of the output variance due to stochastic disturbances. These are assumed to be generated from a dynamic system driven by white noise. In this context the devised approaches are based on the minimum variance concept (Qin, 1998; Harris *et al.*, 1999), which is developed in the discrete-time framework.

8.3.1 Minimum Variance Control

In its basic form, the minimum variance control assumes that the (linear, time-invariant) process is described by the model (Katebi and Ordys, 1996)

$$y(t) = \frac{B(q^{-1})}{A(q^{-1})} q^{-d} u(t) + \frac{C(q^{-1})}{A(q^{-1})} w(t), \qquad (8.1)$$

where $y(t)$ represents the variation of the process variable around a given steady-state operating point, $u(t)$ is the control variable, $w(t)$ is a zero-mean Gaussian white noise, q^{-1} is the backward shift operator, *i.e.*,

$$q^{-1} x(t) = x(t-1), \qquad (8.2)$$

8.3 Stochastic Performance Assessment

and $A(q^{-1})$, $B(q^{-1})$ and $C(q^{-1})$ are nth order polynomials, i.e.,

$$A(q^{-1}) = 1 + a_1 q^{-1} + \cdots + a_n q^{-n}, \tag{8.3}$$

$$B(q^{-1}) = b_0 + b_1 q^{-1} + \cdots + b_n q^{-n}, \quad b_0 \neq 0, \tag{8.4}$$

$$C(q^{-1}) = 1 + c_1 q^{-1} + \cdots + c_n q^{-n}. \tag{8.5}$$

It can be noted that q^{-d} models a d-step delay in the control signal, so that the effects of a change in the control signal appear on the process after d sampling times. The aim of the minimum variance control is to determine the control signal $u(t)$ that minimises the variance of the process output at time $t + d$, given all the information available at time t. Thus, the performance index to minimise is

$$J(t) = E\left\{y(t+d)^2 | Y(t)\right\}, \tag{8.6}$$

where $E\{\cdot|\cdot\}$ is the conditional expectation operator and

$$Y(t) = [u(t-d-1), u(t-d-2), \ldots, y(t), y(t-1), \ldots]. \tag{8.7}$$

For the solution of the problem it is convenient to rewrite the process model as

$$y(t+d) = \frac{B(q^{-1})}{A(q^{-1})} u(t) + \frac{C(q^{-1})}{A(q^{-1})} w(t+d). \tag{8.8}$$

Since the random variables $[w(t+d-1), w(t+d-2), \ldots]$ are assumed independent of the process output $[y(t), y(t-1), \ldots]$ and all the future control inputs are assumed to be zero, the disturbance signal can be separated into causal and noncausal part and therefore Equation (8.8) can be rewritten as

$$y(t+d) = \frac{B(q^{-1})}{A(q^{-1})} u(t) + \left[E(q^{-1}) + z^{-d} \frac{F(q^{-1})}{A(q^{-1})} \right] w(t+d), \tag{8.9}$$

where polynomials $E(q^{-1})$ and $F(q^{-1})$ are defined as

$$E(q^{-1}) = 1 + e_1 q^{-1} + \cdots + e_d q^{-d}, \tag{8.10}$$

$$F(q^{-1}) = f_0 + f_1 q^{-1} + \cdots + f_{n-1} q^{-(n-1)}, \tag{8.11}$$

and they satisfy the following Diophantine equation:

$$A(q^{-1}) E(q^{-1}) + q^{-d} F(q^{-1}) = C(q^{-1}). \tag{8.12}$$

By considering Equations (8.8), (8.9) and (8.12), it can be written

$$y(t+d) = \frac{E(q^{-1})B(q^{-1})}{C(q^{-1})}u(t) + \frac{F(q^{-1})}{C(q^{-1})}y(t) + E(q^{-1})w(t+d). \qquad (8.13)$$

Hence, the performance index (8.6) is:

$$J(t) = E\left\{\left[\frac{E(q^{-1})B(q^{-1})}{C(q^{-1})}u(t) + \frac{F(q^{-1})}{C(q^{-1})}y(t)\right]^2\right\} + E\left\{\left[E(q^{-1})w(t+d)\right]^2\right\}, \qquad (8.14)$$

where the expected value of the cross-product term is zero because $[w(t+d), w(t+d-1), \ldots, w(t+1)]$ are independent from $[y(t), y(t-1), \ldots,]$. Thus, the minimum variance control law is the one that set to zero the first term of the right-hand side of (8.14), i.e.,

$$u(t) = -\frac{F(q^{-1})}{B(q^{-1})E(q^{-1})}y(t). \qquad (8.15)$$

It is worth noting that, in order for the closed-loop to be stable, the process has to be minimum-phase, namely $B(q^{-1})$ has to be stable (see (Åström and Wittenmark, 1997) for the solution of the problem for nonminimum-phase systems).

8.3.2 Assessment of Performance

Based on the concepts expressed in the previous section, the performance achieved by a given feedback controller can be assessed in the sense that it can be evaluated how far it is from the minimum variance performance. In particular, this can be done quite easily if the dead time of the process (expressed as the number d of sampling intervals) is known (see, for example, (Horch, 2000; Björklund, 2003; Ahmed et al., 2006) for techniques for estimating it). Indeed, an autoregressive moving average (ARMA) model of the closed-loop system relating the noise w to the process output y has to be estimated from a selected period of routine data:

$$y(t) = \sum_{i=1}^{n_y} a_i y(t-i) + \sum_{i=1}^{n_w} c_i w(t-i) + w(t), \qquad (8.16)$$

where (n_y, n_w) is the model order (which can be actually selected by a trial and error procedure). The obtained ARMA model has then to be expanded into an impulse response model (by performing a long division) that has to be truncated at the first $d-1$ coefficients (note that the first d terms are invariant with respect to the adopted controller):

$$y(t) = w(t) + \sum_{i=1}^{d-1} \psi_i w(t-i). \qquad (8.17)$$

8.3 Stochastic Performance Assessment

The estimated minimum variance can therefore be determined as

$$\sigma_{MV}^2 = \left(1 + \sum_{i=1}^{d-1} \psi_i^2\right) \sigma_w^2, \tag{8.18}$$

where σ_w^2 is the estimated noise variance. The value of σ_{MV}^2 can then be compared to the estimate of the output variance σ_y^2 which is given by:

$$\sigma_y^2 = \frac{1}{N-1} \sum_{i=1}^{N} (y(i) - \bar{y})^2 \tag{8.19}$$

where $\bar{y}$ is the mean value of the process output. This can be done by means of the so-called Harris Index, which is expressed as (Harris, 1989):

$$HI = \frac{\sigma_y^2}{\sigma_{MV}^2}. \tag{8.20}$$

It can be easily deduced that $HI \in [1, +\infty)$. Determining a value of HI close to one means that the performance achieved by the adopted controller is close to the minimum variance performance, while a large value of HI might indicate that the controller should be retuned. Alternatively, the inverse of the Harris Index

$$\eta = \frac{\sigma_{MV}^2}{\sigma_y^2}. \tag{8.21}$$

or the Normalised Harris Index:

$$NHI = 1 - \eta = 1 - \frac{\sigma_{MV}^2}{\sigma_y^2} = \frac{\sigma_y^2 - \sigma_{MV}^2}{\sigma_y^2} \tag{8.22}$$

can be evaluated. It appears that $\eta \in (0, 1]$ and the less the value of η is, the worse is the performance. On the contrary, $NHI = 0$ means that the optimal performance is achieved, while a value of NHI close to one means that the controller is not performing well (note that $NHI \in [0, 1)$).

As an illustrative example, consider the following process:

$$y(t) = \frac{1 - 0.1q^{-1}}{1 - 0.7q^{-1} + 0.1q^{-2}} q^{-3} u(t) + \frac{1 - 0.2q^{-1} + 0.5q^{-2}}{1 - 0.7q^{-1} + 0.1q^{-2}} w(t). \tag{8.23}$$

By solving the Diophantine equation (8.12), it results:

$$E(q^{-1}) = 1 + 0.5q^{-1} + 0.75q^{-2} \tag{8.24}$$

and

$$F(q^{-1}) = 0.475 - 0.075q^{-1}. \tag{8.25}$$

Hence, the minimum variance control law results to be:

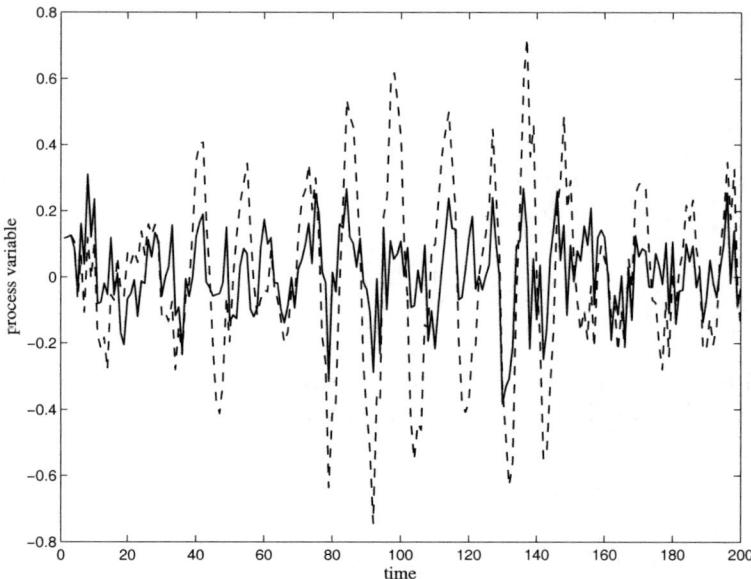

Fig. 8.1. Process variable with a minimum variance controller (solid line) and with a PI controller (dashed line)

$$u(t) = -\frac{0.475 - 0.075q^{-1}}{1 + 0.4q^{-1} + 0.7q^{-2} - 0.075q^{-3}}y(t). \quad (8.26)$$

The result of an experiment with the minimum variance controller is shown in Figure 8.1 (the noise variance has been fixed to 0.01). The application of the performance assessment methodology, where a second-order ARMA model has been estimated, results in $\eta = 0.98$, which confirms that an optimal controller has been adopted.

Conversely, if the (PI) control law

$$u(t) = -\frac{0.3 - 0.1q^{-1}}{1 - q^{-1}}y(t) \quad (8.27)$$

is adopted, the resulting value of η is 0.29 (the process variable obtained is shown again in Figure 8.1), so that it can be deduced that the performance can be improved.

As an alternative methodology, the autocorrelation of the output can be calculated in order to verify if there is significant correlation beyond the time delay (Qin, 1998). In particular, given a time-series $[y(t), y(t+1), \ldots, y(t+N-1)]$, the autocorrelation function is determined by calculating

$$\rho_y(j) = \frac{1}{N}\sum_{i=0}^{N-1-j}(y(t+i) - \bar{y})(y(t+i+j) - \bar{y}) \quad (8.28)$$

where $\bar{y}$ is again the mean value of the time series. The control is roughly achieving the minimum variance if, for $j \geq d$,

$$\rho_y(j) \in \left[-2\sqrt{var[\rho_y(j)]}, +2\sqrt{var[\rho_y(j)]}\right], \qquad (8.29)$$

where

$$var[\rho_y(j)] = \frac{1}{N}\left[1 + 2\sum_{i=1}^{d-1} \rho_y^2(i)\right]. \qquad (8.30)$$

Although it is appealing to determine how far the performance achieved by a given controller is from the minimum variance controller, the approach presents serious drawbacks that decrease its significance in a wide range of practical situations (Qin, 1998). Actually, the theoretical minimum variance σ_{MV}^2 is invariant regardless of the control structure. Thus, it represents a lower bound that could not be achieved by the given control structure (for example, a PID controller) and therefore it is not easy to understand if retuning the controller would improve the performance or if it would be necessary to change the controller structure. In other words, despite a high value of the Harris Index, the performance achieved by a controller could not be improved by simply retuning it. This is particularly true for processes with a significant dead time if a model-based dead-time compensator is not employed in the control system, since the minimum variance control has the same structure of the Smith predictor (Palmor, 1996). By following the same reasoning, the minimum variance performance assessment is not applicable for processes with a varying time delay. In any case, it is worth stressing that it is wise to repeat the calculation of the Harris Index from time to time because of the varying nature of the disturbances and of the process dynamics. This means that an accurate estimation of the dead time of the process has to be performed every time.

From another point of view, even if the achieved stochastic performance is satisfactory, this might not be compatible with the required deterministic performance (for example in the set-point following and load disturbance rejection task) (Thornhill et al., 2003). It is worth noting in this context that the minimum variance controller is based on pole-zero cancellation (see Equation (8.15)) and that there is not a chance to address the set-point task rather than the load disturbance rejection task (or *vice versa*) and to handle the trade-off between aggressiveness, robustness and control effort. These concepts will be further discussed in Section 8.4.

Summarising, the minimum variance performance assessment approach is effective most of all for low-order processes with negligible dead time (such as flow loops), when the rejection of stochastic disturbances is of primary concern.

8.3.3 Assessment of PID Control Performance

In the previous section it has been highlighted that the use of the Harris Index for the performance assessment of PID controllers can be misleading, since, even if the actual variance is far from the minimum one, the performance could not be improved by retuning the controller. Thus, a more realistic performance measure, which takes into account the controller structure, should be adopted.

An iterative solution method for the determination of PID achievable performance bound has been proposed in (Ko and Edgar, 2004). The process output is described by the discrete-time model

$$y(t) = P(q^{-1})q^{-d}u(t) + N(q^{-1})w(t) \qquad (8.31)$$

where $P(q^{-1})q^{-d}$ is the process model with a dead time d and $N(q^{-1})$ is the disturbance model driven by a zero-mean white noise $w(t)$. The PID controller is expressed as (see (1.41), note that here lower letters are adopted for the PID parameters in order to stress that they are scalar quantities):

$$C(q^{-1}) = \frac{k_1 + k_2 q^{-1} + k_3 q^{-2}}{1 - q^{-1}}. \qquad (8.32)$$

By considering Expression (8.32), the process output can be rewritten as

$$y(t) = -\sum_{i=1}^{m} s_i \left(k_1 + k_2 q^{-1} + k_3 q^{-2}\right) y(t-i) + \sum_{i=0}^{\infty} n_i w(t-i) \qquad (8.33)$$

where the s_i ($i = 1, \ldots, m$) represent the process step response coefficients and the n_i ($i = 0, \ldots, \infty$) represent the disturbance impulse response coefficients. In order to obtain the output variance as a function of the PID parameters, it is assumed that a single random shock w_0 is introduced in the closed-loop system at time $t = 0$. In this case the closed-loop response over a finite horizon p can be written as

$$\begin{bmatrix} y_0 \\ y_1 \\ \vdots \\ y_p \end{bmatrix} = \left(I + Sk_1 + FSk_2 + F^2 Sk_3\right)^{-1} \bar{n} w_0 \qquad (8.34)$$

where I is the $(p+1) \times (p+1)$ identity matrix,

$$S = \begin{bmatrix} 0 & \cdots & & & \\ s_1 & 0 & & & \\ s_2 & s_1 & 0 & & \\ \vdots & \vdots & \ddots & \ddots & \\ s_p & s_{p-1} & \cdots & s_1 & 0 \end{bmatrix}, \qquad (8.35)$$

$$\bar{n} = \begin{bmatrix} n_0 \\ n_1 \\ \vdots \\ n_p \end{bmatrix}, \tag{8.36}$$

and F is a $(p+1) \times (p+1)$ matrix defined as:

$$F = \begin{bmatrix} 0 & & & 0 \\ 1 & \ddots & & \\ & \ddots & \ddots & \\ 0 & & 1 & 0 \end{bmatrix}. \tag{8.37}$$

If the random shock occurs at every time instant, the closed-loop system response can be determined as

$$y(t) = \sum_{i=0}^{p} \psi_i w(t-i) \tag{8.38}$$

where

$$\begin{bmatrix} \psi_0 \\ \psi_1 \\ \vdots \\ \psi_p \end{bmatrix} = \left(I + Sk_1 + FSk_2 + F^2 Sk_3\right)^{-1} \bar{n} \tag{8.39}$$

defines the vector of the closed-loop impulse response coefficients. The output variance can be therefore expressed as

$$\sigma_{PID}^2 = \bar{n}^T \left(I + S^T k_1 + (FS)^T k_2 + (F^2 S)^T k_3\right)^{-1} \\ \cdot \left(I + Sk_1 + FSk_2 + F^2 Sk_3\right)^{-1} \bar{n}\sigma_w^2. \tag{8.40}$$

It is worth stressing that the process model is described by means of its step response coefficients. It can be remarked that Expression (8.40) can be used to determine the optimal PID parameters that minimise the output variance. Indeed, the optimal parameters satisfy the following first-order necessary condition:

$$\frac{\partial \sigma_{PID}^2}{\partial k_1} = -2\bar{n}^T (\Gamma^{-1})^T S \Gamma^{-2} \bar{n} = 0 \tag{8.41}$$

$$\frac{\partial \sigma_{PID}^2}{\partial k_2} = -2\bar{n}^T (\Gamma^{-1})^T FS \Gamma^{-2} \bar{n} = 0 \tag{8.42}$$

$$\frac{\partial \sigma_{PID}^2}{\partial k_3} = -2\bar{n}^T (\Gamma^{-1})^T F^2 S \Gamma^{-2} \bar{n} = 0 \tag{8.43}$$

where $\Gamma = I + Sk_1 + FSk_2 + F^2Sk_3$. The second-order derivatives can be expressed as:

$$\frac{\partial^2 \sigma_{PID}^2}{\partial k_1^2} = 2\bar{n}^T(\Gamma^{-2})^T S^T S \Gamma^{-2}\bar{n} + 4\bar{n}^T(\Gamma^{-1})^T S^2 \Gamma^{-3}\bar{n}, \quad (8.44)$$

$$\frac{\partial^2 \sigma_{PID}^2}{\partial k_1 \partial k_2} = 2\bar{n}^T(\Gamma^{-2})^T (FS)^T S \Gamma^{-2}\bar{n} + 4\bar{n}^T(\Gamma^{-1})^T FS^2 \Gamma^{-3}\bar{n}, \quad (8.45)$$

$$\frac{\partial^2 \sigma_{PID}^2}{\partial k_1 \partial k_3} = 2\bar{n}^T(\Gamma^{-2})^T (F^2S)^T S \Gamma^{-2}\bar{n} + 4\bar{n}^T(\Gamma^{-1})^T F^2 S^2 \Gamma^{-3}\bar{n}, \quad (8.46)$$

$$\frac{\partial^2 \sigma_{PID}^2}{\partial k_2^2} = 2\bar{n}^T(\Gamma^{-2})^T (FS)^T (FS) \Gamma^{-2}\bar{n} + 4\bar{n}^T(\Gamma^{-1})^T F^2 S^2 \Gamma^{-3}\bar{n}, \quad (8.47)$$

$$\frac{\partial^2 \sigma_{PID}^2}{\partial k_2 \partial k_3} = 2\bar{n}^T(\Gamma^{-2})^T (F^2S)^T (FS) \Gamma^{-2}\bar{n} + 4\bar{n}^T(\Gamma^{-1})^T F^3 S^2 \Gamma^{-3}\bar{n}, \quad (8.48)$$

$$\frac{\partial^2 \sigma_{PID}^2}{\partial k_3^2} = 2\bar{n}^T(\Gamma^{-2})^T (F^2S)^T (F^2S) \Gamma^{-2}\bar{n} + 4\bar{n}^T(\Gamma^{-1})^T F^4 S^2 \Gamma^{-3}\bar{n}. \quad (8.49)$$

These expressions can be adopted to find the optimal PID settings $\bar{k}_{opt} = [k_{1,opt}, k_{2,opt}, k_{3,opt}]$ by means of Newton's iterative method, namely, by updating the PID parameters by means of the expression:

$$\bar{k}_{new} = \bar{k}_{opt} - \begin{bmatrix} \frac{\partial^2 \sigma_{PID}^2}{\partial k_1^2} & \frac{\partial^2 \sigma_{PID}^2}{\partial k_1 \partial k_2} & \frac{\partial^2 \sigma_{PID}^2}{\partial k_1 \partial k_3} \\ \frac{\partial^2 \sigma_{PID}^2}{\partial k_1 \partial k_2} & \frac{\partial^2 \sigma_{PID}^2}{\partial k_2^2} & \frac{\partial^2 \sigma_{PID}^2}{\partial k_2 \partial k_3} \\ \frac{\partial^2 \sigma_{PID}^2}{\partial k_1 \partial k_3} & \frac{\partial^2 \sigma_{PID}^2}{\partial k_2 \partial k_3} & \frac{\partial^2 \sigma_{PID}^2}{\partial k_3^2} \end{bmatrix}_{\bar{k}=\bar{k}_{old}}^{-1} \cdot \begin{bmatrix} \frac{\partial \sigma_{PID}^2}{\partial k_1} \\ \frac{\partial \sigma_{PID}^2}{\partial k_2} \\ \frac{\partial \sigma_{PID}^2}{\partial k_3} \end{bmatrix}_{\bar{k}=\bar{k}_{old}} . \quad (8.50)$$

When the convergence is obtained, the optimality of the determined PID parameters can be checked by verifying the positive definitiveness of the Hessian matrix.

The best achievable performance bound can then be calculated as:

$$(\sigma_{PID}^2)_{opt} = \bar{n}^T (\Gamma_{opt}^{-1})^T \Gamma_{opt}^{-1} \bar{n} \sigma_w^2, \quad (8.51)$$

where

$$\Gamma_{opt} = I + Sk_{1,opt} + FSk_{2,opt} + F^2 Sk_{3,opt}. \quad (8.52)$$

It is worth noting at this point that the optimal PID parameters and the achievable minimum variance can be derived starting from the knowledge of the process step response coefficients (which can be obtained by means of a standard simple open-loop experiment) and from the impulse response coefficients of the disturbance. These can be obtained by considering the following relationship between the disturbance model impulse response and the closed-loop impulse response:

$$\bar{n} = (I + Sk_1 + FSk_2 + F^2 Sk_3) \cdot \begin{bmatrix} \psi_0 \\ \psi_1 \\ \vdots \\ \psi_p \end{bmatrix}. \tag{8.53}$$

Hence, an estimation of the disturbance impulse response coefficients $\hat{\bar{n}}$ can be derived from the estimation of the closed-loop impulse response $\hat{\psi}$ obtained by means of a time-series modelling of the closed-loop data:

$$\hat{\bar{n}} = (I + Sk_1 + FSk_2 + F^2 Sk_3) \cdot \begin{bmatrix} \hat{\psi}_0 \\ \hat{\psi}_1 \\ \vdots \\ \hat{\psi}_p \end{bmatrix}. \tag{8.54}$$

It can be remarked that the devised methodology can be adopted also for assessing the control performance (and for determining the optimal PID parameters) in the context of deterministic set-point tracking. Indeed, it is recognised that a step change in the set-point can be modelled as

$$\frac{1}{1-q^{-1}} w(t) \tag{8.55}$$

where $w(t)$ is zero except at the time of the set-point change. Thus, in this case the vector of the disturbance impulse response coefficients is simply

$$\bar{n} = \begin{bmatrix} 1 \\ \vdots \\ 1 \end{bmatrix}. \tag{8.56}$$

It is worth also noting that the minimum variance index (8.21) can be calculated directly as (Thornhill et al., 2003)

$$\eta = \frac{\sigma_{MV}^2}{\sigma_y^2} = \frac{\sum_{i=0}^{d-1} y^2(t)}{\sum_{i=0}^{\infty} y^2(t)}. \tag{8.57}$$

As an illustrative example consider again the process (8.23). A PI controller

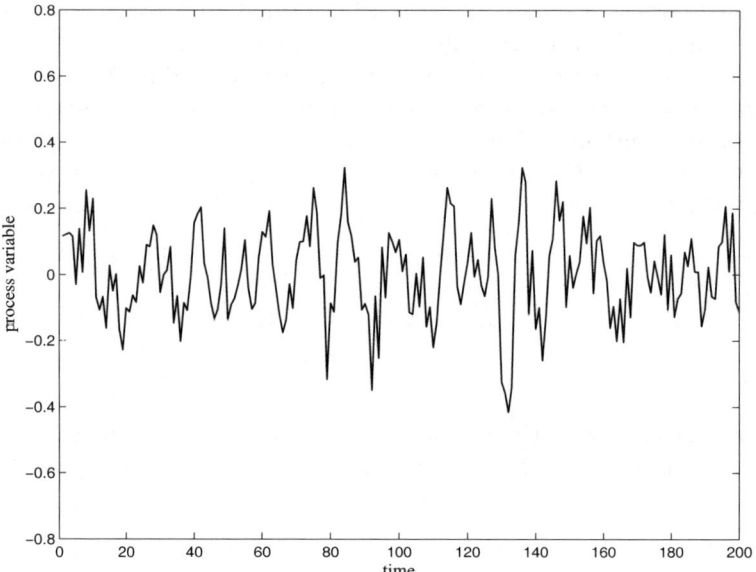

Fig. 8.2. Process variable with a PI controller

$$C(q^{-1}) = \frac{0.22 - 0.2q^{-1}}{1 - q^{-1}} \qquad (8.58)$$

is initially adopted. The process output obtained by setting a noise variance of 0.01 is shown in Figure 8.2 (compare it with Figure 8.1). The performance index (8.21) results to be $\eta = 0.90$, which indicates that the PI controller is well tuned from the regulatory control point of view. However, the closed-loop set-point step response performance is not satisfactory, as it appears from Figure 8.3. By running Newton's iterative method (8.50), the optimal PID parameter is then derived as

$$C(q^{-1}) = \frac{0.7286 - 1.0386q^{-1} + 0.4252q^{-2}}{1 - q^{-1}}. \qquad (8.59)$$

The resulting closed-loop set-point step response is shown in Figure 8.4, where the sensible performance improvement appears.

In order to assess the performance of a PID controller, it is suggested to compare the actual variance to that obtained by the (determined) optimal PID controller (Ko and Edgar, 2004). Thus, similarly to the inverse Harris index, the following performance index can be used:

$$\eta = \frac{(\sigma^2_{PID})_{opt}}{\sigma^2_y}. \qquad (8.60)$$

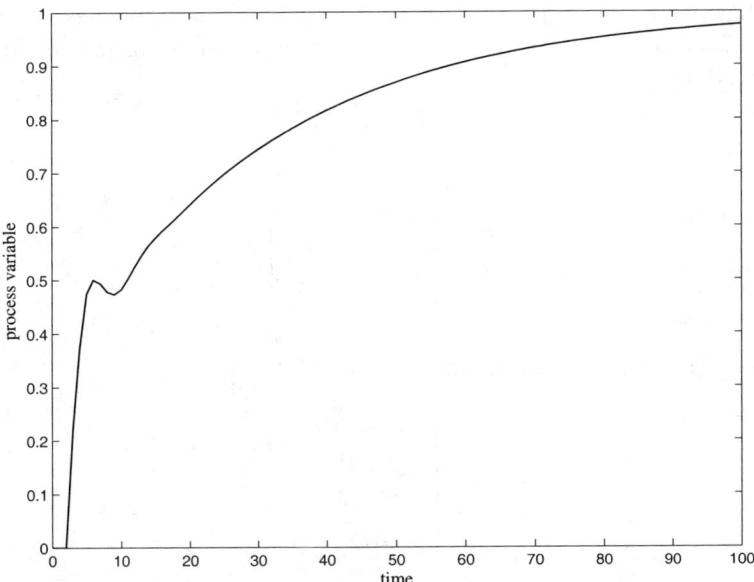

Fig. 8.3. Set-point response with a PI controller

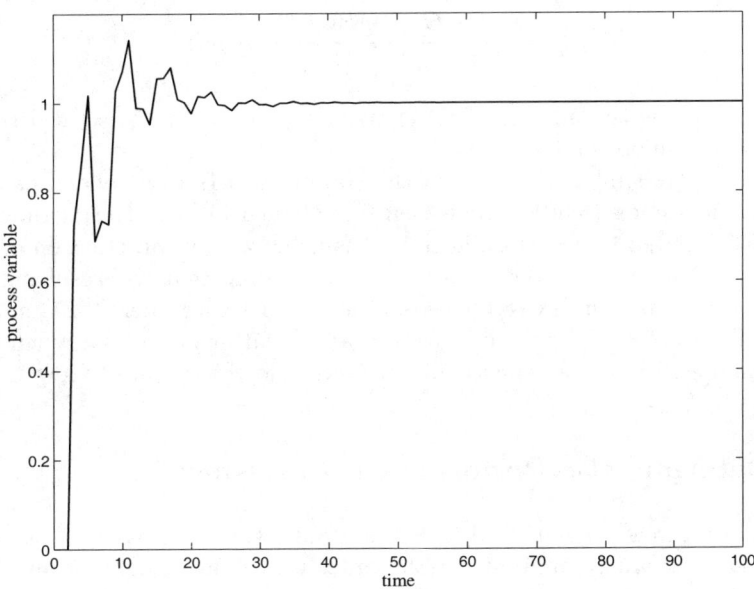

Fig. 8.4. Set-point response with the optimal PID controller

Note that $0 < \eta \leq 1$ and $\eta = 1$ means that the best performance is achieved. Given a time-series of n samples, an estimation of the performance index can be derived as:
$$\hat{\eta} = \frac{\bar{y}_{opt}^T \bar{y}_{opt}}{\bar{y}^T \bar{y}}, \qquad (8.61)$$
where
$$\bar{y} = \begin{bmatrix} y(t) \\ y(t-1) \\ \vdots \\ y(t-n+1) \end{bmatrix} \qquad (8.62)$$
is the vector of the measured output, and
$$\bar{y}_{opt} = \begin{bmatrix} y_{opt}(t) \\ y_{opt}(t-1) \\ \vdots \\ y_{opt}(t-n+1) \end{bmatrix} \qquad (8.63)$$
is the vector of the closed-loop outputs that can be achieved by the optimal PID settings. Once the optimal PID parameters are determined by iteratively applying Equation (8.50), $\bar{y}_{opt}$ can be calculated as
$$\bar{y}_{opt} = \frac{1 + C(q^{-1})P(q^{-1})q^{-d}}{1 + C_{opt}(q^{-1})P(q^{-1})q^{-d}} y(t) \qquad (8.64)$$
where $C(q^{-1})$ is the adopted PID controller and $C_{opt}(q^{-1})$ is the PID controller with the optimal settings.

It is worth stressing again that, in the context of PID controllers, using the performance index (8.60) is more sensible than using the Harris Index, because it is related to a performance that is achievable in practice. In order to understand this fact better, it is worth considering that the resulting value of the performance index (8.60) is 0.32 for the PI controller (8.27) and 0.92 for the PI controller (8.58), indicating that the latter provides a much better performance from the stochastic disturbance rejection point of view.

8.4 Deterministic Performance Assessment

It has been already mentioned in Section 8.3.2 that addressing a stochastic performance (namely, minimising the output variance) might be in conflict with satisfying (deterministic) performance requirements related to the set-point following or the load disturbance rejection task. For example, if the noise dynamics is ARMA, then the minimum variance controller has no integral action (Qin, 1998), so that a null steady-state error is not achieved in general (more precisely, the mean value of the steady-state control error is not null). Indeed, in practical cases the control requirements are often specified in terms of maximum overshoot, settling time, and so on. Further, robustness constraints and control effort should be taken into account. In any case, a trade-off between stochastic and deterministic performance should be considered.

When the deterministic requirements are considered, it is realised that an unsatisfactory performance can be caused by different factors (Patwardhan and Shah, 2002).

Thus, there is the need to integrate different techniques, each of them devoted to deal with a particular situation. Obviously, it is desirable that each technique be based as much as possible on routine operating data and that no process model is required in order to be employed in general. In the following section different functionalities devoted to detect and analyse a particular situation are presented. They will be subsequently exploited in the context of PID performance assessment.

8.4.1 Useful Functionalities

Oscillation Detection

A major source of degradation of the quality of the end product is given by the presence of oscillations in a control loop, which cause also an increased energy consumption. Actually, oscillations can be caused by different reasons and therefore, in addition to detect automatically the occurrence of oscillations, it is important to detect the reason for it, in order to take the appropriate correction. This aspect will be discussed in the following section.

In order to detect a persistent oscillation, a method has been presented in (Hägglund, 1995) (note that it has been successfully tested in industrial environments). It is based on a load disturbance detection procedure which consists of determining the integrated absolute error between two successive zero crossings time instants t_{i-1} and t_i of the control error:

$$IAE = \int_{t_{i-1}}^{t_i} |e(t)| dt. \tag{8.65}$$

If the controller has no integral action, then the difference between the measurement signal and its average value should be adopted instead of the control error.

A load disturbance is detected if the value of IAE exceed a given threshold IAE_{lim}. The value of IAE_{lim} is fixed as

$$IAE_{lim} = \frac{2a}{\omega_u} \tag{8.66}$$

where a is the amplitude of an acceptable oscillation, which should not be detected (a reasonable choice for it is $a = 1\%$), and ω_u is the ultimate frequency of the process. If the value of ω_u is not available, it can be substituted (assuming that the PI(D) controller is properly tuned) with $\omega_i = 2\pi/T_i$ where T_i is the integral time constant. Thus, if it is $IAE > IAE_{lim}$ between two successive zero crossings time instants, then it is concluded that a load disturbance has occurred.

It is worth noting at this point that an alternative load detection procedure has been proposed in (Salsbury, 2005). It is based on a statistical change detection procedure that uses zero crossings information to account for autocorrelation.

Once the load detection procedure has been established, the oscillation detection procedure can be derived by monitoring if the number of load disturbances exceed a threshold limit $n_{lim} = 10$ over a supervision time $T_{sup} = 50T_u$, where T_u is the ultimate period of the process (or, alternatively $T_{sup} = 50T_i$). Thus, if at least n_{lim} load disturbances are detected during the last interval of duration T_{sup}, it is concluded that an oscillation is present.

For a more practical implementation of the procedure, it is suggested to make an exponential weighting of the detections. Thus, the following recursive procedure can be applied at every sampling instant:

1. if a load is detected then $load = 1$ else $load = 0$;
2. $x := \gamma x + load$;
3. if $x \geq n_{lim}$ then conclude that an oscillation is present.

The value of γ can be selected as

$$\gamma = 1 - \frac{\Delta t}{T_{sup}} \qquad (8.67)$$

where Δt is the sampling time.

An alternative strategy is proposed in (Forsman and Stattin, 1999). It consists of considering separately the error signal when it is positive and when it is negative and of determining the periodicity of the two parts in terms of the integrated error between successive zero crossings and in terms of the time intervals between successive zero crossings. By considering the notation depicted in Figure 8.5, the following definitions can be applied:

$$A_i := \int_{t_{2i}}^{t_{2i+1}} |e(\tau)| d\tau \qquad B_i := \int_{t_{2i+1}}^{t_{2i+2}} |e(\tau)| d\tau \qquad i = 0, 1, \ldots, N/2 \quad (8.68)$$

where $N > 20$ is suggested. Then, define

$$h_A(N) := \sharp \left\{ i < N/2 : \alpha < \frac{A_{i+1}}{A_i} < \frac{1}{\alpha} \wedge \gamma < \frac{\delta_{i+1}}{\delta_i} < \frac{1}{\gamma} \right\} \qquad (8.69)$$

and

$$h_B(N) := \sharp \left\{ i < N/2 : \alpha < \frac{B_{i+1}}{B_i} < \frac{1}{\alpha} \wedge \gamma < \frac{\varepsilon_{i+1}}{\varepsilon_i} < \frac{1}{\gamma} \right\} \qquad (8.70)$$

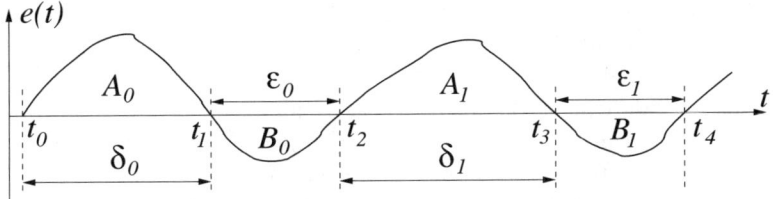

Fig. 8.5. Notation for the alternative method for oscillations detection

where $\sharp S$ denotes the number of elements of the set S and α and γ are tuning parameters to be selected as

$$\alpha \in [0.5, 0.7] \qquad \gamma \in [0.7, 0.8] \qquad \gamma > \alpha. \tag{8.71}$$

The oscillatory behaviour of the control system can be detected by calculating

$$h(N) = \frac{h_A(N) + h_B(N)}{N}. \tag{8.72}$$

In particular, loops having $h > 0.4$ have oscillations and should be better examined. If $h > 0.8$, then the loop is definitively oscillatory. Intermediate situations may arise.

It can be deduced that this method requires the tuning of additional parameters and the result is not always easy to interpret. However, it has the capability to detect different type of oscillations and therefore it can be usefully adopted in practical cases.

Oscillation Diagnosis

As already mentioned, once an oscillation has been detected, it is important to diagnose its cause in order to take the necessary corrective action. Indeed, there are several possible reasons for an oscillatory behaviour of the control loop. For example, an oscillatory load might disturbing the monitored loop. In this case, if the oscillation frequency is low (with respect to the closed-loop bandwidth) it can be efficiently handled by the feedback control system, while if it is high, it should be appropriately filtered in the controller so that it is not transferred to the actuator (see Chapter 2), possibly reducing its life-span. The worst case occurs if the oscillation frequency is close to the critical frequency of the loop transfer function, because in this case it might be amplified by the control system. In this case the controller should be retuned (or the source of disturbance should be eliminated).

Another possible source of an oscillatory control loop is a too aggressively tuned controller. This fact implies that in many cases operators detune the controller when they detect an oscillatory behaviour of the control system. However, it is recognised in the literature that the most common reason for oscillations is an excessive (static) friction in the valve (Hägglund, 1995). In this case detuning the controller is not appropriate and valve maintenance should be performed.

In order to provide valid tools for oscillation diagnosis, various solutions have been proposed in the literature (see, for example, (Thornhill and Hägglund, 1997; Taha *et al.*, 1996)). In particular, based on the above considerations, a significant effort has been provided by researchers in order to detect the presence of stiction in the valves (Choudhury *et al.*, 2005). Different approaches have been followed. For example, the use of a nonlinearity index, based on the determination of the squared bicoherence, has been proposed in (Choudhury *et al.*, 2004) (note that the occurrence of oscillations can be determined also by the presence of hysteresis and dead-band in the loop). Alternatively, in (Horch, 1999), it is proposed to analyse the cross-correlation between the controller output and the process output and in (Horch, 2001) it is suggested to exploit the different probability distributions of the first derivative of the oscillatory process output in case of an aggressive control and in case of the presence of stiction (the second derivative can be used if the process is non self-regulating).

A very simple approach is proposed in (Singhal and Salsbury, 2005). It is based on the fact that an aggressive control results in a sinusoidal control error signal, while for a sticking valve the discontinuous process input signal causes a piecewise exponential control error signal. Thus, a positive (or negative) half period of the oscillation is considered and the ratio of the area before and after the peaks is calculated. If the value of the ratio is close to one it is concluded that the controller is aggressive while if it is greater than one it is concluded that stiction is present.

By following the same idea of evaluating the shape of the oscillation, a technique based on comparing the oscillatory response to a sine wave, to a triangular wave and to the output response of a FOPDT process under relay control (see Section 7.2.2) is presented in (Rossi and Scali, 2005). Note that the last two cases are associated with the presence of stiction.

Finally, a technique based on the analysis of the input-output characteristics of the valve, *i.e.*, of the plot of the valve position *versus* the controller output, is proposed in (Yamashita, 2006). Note that if the valve position is not available, it can be substituted with the corresponding flowrate. In any case, it is recognised that there is not a method that is capable to definitively solve the problem and therefore it is wise to suitably integrate some of them in a monitoring tool (Rossi and Scali, 2005).

Note that a method for the compensation of the static friction of pneumatic control valves has been proposed in (Hägglund, 2002).

Abrupt Load Disturbance Detection

In addition to detect the occurrence of a load disturbance (see the previous sections), it is often necessary to verify if its dynamics is sufficiently exciting for the control loop. This is of particular concern if an adaptive control law is implemented (Hägglund and Åström, 2000), but it is useful also in the performance assessment context, as it will be shown in the next sections (see also Section 5.6).

In order to detect an abrupt (*i.e.*, step-like) load disturbance, an idea derived from the analysis presented in (Hägglund and Åström, 2000) can be applied. In particular, the control variable and process variable signals can be high-pass filtered, according to the expressions (for simplicity, the Laplace transform of the signals is employed):

$$U_{hp}(s) = \frac{s}{s+\omega_{hp}}U(s) \qquad Y_{hp}(s) = \frac{1}{K}\frac{s}{s+\omega_{hp}}Y(s) \qquad (8.73)$$

where K is the process gain (which can be derived by considering the steady-state values of the signals) and its use is for an appropriate scaling of the signals. The frequency ω_{hp} can be chosen to be inversely proportional to the integral time constant T_i. Then, the load disturbance is considered to be abrupt if the obtained signals exceed a given threshold. This can be fixed as a 3% of the amplitude of the disturbance, which can be evaluated by considering the initial and the final steady-state values of the control variable.

Aggressive Controller Detection

The detection of an aggressive controller can be made through the detection of an oscillatory behaviour of a control loop in the presence of an abrupt set-point change or load disturbance. For this purpose, a statistically-based approach has been proposed in (Miao and Seborg, 1999). It consists of calculating the autocorrelation of either the controlled variable or the control error. The sample autocorrelation coefficients are determined as:

$$\rho_k = \frac{\sum_{t=1}^{N-k}(z(t)-\bar{z})(z(t+k)-\bar{z})}{\sum_{t=1}^{N}(z(t)-\bar{z})^2}, \qquad (8.74)$$

where $z(t)$, $t = 1, \ldots, N$ is the evaluated time series data and $\bar{z}$ is the sample mean for the N samples. An oscillation index is then calculated as the decay ratio of the obtained autocorrelation function. It is defined as

$$\mathcal{R} = \frac{a}{b}, \qquad (8.75)$$

where a is the distance from the first maximum to the straight line connecting the first two minima and b is the distance from the first minimum to the straight line that connects the first coefficient of the autocorrelation function to the first maximum (see Figure 8.6). If less than two minima exist in the autocorrelation function, the value of $\mathcal{R}$ is simply set to zero.

It is worth stressing that the algorithm requires a sufficiently long data collection period in order to have at least five cycles of (damped) oscillations. By

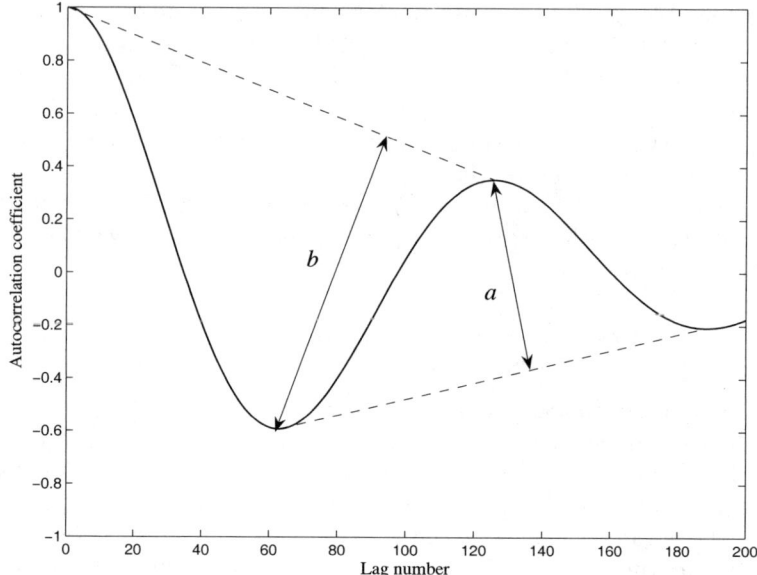

Fig. 8.6. Calculation of the oscillation index $\mathcal{R} = a/b$

considering that it is of interest to detect oscillations with a frequency around the ultimate frequency of the process (as already mentioned, low-frequency oscillations are compensated by the feedback controller while high-frequency oscillations can be easily filtered), the data collection period can be initially set to $50T_i$, where T_i is the integral time constant of the PI(D) controller. If in this period a sufficient number of oscillations is not detected, then the period should be increased to $250T_i$. Note also that the collected data should be appropriately filtered to remove the measurement noise.

An alternative index, called the Area Index (AI), whose aim is to estimate a generalised damping index of the closed-loop system has been proposed in (Visioli, 2006). It is based on the analysis of the control signal $u(t)$ that compensates for an abrupt load disturbance d occurring on the process. In particular, the new steady-state value achieved by the control signal after the transient load disturbance response is denoted as $\bar{u}$. The time instant in which the step load disturbance occurs is denoted as t_0 (note that the value of t_0 does not need to be known) and $t_1, \ldots, t_{n-1}$ are the subsequent time instants in which $u(t) = \bar{u}$. Finally, the time instant in which the transient response ends and the manipulated variable attains its steady-state value $\bar{u}$ is denoted as t_n. From a practical point of view, the value of t_n can be selected as the minimum time after that the control signal $u(t)$ remains within a one percent range of $\bar{u}$. The area delimited by the function $u(t)$ and $\bar{u}$ between two consecutive time instants t_i and t_{i+1} are defined as:

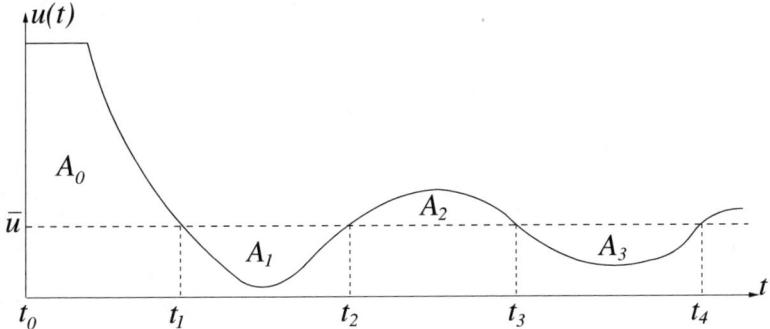

Fig. 8.7. Significant parameters for determining the Area Index

$$A_i := \int_{t_i}^{t_{i+1}} |u(t) - \bar{u}| dt. \tag{8.76}$$

The introduced notation is depicted in Figure 8.7. The Area Index AI is calculated by first eliminating the area A_0, i.e., the area between the time instant in which the step load disturbance occurs and the first time instant in which it is $u(t) = \bar{u}$. Then, the the ratio between the maximum value of the determined areas and their sum is determined, by taking into account that the last area A_{n-1} has to be excluded from the computation of the maximum area. In case during the overall transient response we have just once or we never have $u(t) = \bar{u}$, the Area Index is simply set to one. Formally, the Area Index is therefore defined as:

$$AI := \begin{cases} 1 & \text{if } n < 3 \\ \dfrac{\max\{A_1, \ldots, A_{n-2}\}}{\sum_{i=1}^{n-1} A_i} & \text{elsewhere} \end{cases}. \tag{8.77}$$

From Formula (8.77) it can be trivially deduced that the value of AI is always in the interval $(0, 1]$. The significance of the devised index can also be evaluated by performing the following analysis. Consider the transfer function $T(s)$ from the load disturbance signal (acting at the process input) to the manipulated variable (i.e., the controller output) in a standard unitary-feedback control system:

$$T(s) := -\frac{C(s)P(s)}{1 + C(s)P(s)} \tag{8.78}$$

and assume that $T(s)$ has a pair of complex conjugate dominant poles, i.e., it can be well-approximated by the following transfer function (note that this is not always the case as it will be discussed in Section 8.4.3):

$$\tilde{T}(s) := -\frac{1}{T_1^2 s^2 + 2\xi T_1 s + 1}. \tag{8.79}$$

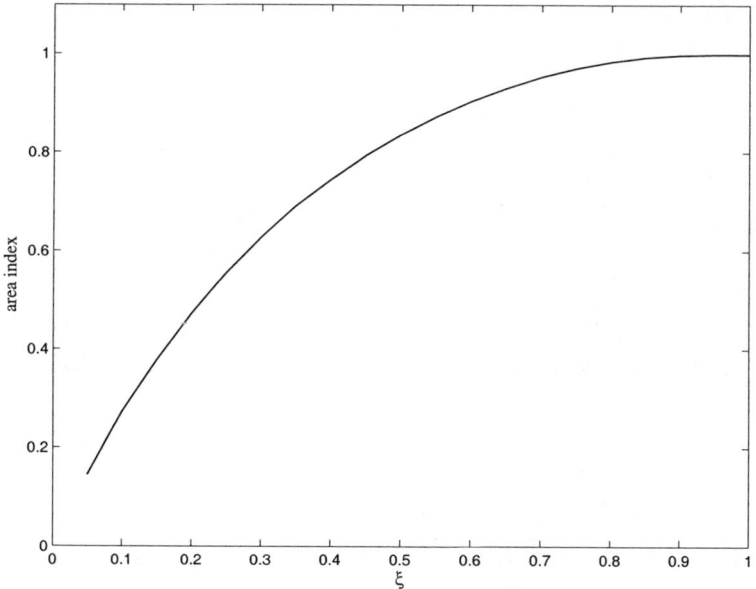

Fig. 8.8. Dependence of the Area Index on the damping factor ξ

If the Area Index AI is calculated by considering the step response of $\tilde{T}(s)$ with different values of T_1 and $\xi \in (0,1]$, it results that the value of AI is independent of the value of T_1 and depends only on the value of the damping factor ξ. The relation between ξ and AI is plotted in Figure 8.8. It appears that the more closely the value of AI approaches zero the more the control loop is oscillatory, whereas the more closely the value of AI approaches one, the more the control loop is sluggish.

It has to be noted that when the technique has to be applied in practical cases, noise has to be considered. As the Area Index is determined off-line, a standard filtering procedure can be applied before calculating the different areas. Alternatively, it is sufficient to discard from the analysis those areas A_i whose value is less than a predefined threshold (because they are actually due to the noise). This threshold can be determined by considering the control signal for a sufficiently long time interval when the process is at an equilibrium point and by determining the maximum area between two consecutive crossings with respect to its steady state value (the latter can be calculated as the mean value of the control signal itself in the considered time interval). Indeed, this procedure is actually similar to the one based on the concept of noise band (Åström *et al.*, 1993). In any case, as the overall procedure is based on the calculus of integrals, it is inherently robust to the noise.

It is worth stressing again that both the oscillation index $\mathcal{R}$ and the area index AI relies on the estimation of an abrupt load detection procedure like that described in the previous section. Note that in this context a load disturbance is considered to be abrupt when its dynamics is much faster than the dynamics of the closed-loop system.

Sluggish Controller Detection

Very often in practical cases there is not a sufficient time to optimise the controllers during the installation of the plant. Thus, the parameters of the controllers are usually tuned in a very conservative way in order to avoid problems with possible changes of the operating conditions.
A methodology to (automatically) detect sluggish control loops has been presented in (Hägglund, 1999) and further discussed in (Hägglund, 2005; Kuehl and Horch, 2005). It is based on the fact that, in the presence of an abrupt stepwise load disturbance, a sluggish response is characterised by the fact that the first time derivative of the manipulated variable and of the process output signals have the same sign for a large period. Thus, it is sensible to apply to the transient response the following calculation:

$$t_{pos} = \begin{cases} t_{pos} + \Delta t & \text{if } \Delta u \Delta y > 0 \\ t_{pos} & \text{if } \Delta u \Delta y \leq 0 \end{cases} \quad (8.80)$$

$$t_{neg} = \begin{cases} t_{neg} + \Delta t & \text{if } \Delta u \Delta y < 0 \\ t_{neg} & \text{if } \Delta u \Delta y \geq 0 \end{cases} \quad (8.81)$$

where Δt is the sampling time and Δu and Δy are the increments of the manipulated variable and of the process output respectively. Then, the so-called Idle Index can be determined as

$$II = \frac{t_{pos} - t_{neg}}{t_{pos} + t_{neg}}. \quad (8.82)$$

The Idle Index can be calculated also recursively by applying the following algorithm at every sampling instant:

if $\Delta u \Delta y > 0$ then $s = 1$
  else if $\Delta u \Delta y < 0$ then $s = -1$
  else $s = 0$
if $s \neq 0$ then $II = \gamma II + (1 - \gamma)s$.

Parameter γ determines the time horizon in the filter and can be related to the supervision time $T_{sup} = t_{pos} + t_{neg}$ of off-line calculation by means of the following relation:

$$\gamma = 1 - \frac{\Delta t}{T_{sup}}. \quad (8.83)$$

Since it is based on the increments of the signals, the procedure is quite sensitive to the measurement noise. For this reason it is necessary to appropriately filter the two signals (note that if the off-line procedure is applied a noncausal filter can be adopted). Different methods in this context have been discussed in (Kuehl and Horch, 2005). For example, the use of a low-pass filtering strategy with triggered reinitialisation is proposed. It consists of filtering the signals by a standard low-pass filter, but when the deviation between the input and output of the filter exceeds a given threshold ε, then the filter is reinitialised with the current signal value (this is done in order to avoid to soften too much the signals and therefore to avoid an incorrect calculation of the Idle Index). The value of ε can be set in the range $(0.4\sigma, 0.6\Delta d)$, where σ denotes the standard deviation of the noise and Δd is the typical size of load disturbances. Alternatively, a linear regression approach can be employed, namely, a polynomial is fitted to the data in the least-squares sense (a sensible polynomial order is ten), and an approach based on wavelet analysis (Daubechies, 1992) can be considered as well. Finally, quantisation of the already filtered signals is suggested.

In any case, it is evident that the value of II is always in the interval $[-1, +1]$ and a positive value close to one indicates that the control loop is sluggish. The problem associated with the use of the Idle Index is that a negative value close to -1 might be obtained both from a well-tuned loop and from an oscillatory loop. Thus, an oscillation detection technique has to be used in conjunction with the Idle Index approach.

8.4.2 Optimal Performance for Single-loop Systems

The achievable optimal performance in terms of integrated absolute error (IAE) for the set-point response has been investigated in (Huang and Jeng, 2002). In particular, a general controller transfer function

$$C(s) = \frac{K_c(a_m s^m + a_{m-1} s^{m-1} + \cdots + a_1 s + 1)}{s(b_n s^n + b_{n-1} s^{n-1} + \cdots + b_1 s + 1)} \qquad (8.84)$$

and a general process transfer function

$$P(s) = \frac{K(\tau_m s^m + \tau_{m-1} s^{m-1} + \cdots + \tau_1 s + 1)}{T_n s^n + T_{n-1} s^{n-1} + \cdots + T_1 s + 1} e^{-Ls} \qquad (8.85)$$

are considered. Then, the loop transfer function $L(s) := C(s)G(s)$ that minimises the performance index

$$J = \int_0^\infty |e(t)| dt \qquad (8.86)$$

is searched by using the simplex method. If a first-order loop transfer function is considered, the optimal transfer function results to be:

$$L^*(s) = \frac{0.76(1 + 0.47Ls)}{Ls} \qquad (8.87)$$

It is worth noting that a biproper transfer function has been considered since the presence of an additional (high-frequency) filter that have to be used to make the controller proper does not influence significantly the result. The corresponding optimal value of the integrated absolute error is $J^* = 1.377L$. Similarly, if a second-order loop transfer function is considered, the optimisation method results in

$$L^*(s) = \frac{0.83(1 + 0.70Ls + 0.18L^2s^2)}{Ls(1 + 0.30Ls)} \tag{8.88}$$

and the corresponding value of J^* is $1.314L$. Then, for a third-order transfer function it is

$$L^*(s) = \frac{0.84(1 + 1.71Ls + 0.97L^2s^2 + 0.25L^3s^3)}{Ls(1 + 1.26Ls + 0.41L^2s^2)} \tag{8.89}$$

and $J^* = 1.310L$. By considering that the value of the optimal integrated absolute error does not decrease by increasing again the order of the loop transfer function, it is concluded that the minimum integrated absolute error achievable by a simple unitary feedback loop is $J^* = 1.31L$.

In case the process has a FOPDT or a SOPDT dynamics, the optimal loop transfer function that can be obtained is the one shown in Expression (8.87) and therefore the practically achievable optimal integrated absolute error is $J^* = 1.38L$.

By applying similar reasonings, it can be found that if the controller is restricted to be of PI type, then the optimal performance for a FOPDT process results to be

$$J^*_{PI} = \begin{cases} L(2.1038 - 0.6023e^{-1.0695L/T}) & \text{for } L/T \leq 5 \\ 2.1038 & \text{for } L/T > 5 \end{cases} \tag{8.90}$$

where T is the process time constant. For a SOPDT transfer function, it is

$$J^*_{PI} = \begin{cases} L(\alpha(\xi)L^2/T^2 + \beta(\xi)L/T + \gamma(\xi)) & \text{for } \xi \leq 2.0 \\ L(-0.0173L^2/T_2^2 + 1.7749L/T_2 + 2.3514) & \text{for } \xi > 2.0 \end{cases} \tag{8.91}$$

where ξ is the damping factor, T is the process time constant if Expression (7.43) is considered, $T_2 \leq T_1$ is the process time constant in Expression (7.42), and

$$\alpha(\xi) = \begin{cases} 0.7444\xi^3 - 1.4975\xi^2 + 1.0202\xi - 0.2525 & \text{for } \xi \leq 0.7 \\ 0.0064\xi - 0.0203 & \text{for } 0.7 < \xi \leq 2.0 \end{cases} \tag{8.92}$$

$$\beta(\xi) = 1.1193\xi^{-0.9339} \tag{8.93}$$

$$\gamma(\xi) = \begin{cases} -18.4675\xi^2 + 17.9592\xi - 2.7222 & \text{for } \xi \leq 0.5 \\ -0.0995\xi^2 + 0.4893\xi + 1.4712 & \text{for } 0.5 < \xi \leq 2.0. \end{cases} \tag{8.94}$$

Conversely, if a PID controller is considered, the following values result if the process is of FOPDT type:

$$J^*_{PID} = \begin{cases} L(1.38 - 0.1134e^{-1.5541L/T}) & \text{for } L/T \leq 3 \\ 1.38 & \text{for } L/T > 3 \end{cases} \quad (8.95)$$

For a SOPDT transfer function, it is

$$J^*_{PID} = \begin{cases} L(2.1038 - \lambda(\xi)e^{-\mu(\xi)LT}) & \text{for } \xi \leq 1.1 \\ L(2.1038 - 0.6728e^{-1.2024LT_2}) & \text{for } \xi > 1.1 \end{cases} \quad (8.96)$$

where

$$\lambda(\xi) = 0.4480\xi^2 - 1.0095\xi + 1.2904 \quad (8.97)$$

$$\mu(\xi) = 6.1998e^{-3.8888\xi} + 0.6708. \quad (8.98)$$

By means of the previous expressions, it is easy to evaluate the performance of a given (PID) controller by comparing the obtained integrated absolute error with the optimal achievable one.

It is also worth stressing that from the previous analysis, it can be deduced that a PID controller is capable to provide virtually the optimal performance for FOPDT processes (*i.e.*, its efficiency is very close to 100%), while for SOPDT processes it might be worth to considering a general structure controller (the efficiency of the PID control is mostly around 65%). Further, it appears that the use of the derivative action allows to significantly improve the performance also for FOPDT processes (see Section 1.8).

8.4.3 PID Tuning Assessment

In practical cases it is very useful to assess the tuning of a PI(D) controller in order to verify if a retuning of the controller should be performed. Obviously, the assessment should be done with respect to the control specifications, since, for example, achieving a satisfactory performance in the load disturbance rejection task requires in general a different tuning from that required to achieve a satisfactory performance in the set-point following task. Further, various kinds of operating data could be available for assessing the performance and therefore there is the need of different techniques to be applied in different contexts. Few of them are presented hereafter.

Assessment Based on Set-point Response Data

The methodology proposed in (Swanda and Seborg, 1999) is based on the evaluation of set-point response data and it is related with the set-point response performance of a PI controller.

The rationale of the technique is to compare the achieved performance with that of a PI controller tuned with the Internal Model Control tuning rule

Table 8.1. Classification of a PI controller set-point following performance

Class	T_s	IAE_d	overshoot
High performance	≤ 4.6	≤ 2.8	*
Excessively sluggish	> 13.3	> 6.3	$\leq 10\%$
Poorly tuned	> 13.3	> 6.3	$> 10\%$

based on a FOPDT process model (see (7.5.1)).

In this context, two indices are evaluated, namely, the dimensionless settling time T_s and the dimensionless integrated absolute error IAE_d. They are defined as

$$T_s := \frac{t_s}{L} \quad (8.99)$$

where L is the process dead time and t_s is the measured settling time defined as the time the process output takes to attain and remain inside a band whose width is equal to $\pm 10\%$ of the amplitude A of the step input, and

$$IAE_d := \frac{IAE}{|A|L} \quad (8.100)$$

where IAE is the measured integrated absolute error. Based on the benchmark values of T_s and IAE_d obtained with the IMC design (high-order models reduced to a FOPDT form have been also taken into account), Table 8.1 has been established to classify the performance of the adopted PI controller (note that also the obtained overshoot is adopted as a performance index). It is worth stressing that the method is useful also in those situations where the a high performance results but the control requirements are not met. In this case it can be deduced that a PI controller is not sufficient and a more complex controller has to be considered.

Relay-feedback-based Approach

A method to assess the set-point following performance of a PI controller based on the data collected during a relay-feedback experiment (see Chapter 7) has been presented in (Thyagarajan and Yu, 2003). It is based on the evaluation shape of the process output when a relay-feedback is connected in series with a standard PI controller (see Figure 8.9). If the process has a FOPDT dynamics, three situations can arise. If the integral time constant is (almost) equal to the process time constant, then the relay is actually applied to an integral process whose transfer function is

$$C(s)P(s) = \frac{K_p(T_i s + 1)}{T_i s} \frac{K}{Ts + 1} e^{-Ls} = \frac{KK_p}{T_i s} e^{-Ls} \quad (8.101)$$

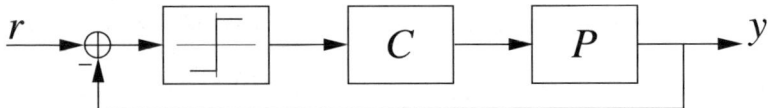

Fig. 8.9. Scheme for the relay-feedback-based approach for PI performance assessment

where K, T and L are respectively the gain, the time constant and the dead time of the process, and K_p and T_i are the proportional gain and integral time constant of the PI controller. The output response is therefore triangular in shape (namely, a series of upward and downward ramps), as shown in Figure 8.10. The half period of the relay response is equal to the dead time of the process (*i.e.*, $L = T_u/2$) and the slope of the process output corresponds to $\bar{K} := KK_p/T_i$ (its peak value A is therefore equal to $\bar{K}L$). Conversely, if the process time constant is less than the integral time constant, the relay feedback response exhibits a convex rise and fall shape (see Figure 8.11). Finally, if the process time constant is greater than the integral time constant, the relay feedback response exhibits a concave rise and fall shape. However, depending on the sharpness of the shape, two cases can be conveniently distinguished. If the ratio T_i/T is greater than a critical value $(T_i/T)_c$, then a sharp peak can be observed (see Figure 8.12), otherwise, a rounded peak occurs (see Figure 8.13). The values of $(T_i/T)_c$ depending on the ratio L/T can be evaluated by means of plot reported in (Thyagarajan *et al.*, 2003).

At this point, it has to be highlighted that the minimum integrated absolute error in the set-point step response is obtained when the integral time constant is equal to the process time constant (*i.e.*, $T_i = T$) and the value of $\bar{K}/L$ is equal to 1.68. This fact, together with the analytical expressions of the output curves obtained in the different above mentioned cases, can be exploited to establish a procedure that assess the controller performance and retune the controller in order to achieve the minimum integrated absolute error performance. Thus, if the shape of the output curve is a triangle, this means that it is already $T_i = T$ and therefore the proportional gain can be optimised by applying the following steps:

1. Estimate the dead time L of the process by measuring the half period of the relay response and measure the peak amplitude A.
2. Calculate $\bar{K}_{old} = L/A$.
3. Calculate $\bar{K}_{new} = 1.68L$.
4. Set the new value of the proportional gain $K_{p,new}$ to $K_{p,old}\bar{K}_{old}/\bar{K}_{new}$, where $K_{p,old}$ is the previously adopted value.

On the contrary, if the shape of the process output is similar to that of Figure 8.11, it means that $T_i > T$ and therefore the following procedure has to be applied.

1. Estimate the dead time L of the process by measuring the interval between a relay commutation and the successive peak amplitude of the process output. Measure also the value of A and of the ultimate period T_u.
2. The values of $\bar{K}$ ($\bar{K} = \bar{K}_{old}$) and of T can be derived by solving the following two equations (note that the value of T_i is known):

$$\frac{T_u}{4\bar{K}} - \left(\frac{T_i - T}{\bar{K}}\right)\left(1 - \frac{2}{1 + e^{-\frac{T_u}{2T}}}\right) - A = 0, \qquad (8.102)$$

$$\frac{T_u/4 - L}{\bar{K}} - \left(\frac{T_i - T}{\bar{K}}\right)\left(1 - \frac{2e^{-\frac{T_u/2 - L}{T}}}{1 + e^{-\frac{T_u}{2T}}}\right) - A = 0. \qquad (8.103)$$

3. Set the new value of the integral time constant $T_{i,new}$ to T.
4. Calculate $\bar{K}_{new} = 1.68L$.
5. Set the new value of the proportional gain $K_{p,new}$ to

$$\frac{(K_{p,old})(T_{i,new})(\bar{K}_{old})}{(T_{i,old})(\bar{K}_{new})}.$$

The initial estimates for $\bar{K}$ and T in the numerical procedure that has to be applied to solve Equations (8.102) and (8.103) can be taken as $\bar{K} = L/A$ and $T = 0.8 T_i$.

The same procedure has to be applied if the shape of the relay output is similar to that of Figure 8.12 (i.e., $T_i < T$) and it is therefore found that $(T_i/T)_c \le (T_i/T)$. The only difference is that the initial estimate of T has to be set to $1.2 T_i$.

Finally, if the relay output shape is similar to that shown in Figure 8.13 and therefore $(T_i/T)_c > (T_i/T)$, the following algorithm has to be applied.

1. Estimate the (approximate) dead time L^* of the process by measuring again the time elapsed from zero to peak value of the process output. Measure also the value of A and of the ultimate period T_u.
2. The values of $\bar{K}$ ($\bar{K} = \bar{K}_{old}$) and of T can be derived by solving the following two equations (note that the value of T_i is known):

$$-\frac{T_u}{4\bar{K}} - \frac{T}{\bar{K}}\left(1 + \ln\left(\frac{2(T - T_i)}{T(1 + e^{-\frac{T_u}{2T}})}\right)\right) + \frac{T - T_i}{\bar{K}} - A = 0, \qquad (8.104)$$

$$\frac{T_u/4 - L}{\bar{K}} - \left(\frac{T_i - T}{\bar{K}}\right)\left(1 - \frac{2e^{-\frac{T_u/2 - L}{T}}}{1 + e^{-\frac{T_u}{2T}}}\right) - A = 0. \qquad (8.105)$$

3. Set the new value of the integral time constant $T_{i,new}$ to T.
4. Determine L from the following equation:

$$L = L^* - T\left(\ln\left(\frac{2(T - T_i)}{T(1 + e^{-\frac{T_u}{2T}})}\right)\right). \qquad (8.106)$$

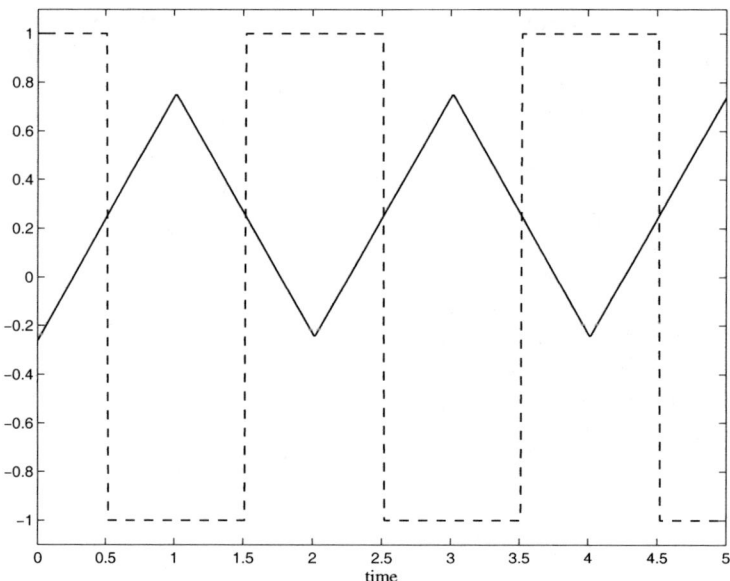

Fig. 8.10. Process output (solid line) and relay output (dashed line) when the process time constant is equal to the integral time constant

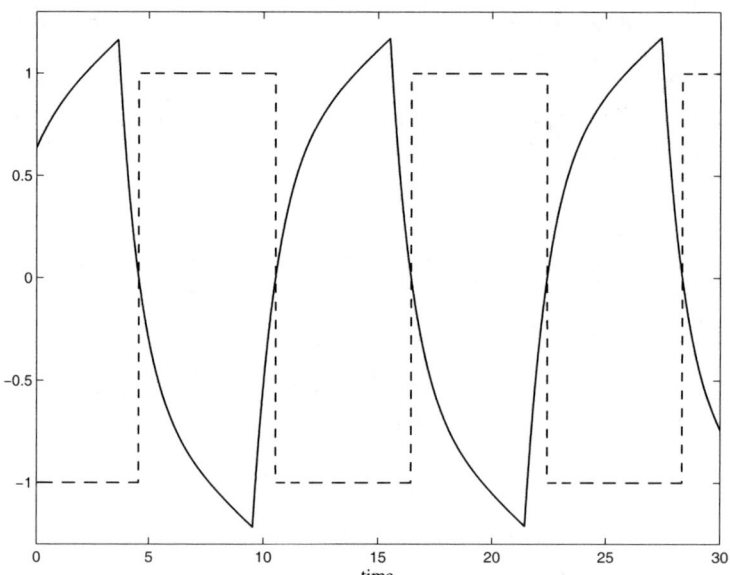

Fig. 8.11. Process output (solid line) and relay output (dashed line) when the process time constant is less than the integral time constant

8.4 Deterministic Performance Assessment   239

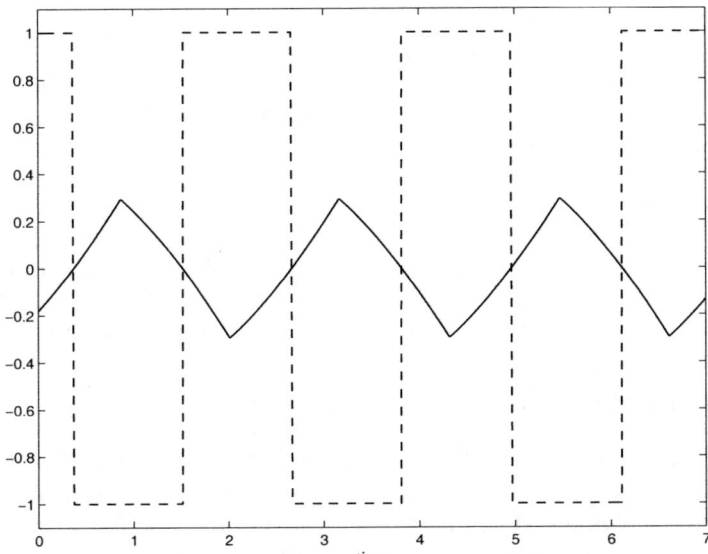

Fig. 8.12. Process output (solid line) and relay output (dashed line) when the process time constant is greater than the integral time constant (and $(T_i/T) > (T_i/T)_c$)

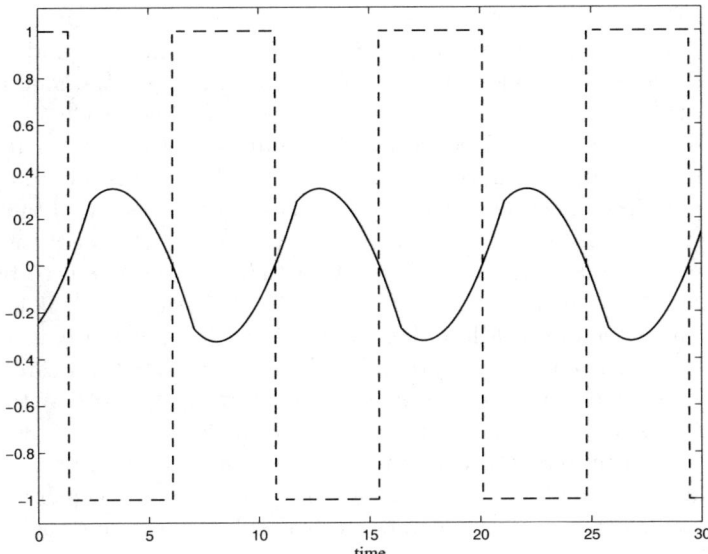

Fig. 8.13. Process output (solid line) and relay output (dashed line) when the process time constant is greater than the integral time constant (and $(T_i/T) < (T_i/T)_c$)

5. Calculate $\bar{K}_{new} = 1.68L$.
6. Set the new value of the proportional gain $K_{p,new}$ to

$$\frac{(K_{p,old})(T_{i,new})(\bar{K}_{old})}{(T_{i,old})(\bar{K}_{new})}.$$

In (Thyagarajan et al., 2003) it has been shown that the methodology is effective also for higher-order processes (although the achieved performance is obviously suboptimal). However, being based on the use of a relay-feedback test, it is sensible to the measurement noise and therefore the noise filtering procedure that has to be adopted in practical cases has to be selected carefully.

Assessment of Load Disturbance Rejection Performance

An approach to assess the load disturbance rejection performance of a PI controller (in terms of the integrated absolute error IAE) has been proposed in (Visioli, 2006). It is based on the simultaneous evaluation of the Area Index and of the Idle Index when a stepwise load disturbance occurs. As a result, guidelines on how to improve the controller tuning are given.

Actually, it is evident that a well-tuned controller gives a low value of the Idle Index II and at the same time a medium value of the Area Index AI, as this means that the control loop is neither sluggish nor oscillating. However, evaluating the values of the two indexes can give indications on how to improve the tuning. Guidelines in this context have been derived from the analysis of different processes with a FOPDT transfer function and with different normalised dead times. In particular, having applied the methodology presented in (Silva et al., 2002), the set of stabilising PI controllers have been determined and for each PI controller determined in this way, a unit step load disturbance response have been simulated and the corresponding values of AI, II and IAE have been computed. Based on the results obtained, the rules presented in Table 8.2 have been devised in order to assess the tuning of the PI parameters. The value of the Area Index is considered to be low if it is less than 0.35, medium if it is $0.35 < AI < 0.7$ and high if it is greater than 0.7. The value of the Idle Index is considered to be low if it is less than -0.6, medium if it is $-0.6 < II < 0$ and high if it is greater than zero.

Although these rules might appear somewhat intuitive, it is worthy to discuss two of them in some detail. First, the case when the value of AI is low and the value of II is medium/high is examined, because these seem to be two results that indicate an oscillatory loop from one side (AI) and a sluggish loop from another side (II). The situation can be evaluated by considering the following process

$$P(s) = \frac{1}{10s + 1}e^{-5s} \tag{8.107}$$

controlled by a PI controller whose parameters are $K_p = 1.81$ and $T_i = 20$ (note that the parameters that provide the minimum IAE of 6.11 are

Table 8.2. Rules for the assessment of the PI tuning. (*): an additional test is useful (see the text).

Value of AI	Value of II	Tuning assessment
high	high	K_p too low, T_i too high
high	low	K_p too low
medium/high	medium	K_p too low, T_i too low
medium	low	K_p ok, T_i ok
low	medium/high	T_i too high
low	low	K_p too high and/or T_i too low (*)

$K_p = 1.81$ and $T_i = 10.36$ with corresponding values of $AI = 0.61$ and $II = -0.71$). The unit step load disturbance response and the corresponding control variable are plotted in Figure 8.14. The resulting values of the Area Index and of the Idle Index are $AI = 0.14$ and $II = -0.21$ respectively (while $IAE = 11.03$). It appears that in this case the low value of AI is not associated to an oscillatory loop but to a control loop in which the dynamics of the complementary sensitivity function (see (8.79)) is not dominated by a pair of complex conjugate poles.

Thus, although in this case the Area Index is not indicative of the damping factor of the closed-loop system, it gives the important information that the value of the integral time constant is too high. It has to be noted that this conclusion cannot be drawn easily if a technique that reveals an oscillatory behavior of the manipulated variable (for example, by considering the autocorrelation function) is employed only. For example, the method proposed in (Miao and Seborg, 1999) gives almost the same oscillation index $\mathcal{R}$ in the two cases of $T_i = 10.36$ and $T_i = 20$ (it results respectively $\mathcal{R} = 0.37$ and $\mathcal{R} = 0.39$ if the control variable is analysed and $\mathcal{R} = 0.08$ and $\mathcal{R} = 0$ if the process output is analysed; that is, no significant oscillation is actually detected in both cases).

The second case that is worthy to be discussed is when both values of AI and II are low. This means that the control loop is too oscillatory and this fact is motivated by a high value of the proportional gain of the controller and/or by a low value of the integral time constant. In order to provide a possible additional information on the value of T_i it is useful to calculate another simple index related to the process output signal. This fact is explained by the results shown in Figures 8.15 and 8.16 where again the process modelled by transfer function (8.107) has been considered. In the first case the PI parameters are $K_p = 3$ and $T_i = 20$ and therefore the oscillatory response is caused by a too high value of the proportional gain.

The resulting indexes are $AI = 0.19$ and $II = -0.9$ (the resulting integrated absolute error is $IAE = 9.75$). In the second case the PI parameters are $K_p = 2.2$ and $T_i = 6.5$ and therefore the oscillatory response is caused by both a too high value of the proportional gain and a too low value of the integral time constant. The resulting indexes are $AI = 0.23$ and $II = -0.64$ (the corresponding integrated absolute error is $IAE = 14.02$).

It appears that the two considered indexes are not sufficient to distinguish the two situations. However, a look at the process output functions suggests to calculate a new index (called Output Index OI), namely, the ratio between the sum of the negative areas with respect to the final steady-state value and the sum of all the areas with the exception of the first one (note that a positive step load disturbance has been here assumed without loss of generality). In case the process output does never intersect its steady-state value, it has to be set simply $OI = 0$. This choice is motivated by the fact that when both K_p and T_i are high, the dynamics of the transfer function from the load disturbance to the process output is not dominated by a pair of complex conjugate poles only. The resulting values of OI are 0.26 and 0.56 for the first and second case respectively.

Summarising, when both the values of the Area Index and of the Idle Index are low it is convenient to evaluate the devised Output Index. In case $OI < 0.35$ it can be concluded that both the proportional gain and integral time constant values are too high. Otherwise, the oscillatory response is caused by a too high value of K_p and/or a too low value of T_i.

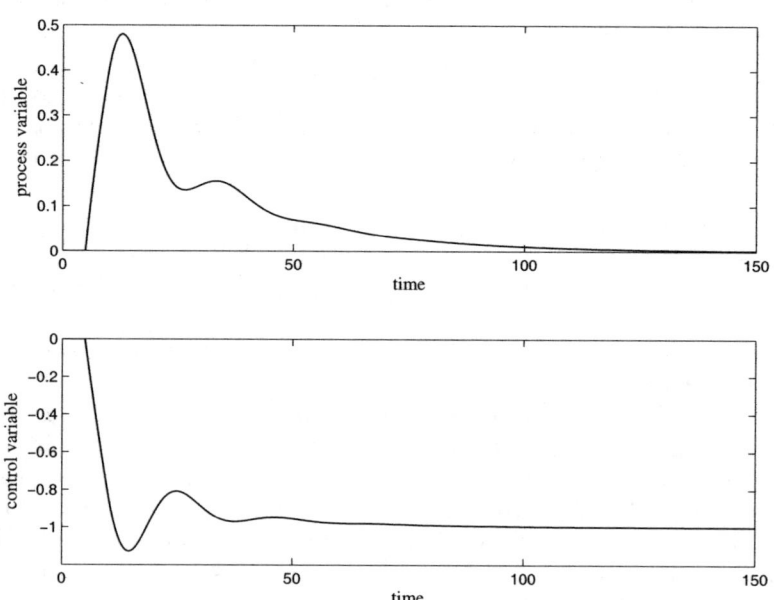

Fig. 8.14. Example of a load disturbance response for a too high value of T_i

8.4 Deterministic Performance Assessment   243

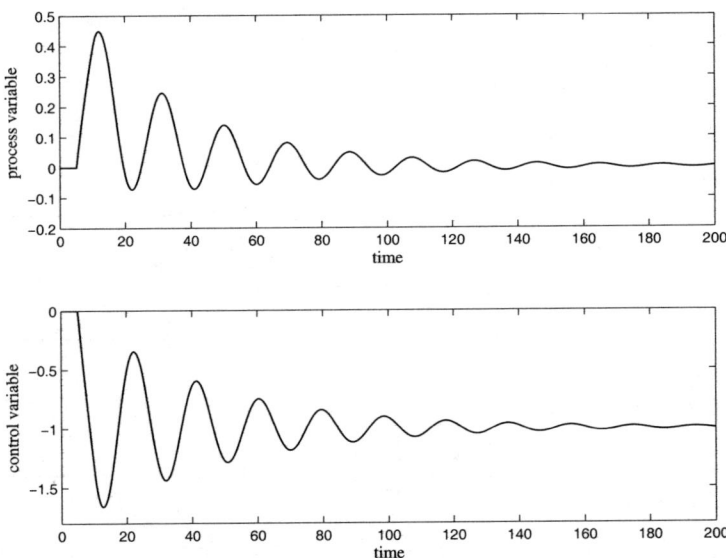

Fig. 8.15. Example of a load disturbance response for a too high values of K_p and T_i

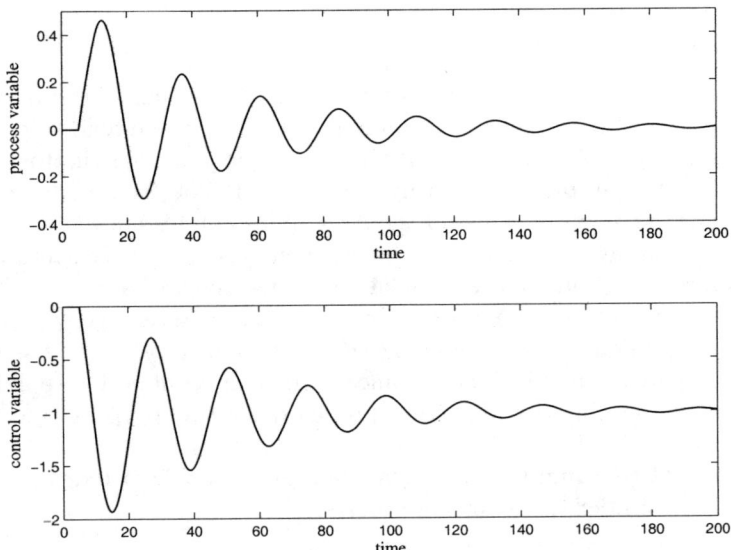

Fig. 8.16. Example of a load disturbance response for a too high value of K_p and a too low value of T_i

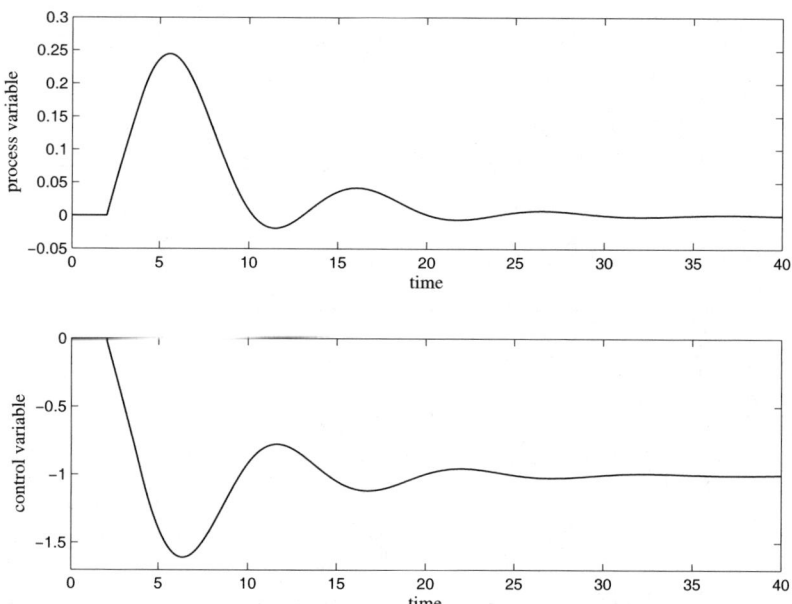

Fig. 8.17. Load disturbance response for $P_1(s)$ with $K_p = 4.61$ and $T_i = 6.06$ ($IAE = 1.42$, $AI = 0.61$, $II = -0.81$)

It is worth noting at this point that, with the decreasing of the normalised dead time, the PI parameters that minimise the IAE value tend to produce a more oscillatory control variable. Applications where a too oscillatory control variable is not desirable can be easily handled by the devised methodology, as the range of the medium values of the Area Index can be suitably modified to address the operator specifications. Further, it has to be taken into account that, as already mentioned, an oscillatory response can be caused either by unsuitable controller parameters or by the excessive presence of stiction in the actuators. Thus, before applying the devised methodology it is wise to determine if valves require maintenance. This can be done by applying one (or more) of the different algorithm proposed for this purpose (see Section 8.4.1).

As an illustrative example of the combined use of the Area Index and of the Area Index, consider the following process:

$$P_1(s) = \frac{1}{10s+1} e^{-2s} \tag{8.108}$$

The PI parameters that minimise the IAE index are $K_p = 4.61$ and $T_i = 6.06$ and the corresponding indexes are $IAE = 1.42$, $AI = 0.61$ and $II = -0.81$. Thus, according to Table 8.2, the proposed method suggests correctly that the controller is well tuned. The unit step load disturbance response, together

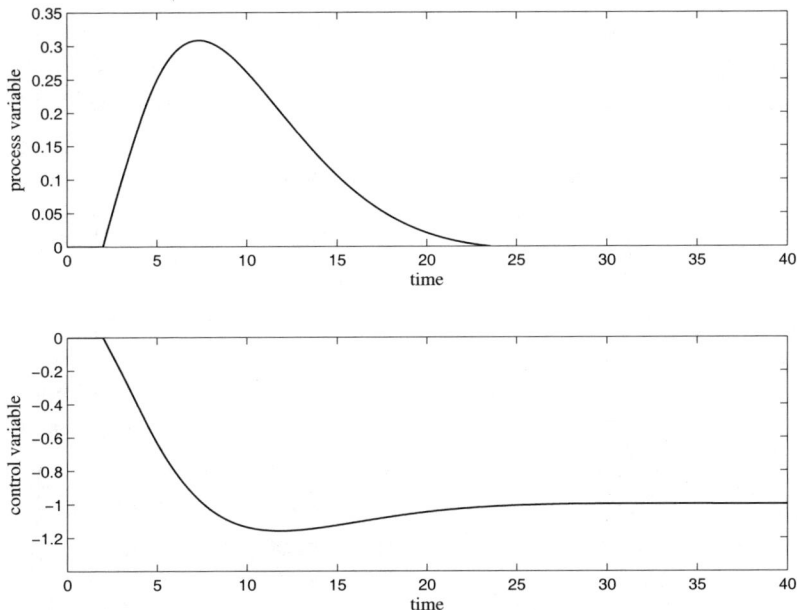

Fig. 8.18. Load disturbance response for $P_1(s)$ with $K_p = 2$ and $T_i = 6.06$ ($IAE = 2.02$, $AI = 0.97$, $II = -0.68$)

with the corresponding manipulated variable signal is plotted in Figure 8.17. Then, it has been fixed $K_p = 2$ (keeping the same value as before of the integral time constant). The performance obtained is shown in Figure 8.18 and the calculated indexes are $AI = 0.97$ and $II = -0.68$. Thus, the too low value of the proportional gain is recognised by the devised technique.

As another example, the following fourth-order process has been considered:

$$P_2(s) = \frac{1}{(s+1)^4} \qquad (8.109)$$

Note that (8.109) can be approximated by a FOPDT transfer function with a time constant $T = 2.1$ and a dead time $L = 1.9$. The optimal tuning is $K_p = 1.65$ and $T_i = 4.15$, which implies a minimum IAE of 2.79 and $AI = 0.36$ and $II = -0.80$.

The control system response to a unit step load disturbance is plotted in Figure 8.19. By decreasing both values of the proportional gain and of the integral time constant to $K_p = 1.2$ and $T_i = 2$ the results shown in Figure 8.20 are obtained. The corresponding values of the considered indexes are $IAE = 4.07$, $AI = 0.40$ and $II = -0.55$. From Table 8.2 it results that both the PI parameters are too low.

It should be noted that whereas the technique proposed confirms that the

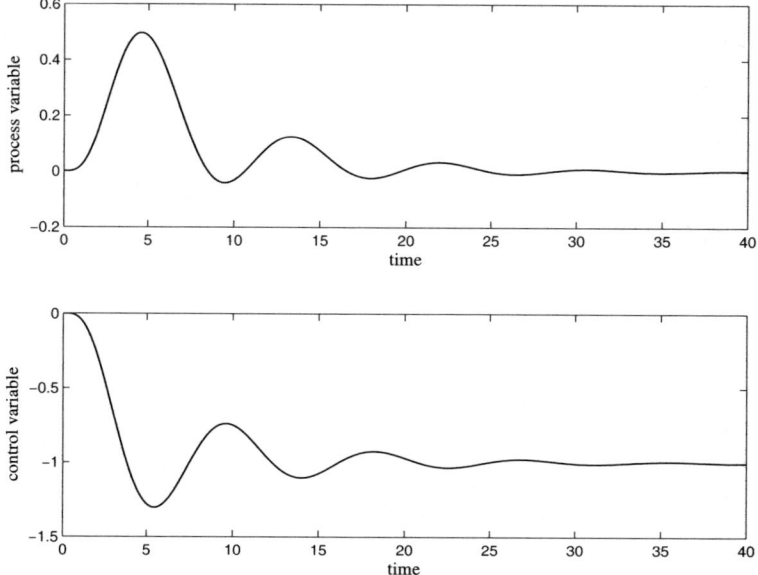

Fig. 8.19. Load disturbance response for $P_2(s)$ with $K_p = 1.65$ and $T_i = 4.15$ ($IAE = 2.79$, $AI = 0.36$, $II = -0.80$)

tuning is good the achieved integrated absolute error value is not far from the optimum. Note also that, being the overall technique based on the closed-loop systems, the effects of model uncertainties (*i.e.*, the discrepancy between the actual dynamics and the FOPDT model on which the method is based) are reduced.

Experimental results for a level control task have been obtained by means of the laboratory equipment described in Section A.1.

A single tank has been employed and a second inflow (driven by a second pump) has been adopted as a disturbance input. In particular, when the system is at the steady state with the process output sensor at 3 V, the second pump is activated by applying a step signal from 0 to 1.8 V. A time delay of 10 s has been added via software to the plant input in order to increase the normalised dead time of the system. Three experiments are presented hereafter. In the first experiment, the PI parameters have been set to $K_p = 0.5$ and $T_i = 50$. The load response is plotted in Figure 8.21. The abrupt load response detection method described in Section 8.4.1, based on a high-pass filtering of the process variable and of the control variable, has been applied, giving the result shown in Figure 8.22. It appears that a sufficiently abrupt load change has been detected, so that the proposed method can be applied. The calculated indexes are $AI = 1$ and $II = 0.22$ (the integrated absolute error is 106.5) and therefore Table 8.2 suggests to increase the proportional

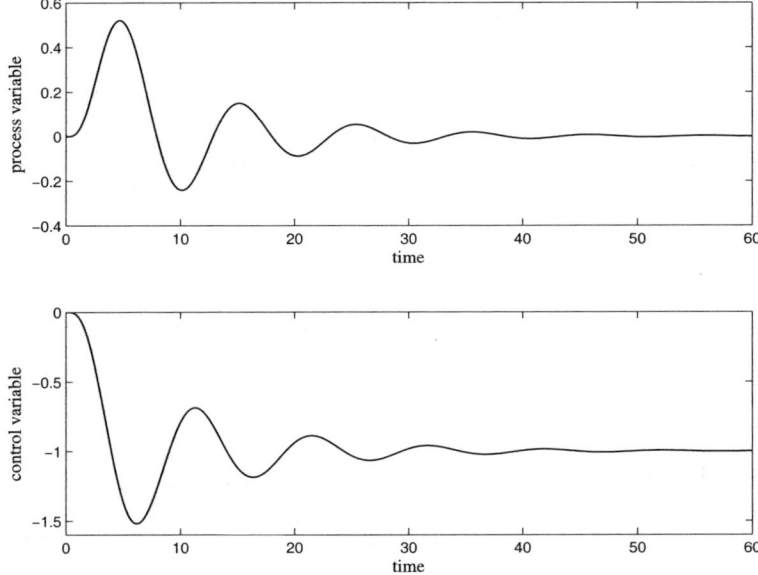

Fig. 8.20. Load disturbance response for $P_2(s)$ with $K_p = 1.2$ and $T_i = 2$ ($IAE = 4.07$, $AI = 0.40$, $II = -0.55$)

gain value and to decrease the integral time constant. With the values of the PI parameters modified to $K_p = 1$ and $T_i = 25$ the response shown in Figure 8.23 has been obtained. In this case it results $AI = 0.25$ and $II = -0.73$ (and $IAE = 49.17$). The oscillatory response is detected by the low value of the Area Index and according to Table 8.2 the value of the proportional gain has to be decreased. This fact is confirmed by the third experiment, where $K_p = 0.8$ and $T_i = 25$ have been selected. The corresponding response is plotted in Figure 8.24. It is $AI = 0.40$ and $II = -0.69$, indicating that the PI controller is well tuned, as it is ascertained also by the obtained value of $IAE = 34.83$. The experimental results are summarised in Table 8.3 and the effectiveness of the devised methodology can be deduced.

Table 8.3. Summary of the experimental results for the method for the PI controller tuning assessment

K_p	T_i	AI	II	IAE
0.5	50	1	0.22	106.5
1	25	0.25	-0.73	49.17
0.8	25	0.40	-0.69	34.83

248   8 Performance Assessment

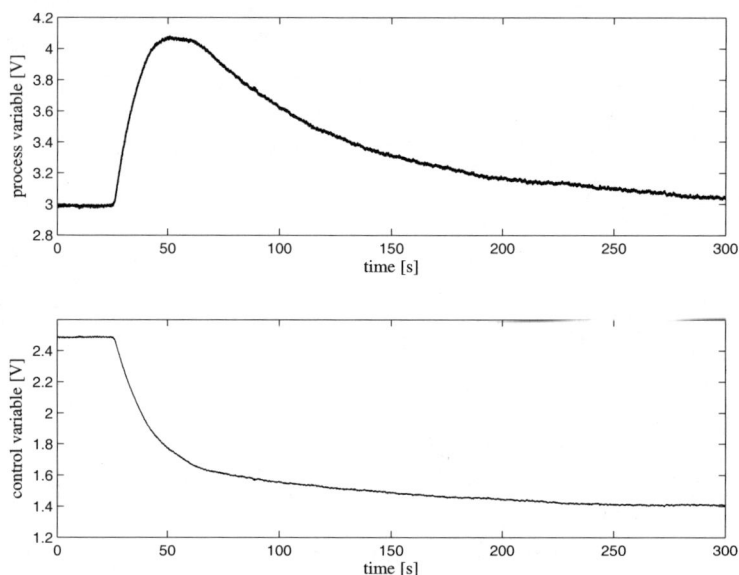

Fig. 8.21. Experimental load disturbance response with $K_p = 0.5$ and $T_i = 50$ ($IAE = 106.5$, $AI = 1$, $II = 0.22$)

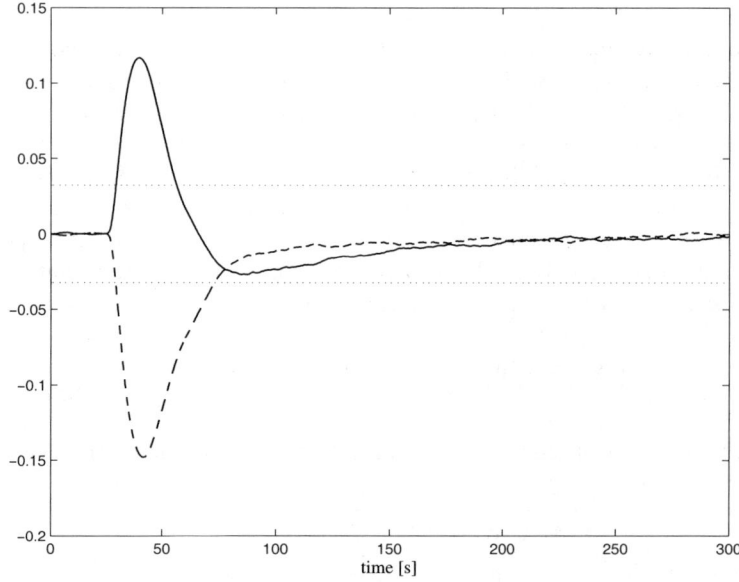

Fig. 8.22. Result of the evaluation of an abrupt load change in the case of the experimental result for $K_p = 0.5$ and $T_i = 50$. Solid line: filtered process variable; dashed line: filtered control variable; dotted line: threshold.

8.4 Deterministic Performance Assessment   249

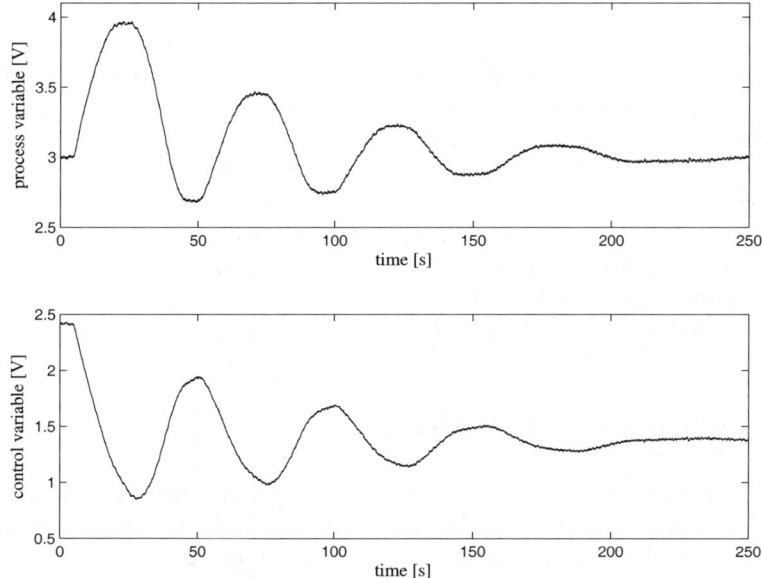

Fig. 8.23. Experimental load disturbance response with $K_p = 1$ and $T_i = 25$ ($IAE = 49.17$, $AI = 0.25$, $II = -0.73$).

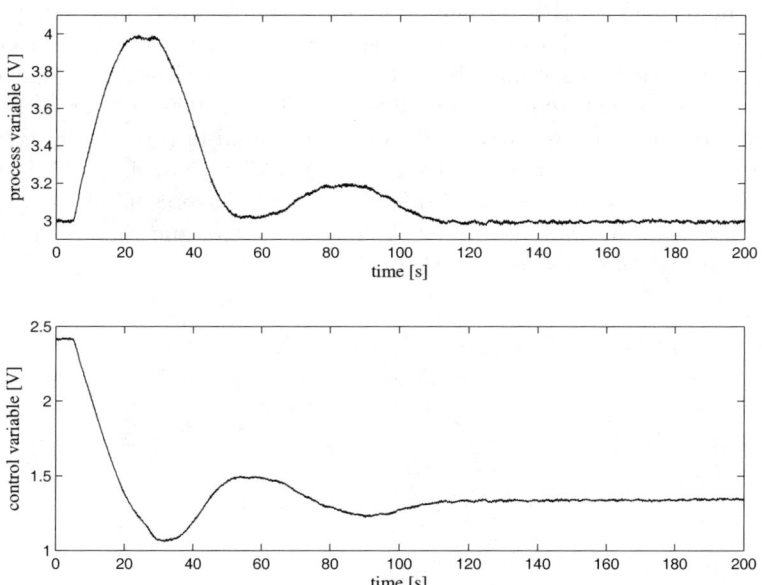

Fig. 8.24. Experimental load disturbance response with $K_p = 0.8$ and $T_i = 25$ ($IAE = 34.83$, $AI = 0.40$, $II = -0.69$).

Discussion

From the presented analysis, it appears that there are different methodologies for different performance requirements. In a practical context, the user should select the appropriate technique for a given application.
In any case, it is worth highlighting that the most relevant results deal with PI controllers and assume that the process is described by a FOPDT dynamics (although they are robust with respect to modelling uncertainties). Thus, there is still the need to devise effective methods that are capable to provide the tuning assessment for PID controllers (where the derivative action is employed) and for processes with integral and oscillatory dynamics.

8.5 Conclusions and References

The problem of how assess the performance of a given control system has been addressed in this chapter. It has been stressed that different methodologies are available for different performance requirement. Indeed, it is believed that a performance assessment package should actually consist of a set of functions, each dealing with a particular situation. Thus, the role of each method in a given situation should be clearly outlined.
The literature presents a large number of contributions on the performance assessment problem and the research effort is continuously increasing in the recent years. In this chapter just some ideas on how the problem can be tackled in practical cases have been presented. For a closer investigation of the topic, an excellent review can be found in (Jelali, 2006). Other reviews for the minimum variance approach have been presented in (Qin, 1998; Harris *et al.*, 1999). An industrial perspective can be found in (Kozub, 2002; Paulonis and Cox, 2003). Advanced methodologies have been presented in a special issue of the International Journal of Adaptive Control and Signal Processing published on September-November, 2003.

9

Control Structures

9.1 Introduction

One of the reasons for the great success of PID controllers is that they can be employed also as a basic component for more advanced control systems so that (relatively) complex control tasks can be addressed by still exploiting the available know-how. This chapter focuses on two control structures widely applied in industry, namely (series) cascade control and ratio control, which are still the subject of new investigations in order to find methodologies that allow the improvement of the performance and/or to simplify the overall control system design.

9.2 Cascade Control

9.2.1 Generalities

In process control applications, the rejection of load disturbances is often of main concern. In order to improve the performance for this task, the implementation of a cascade control system can be considered. In a cascade control scheme the process has one input and two (or more) outputs. Indeed, in order to provide an effective disturbance rejection, an additional sensor is employed so that the fast dynamics of the process is separated as much as possible from the slow dynamics (*i.e.*, that with the slowest poles and the nonminimum-phase part).

The typical series cascade control system is shown in Figure 9.1. For the sake of simplicity, here only two nested loops are considered but the approach can be generalised to more loops. The process transfer function is denoted by $P(s) = P_2(s)P_1(s)$, y_1 is the primary output, y_2 is the secondary output, C_2 is the secondary (or slave) controller and C_1 is the primary (or master) controller (it appears that the output signal of the master controller serves as the set-point for the slave controller). Analogously, the inner loop is de-

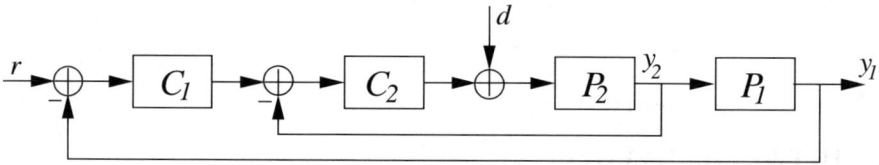

Fig. 9.1. Typical cascade control scheme

nominated as the secondary loop, while the outer loop is denominated as the primary loop.

Intuitively, if $P_1(s)$ represents the slow dynamics of the process and $P_2(s)$ represents the fast dynamics, the effectiveness of the cascade control system is due to the fact that disturbances affecting the (fast) secondary loop are effectively compensated before they affect the main process output y_1. Formally, the transfer function from the load disturbance d to the process variable y_1 is

$$T(s) := \frac{P_1(s)P_2(s)}{1 + C_2(s)P_2(s) + C_1(s)C_2(s)P_1(s)P_2(s)}. \tag{9.1}$$

The characteristic equation is therefore

$$1 + C_2(s)P_2(s) + C_1(s)C_2(s)P_1(s)P_2(s) = 0, \tag{9.2}$$

while, if a conventional (single-loop) feedback control is employed, the characteristic equation is

$$1 + C(s)P_1(s)P_2(s) = 0, \tag{9.3}$$

where $C(s)$ is the single-loop controller. When the dynamics of the secondary loop is faster than the dynamics of the primary loop, the cascade control system has improved stability characteristics and therefore a higher gain in the primary loop can be adopted.

Based on this fact, it appears that the improvement in the cascade control performance is more significant when disturbances act in the inner loop and when the secondary sensor is placed in order to separate as far as possible the fast dynamics of the process from the slow dynamics (Krishnaswami et al., 1990). Actually, when the secondary process exhibits a significant dead time or there is an unstable (positive) zero, the use of cascade control is not useful in general (taking into account the additional cost due to the secondary sensor and to the secondary controller). As an additional advantage, the nonlinearities of the process in the inner loop are handled by that loop and therefore they are removed from the more important outer loop.

In this context, the parameters of the overall control system should be selected in order to provide a tight tuning of the inner loop (with respect to the outer one). Note that the presence of an integrator in the inner loop is not strictly necessary since the null steady-state error can be assured by the outer loop. It is worth stressing that if integral action is employed both in the master and in the slave controller, the integrator windup should be carefully handled. In particular, the saturation of the actuator requires that an anti-windup strategy (see Chapter 3) is implemented for the secondary controller. However, when the secondary controller attains a limit, the primary controller acts in open-loop at the same time. A typical approach is therefore to stop the integration of the primary controller when the output of the secondary controller attains its limits. This solution prevents the master controller from unnecessarily increasing its output and therefore forcing the primary controller to be more saturated. However, a better solution is, when the output of the secondary controller attains its limit, to use the secondary process output as a tracking signal for the primary controller. This solution is effective also in providing a bumpless transfer when the secondary controller switches from manual to automatic mode.

The design of the overall cascade control system is usually performed by first tuning the secondary controller, based on the secondary process transfer function (the primary loop is placed in manual mode). Then, the primary controller is tuned on the basis of the closed-loop transfer function of the secondary loop in series with the primary process transfer function (which contains the dominant dynamics, because of the tight tuning of the secondary loop). It appears that the design is performed sequentially and therefore it is more time-consuming than the design of a classical single-loop controller. There is, therefore, the need to have automatic tuning functionalities that are potentially able to provide a simultaneous tuning of the two controllers.

9.2.2 Relay Feedback Sequential Auto-tuning

A technique based on the relay feedback for the automatic tuning of a cascade controller has been proposed in (Hang *et al.*, 1994). It basically consists of applying the standard relay feedback approach (see Chapter 7) first to the secondary loop (with the primary loop placed in manual mode) and then to the primary loop (with the secondary feedback controller already tuned). Actually, any tuning rule based on the ultimate gain and the ultimate frequency of the process can be applied in this context.

Remarkably, the ratio of the ultimate frequencies obtained in these two steps can be adopted to assess whether a cascade controller is worth being applied (since it indicates the ratio of the speeds of the loops). Further, a refined tuning of the secondary controller can be performed (in closed-loop) by applying again the relay feedback controller to the secondary loop with the primary loop closed (*i.e.*, with the PID controller previously tuned that acts as a primary controller). The same refinement can be performed also on the master controller.

9.2.3 Relay Feedback Simultaneous Auto-tuning

The method presented in the previous section has the disadvantage that a sequential (and therefore time-consuming) tuning is actually performed. A

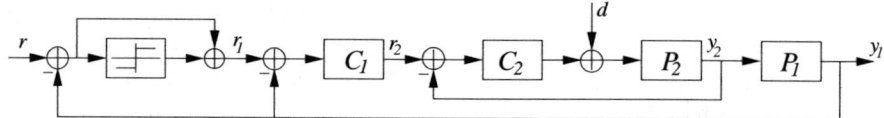

Fig. 9.2. Scheme for the on-line tuning of the cascade controller

technique that allows the achievement of a simultaneous tuning of the two controllers has been proposed in (Tan et al., 2000). Therein it is assumed that the two controllers are already (roughly) tuned (mainly in order to stabilise the process). Then, the scheme shown in Figure 9.2 is applied in order to tune on-line the two controllers simultaneously. In particular, denote by $C_{1,0}(s)$ and $C_{2,0}(s)$ the transfer functions of the two initial controllers and by ω_u the ultimate frequency obtained from the experiment. A Fourier or Spectral analysis, with an appropriate weighting window, is then applied to the signals r_2 and y_2 in order to determine $T_{r_2 y_2,0}(j\omega_u)$, where $T_{r_2 y_2}$ denotes the closed-loop transfer function of the inner loop. Then, the frequency response of the secondary process at $\omega = \omega_u$ can be derived as

$$P_2(j\omega_u) = \frac{T_{r_2 y_2,0}(j\omega_u)}{C_{2,0}(j\omega_u)\left(1 - T_{r_2 y_2,0}(j\omega_u)\right)}. \tag{9.4}$$

A desired frequency response $\bar{T}_{r_2 y_2}(j\omega_u)$ has now to be specified by the user. This can be done easily by considering a prototype first-order-plus-dead-time (FOPDT) transfer function

$$\bar{T}_{r_2 y_2}(s) = \frac{1}{T_2 s + 1} e^{-L_2 s}. \tag{9.5}$$

The (desired) values of the dead time and of the time constant can be selected, starting from the relay feedback experiment, assuming that

$$T_{r_2 y_2,0}(j\omega_u) = \gamma + j\theta. \tag{9.6}$$

They can be fixed as

$$T_2 = \frac{0.5}{\omega_u}\sqrt{\frac{1 - \gamma^2 + \theta^2}{(\gamma^2 + \theta^2)}} \tag{9.7}$$

and

$$L_2 = \frac{1}{\omega_u}\arccos(\gamma - \theta\omega_u T_2). \tag{9.8}$$

Once $T_{r_2 y_2}(j\omega_u)$ has been determined, the frequency response of the new secondary controller at $\omega = \omega_u$ can be calculated as

$$C_2(j\omega_u) = \frac{\bar{T}_{r_2 y_2}(j\omega_u)}{P_2(j\omega_u)\left(1 - \bar{T}_{r_2 y_2}(j\omega_u)\right)}. \tag{9.9}$$

9.2 Cascade Control

Starting from this expression and denoting $C_2(j\omega_u)$ as $\alpha_2 + j\beta_2$, the parameters of the PID secondary controller (in ideal form) are then determined as:

$$K_{p2} = \alpha_2, \tag{9.10}$$

$$T_{i2} = -\frac{\alpha_2}{\beta_2 \omega_u}, \tag{9.11}$$

$$T_{d2} = 0.25 T_i. \tag{9.12}$$

If a PI controller is employed, it can be simply set $T_{d2} = 0$, by keeping the same values of the proportional gain and of the integral time constant.

A similar reasoning is applied in order to design the primary controller. After having applied a Fourier or Spectral analysis to the signals r_1 and y_1 so that $T_{r_1 y_1, 0}(j\omega_u)$ is derived, the frequency response of the primary process at $\omega = \omega_u$ can be determined as:

$$P_1(j\omega_u) = \frac{T_{r_1 y_1, 0}(j\omega_u)}{T_{r_2 y_2, 0}(j\omega_u) C_{1,0}(j\omega_u) \left(1 - T_{r_1 y_1, 0}(j\omega_u)\right)}. \tag{9.13}$$

Then, the frequency response prototype of the primary loop at $\omega = \omega_u$, denoted as $\bar{T}_{r_1 y_1}(j\omega_u)$ is determined as for the secondary loop (see (9.6)–(9.8)). Finally, the frequency response of the new primary controller at $\omega = \omega_u$ is

$$C_1(j\omega_u) = \frac{\bar{T}_{r_1 y_1}(j\omega_u)}{P_1(j\omega_u) \bar{T}_{r_2 y_2}(j\omega_u) \left(1 - \bar{T}_{r_1 y_1}(j\omega_u)\right)}. \tag{9.14}$$

The PID controller parameters are finally determined by denoting $C_1(j\omega_u)$ as $\alpha_1 + j\beta_1$ and calculating

$$K_{p1} = \alpha_1, \tag{9.15}$$

$$T_{i1} = -\frac{\alpha_1}{\beta_1 \omega_u}, \tag{9.16}$$

$$T_{d1} = 0.25 T_i. \tag{9.17}$$

As for the slave controller, a PI controller can be adopted simply by fixing $T_{d1} = 0$.

In order to illustrate the methodology, the following example is provided. Consider the processes

$$P_1(s) = \frac{1}{(5s+1)^2} e^{-4s}, \tag{9.18}$$

$$P_2(s) = \frac{1}{s+1} e^{-0.2s}. \tag{9.19}$$

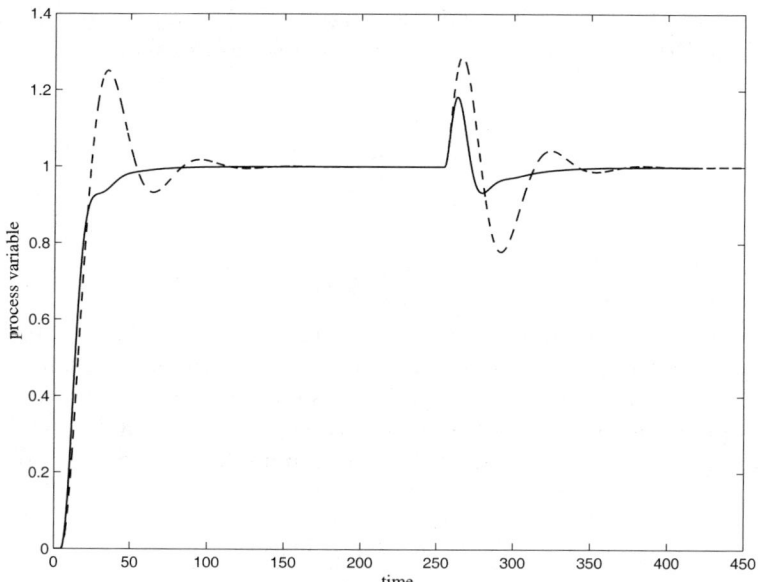

Fig. 9.3. Process variable before (dashed line) and after (solid line) the relay feedback simultaneous auto-tuning

Initially, the tuning of the two controllers is

$$K_{p1} = 1, \quad T_{i1} = 12, \quad T_{d1} = 0, \tag{9.20}$$

and

$$K_{p2} = 0.5, \quad T_{i2} = 4, \quad T_{d2} = 0. \tag{9.21}$$

After the application of the methodology, the parameters of the PID controllers are determined as (note that the derivative action of the secondary controller is not adopted):

$$K_{p1} = 1.18, \quad T_{i1} = 18.99, \quad T_{d1} = 4.75, \tag{9.22}$$

and

$$K_{p2} = 0.56, \quad T_{i2} = 2.14, \quad T_{d2} = 0. \tag{9.23}$$

The process variable before and after the refinement of the tuning is plotted in Figure 9.3. In particular, the experiment consists of applying in both cases a unit step in the set-point signal at time $t = 0$ and a load disturbance unit step at time $t = 250$. It appears that the performance has been improved significantly both with respect to the set-point following and to the load disturbance rejection task. However, it has to be stressed again that a (rough) tuning of the two controllers has to be performed before applying the technique.

9.2.4 Simultaneous Identification Based on Step Response

A technique based on the step response for the simultaneous identification of the primary and of the secondary process has been presented in (Visioli and Piazzi, 2006). It consists of applying a step signal to the process $P(s)$. A FOPDT transfer function of the fast dynamics of the process can be estimated by evaluating the step response of $P_2(s)$, for example by applying the area method (see Section 7.2.1).

At the same time, a model for the slow dynamics of the process can be estimated by considering its input signal y_2 and its output signal y_1 and by applying a least-squares procedure, such as the one proposed in (Sung et al., 1998) which is based on the integrated input and output signals and therefore it is inherently robust to measurement noise (see again Section 7.2.1). The obtained (possibly high-order) model can then be reduced if a tuning rule that requires a FOPDT or a SOPDT model of the primary process is employed (see the next sections). It has to be noted that, because of the different dynamics of P_2 and P_1, the step response of P_2 is indeed a sufficiently exciting signal to be adopted as an input signal for the least-squares based estimation of $P_1(s)$.

9.2.5 Simultaneous Tuning of the Controllers

A methodology for the simultaneous tuning of the two controllers in the cascade control system has been proposed in (Lee et al., 1998a). It is based on the Internal Model Control (IMC) design methodology (Morari and Zafiriou, 1989) and on the reduction of the controllers obtained by means of a Maclaurin series expansion. In particular, the secondary process transfer function can be written as

$$P_2(s) = P_{2m}(s) P_{2a}(s) \tag{9.24}$$

where $P_{2a}(s)$ is the all-pass portion of the transfer function containing all the nonminimmum phase dynamics ($P_{2a}(0) = 1$). Then, the desired inner loop transfer function is specified as

$$\bar{T}_{r_2 y_2}(s) = \frac{P_{2a}(s)}{(\lambda_2 s + 1)^{n_2}} \tag{9.25}$$

where λ_2 is the user-chosen time constant of the IMC filter and the value of n_2 is selected in order to make the resulting controller proper. The secondary controller can then be determined as

$$C_2(s) = \frac{P_{2m}^{-1}(s)}{(\lambda_2 s + 1)^{n_2} - P_{2a}(s)}. \tag{9.26}$$

In order to approximate the controller obtained to a PID controller, the same procedure already presented in Section 7.5.3 can be used, i.e., Expression (9.26) can be rewritten as

$$C_2(s) = \frac{k(s)}{s} \tag{9.27}$$

and expanding $C_2(s)$ in a Maclaurin series in s:

$$C_2(s) = \frac{1}{s}\left[k(0) + k'(0)s + \frac{k''(0)}{2}s^2 + \cdots\right]. \tag{9.28}$$

Expression (9.28) is indeed a PID controller in ideal form with

$$\begin{aligned} K_{p2} &= k'(0) \\ T_{i2} &= \frac{k'(0)}{k(0)} \\ T_{d2} &= \frac{k''(0)}{2k'(0)}. \end{aligned} \tag{9.29}$$

An analogous procedure can be adopted for the design of the primary controller. By assuming that the transfer function of the inner loop is (9.25), the process model of the outer loop can be expressed as

$$P_{12}(s) = P_1(s)\frac{P_{2a}(s)}{(\lambda_2 s + 1)^{n_2}} \tag{9.30}$$

and it can be rewritten as

$$P_{12}(s) = P_{12m}(s)P_{12a}(s), \tag{9.31}$$

where again $P_{12m}(s)$ contains the invertible part of the model and $P_{12a}(s)$ contains the nonminimum-phase part in all-pass form. Then, the desired outer loop transfer function is specified as

$$\bar{T}_{r_1 y_1}(s) = \frac{P_{12a}(s)}{(\lambda_1 s + 1)^{n_1}} \tag{9.32}$$

and therefore the primary controller transfer function is determined as

$$C_1(s) = \frac{P_{12m}^{-1}(s)(\lambda_2 s + 1)^{n_2}}{P_{2a}(s)\left((\lambda_1 s + 1)^{n_1} - P_{12a}(s)\right)}. \tag{9.33}$$

Finally, the controller transfer function can be reduced to a PID form by applying again the Maclaurin series expansion.

An interesting case that is worth analysing is when both processes are described by a FOPDT transfer function, namely,

$$P_1(s) = \frac{K_1}{T_1 s + 1}e^{-L_1 s}, \qquad P_2(s) = \frac{K_2}{T_2 s + 1}e^{-L_2 s}. \tag{9.34}$$

In this case, it is

$$\bar{T}_{r_2y_2}(s) = \frac{e^{-L_2 s}}{\lambda_2 s + 1} \tag{9.35}$$

and

$$\bar{T}_{r_1y_1}(s) = \frac{e^{-(L_1+L_2)s}}{\lambda_1 s + 1}. \tag{9.36}$$

An explicit tuning rule can be therefore derived for both the primary and the secondary controller:

$$\begin{aligned} K_{p2} &= \frac{T_2 + \dfrac{L_2^2}{2(\lambda_2 + L_2)}}{K_2(\lambda_2 + L_2)} \\ T_{i2} &= T_2 + \frac{L_2^2}{2(\lambda_2 + L_2)} \\ T_{d2} &= \frac{L_2^2}{6(\lambda_2 + L_2)}\left(3 - \frac{L_2}{T_2 + \dfrac{L_2^2}{2(\lambda_2 + L_2)}}\right) \end{aligned} \tag{9.37}$$

and

$$\begin{aligned} K_{p1} &= \frac{T_1 + \lambda_2 + \dfrac{(L_1+L_2)^2}{2(\lambda_1 + L_1 + L_2)}}{K_1(\lambda_1 + L_1 + L_2)} \\ T_{i1} &= T_1 + \lambda_2 + \frac{(L_1+L_2)^2}{2(\lambda_1 + L_1 + L_2)} \\ T_{d1} &= \frac{\lambda_2 T_1 - \dfrac{(L_1+L_2)^3}{6(\lambda_1 + L_1 + L_2)}}{T_1 + \lambda_2 + \dfrac{(L_1+L_2)^2}{2(\lambda_1 + L_1 + L_2)}} + \frac{(L_1+L_2)^2}{2(\lambda_1 + L_1 + L_2)}. \end{aligned} \tag{9.38}$$

Similarly, tuning rules can be derived also by assuming that the two processes are described by SOPDT transfer functions (Lee et al., 1998a).

In any case, it can be deduced that the design phase involves the selection of the two time constants λ_1 and λ_2. In principle, they allow to handle the trade-off between aggressiveness and robustness. Actually, since the Maclaurin series expansion is eventually adopted to determine the two PID controllers, the conclusions drawn in Section 7.5.5 should be considered. However, based on many simulations, in (Lee et al., 1998a) the suggestion is to set

$$\lambda_2 = 0.5 L_2 \tag{9.39}$$

and

$$\lambda_1 = 0.5(L_1 + L_2). \tag{9.40}$$

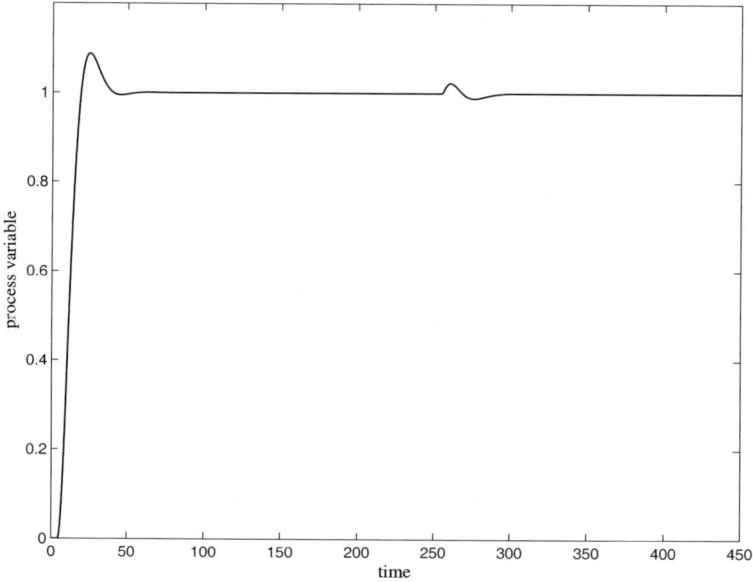

Fig. 9.4. Process variable obtained with the simultaneous tuning of the controllers

The following simulation result is provided in order to evaluate technique. Consider again the process whose dynamics is described by the series of the two transfer functions (9.18)–(9.19). If the method described in Section 9.2.4 is applied in order to identify simultaneously the two parts of the process, two FOPDT transfer functions are derived. In particular, we obtain $K_2 = 1$, $T_2 = 0.99$ and $L_2 = 0.21$ for the secondary process and $K_1 = 1$, $T_1 = 7.55$ and $L_1 = 6.88$ for the primary process. The resulting filter time constants are therefore, according to Expressions (9.39)–(9.40), $\lambda_2 = 0.11$ and $\lambda_1 = 3.55$. By applying the tuning rules (9.37)–(9.38), we obtain $K_{p2} = 3.31$, $T_{i2} = 1.06$, $T_{d2} = 0.07$, $K_{p1} = 0.94$, $T_{i1} = 10.02$, and $T_{d1} = 1.89$. When a unit step in the set-point signal is applied at time $t = 0$ and a load disturbance unit step is applied at time $t = 250$, the resulting process variable is plotted in Figure 9.4. The effectiveness of the overall methodology is apparent.

9.2.6 Tuning of the General Cascade Control Structure

The approach presented in the previous section has been generalised in (Lee et al., 2002), where also integral and unstable processes are considered. The scheme devised exploits the presence of a set-point filter both in the primary and in the secondary controller (see Figure 9.5). In particular, this is adopted for unstable and integrating processes and for stable processes with poles near zero, while for normal processes the same approach of the previous section has to be employed (and therefore the set-point filters are not used). Whether a stable process is considered a normal process or a process with a pole near zero depends on the design sense.

Restricting the analysis again to the case where both processes are described by FOPDT transfer functions (9.34), explicit tuning rules for the secondary and primary PID controllers result. They are (9.37) and (9.38) when normal stable processes are considered, while when stable processes with poles near zero are addressed:

$$K_{p2} = \frac{T_2 + \alpha - \dfrac{\lambda_2^2 + L_2\alpha - \frac{1}{2}L_2^2}{2\lambda_2 + L_2 - \alpha}}{K_2(2\lambda_2 + L_2 - \alpha)}$$

$$T_{i2} = T_2 + \alpha - \frac{\lambda_2^2 + L_2\alpha - \frac{1}{2}L_2^2}{2\lambda_2 + L_2 - \alpha} \qquad (9.41)$$

$$T_{d2} = \frac{T_2\alpha - \dfrac{\frac{1}{6}L_2^3 - \frac{1}{2}L_2^2\alpha}{2\lambda_2 + L_2 - \alpha}}{T_2 + \alpha - \dfrac{\lambda_2^2 + L_2\alpha - \frac{1}{2}L_2^2}{2\lambda_2 + L_2 - \alpha}} - \frac{\lambda_2^2 + L_2\alpha - \frac{1}{2}L_2^2}{2\lambda_2 + L_2 - \alpha}$$

and

$$K_{p1} = \frac{T_1 + \lambda_2 + \beta - \dfrac{\lambda_1^2 + (L_1 + L_2)\beta - \frac{1}{2}(L_1 + L_2)^2}{2\lambda_1 + L_1 + L_2 - \beta}}{K_1(2\lambda_1 + L_1 + L_2 - \beta)}$$

$$T_{i1} = T_1 + \lambda_2 + \beta - \frac{\lambda_1^2 + (L_1 + L_2)\beta - \frac{1}{2}(L_1 + L_2)^2}{2\lambda_1 + L_1 + L_2 - \beta} \qquad (9.42)$$

$$T_{d1} = \frac{\lambda_2 T_1 + \lambda_2\beta + T_1\beta - \dfrac{\frac{1}{6}(L_1+L_2)^3 - \frac{1}{2}(L_1+L_2)^2\beta}{2\lambda_1 + L_1 + L_2 - \beta}}{T_1 + \lambda_2 + \beta - \dfrac{\lambda_1^2 + (L_1+L_2)\beta - \frac{1}{2}(L_1+L_2)^2}{2\lambda_1 + L_1 + L_2 - \beta}} - \frac{\lambda_1^2 + (L_1+L_2)\beta - \frac{1}{2}(L_1+L_2)^2}{2\lambda_1 + L_1 + L_2 - \beta}$$

where

$$\alpha = T_2\left[1 - \left(1 - \frac{\lambda_2}{T_2}\right)^2 e^{-\frac{L_2}{T_2}}\right] \qquad (9.43)$$

and

$$\beta = T_1\left[1 - \left(1 - \frac{\lambda_1}{T_1}\right)^2 e^{-\frac{L_1+L_2}{T_1}}\right]. \qquad (9.44)$$

The set-point filters are selected as

$$F_2(s) = \frac{1}{\alpha s + 1} \qquad (9.45)$$

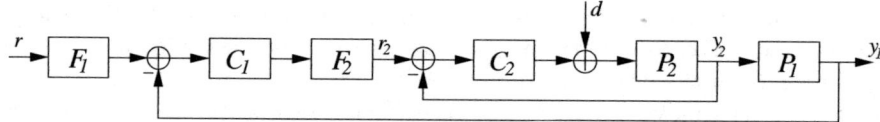

Fig. 9.5. General cascade control structure

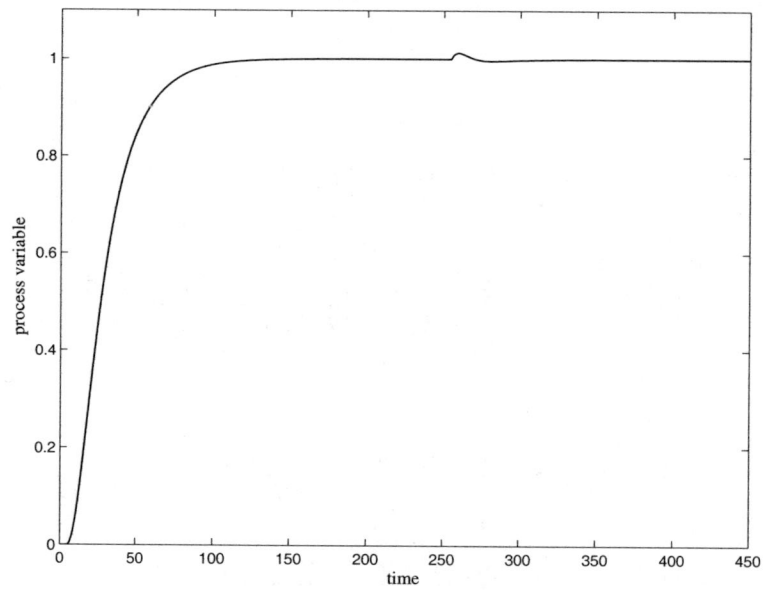

Fig. 9.6. Process variable obtained with the general cascade control structure

and
$$F_1(s) = \frac{1}{\beta s + 1}. \qquad (9.46)$$
Similarly to the case of normal processes, the closed-loop time constants λ_1 and λ_2 can be chosen as
$$\frac{\lambda_2}{L_2} = 0.5 \div 1 \qquad (9.47)$$
and
$$\frac{\lambda_1}{L_1 + L_2} = 0.5 \div 1. \qquad (9.48)$$
The same process (9.18)–(9.19) of the previous sections are used to illustrate the methodology. The technique described in Section 9.2.4 is applied again for the simultaneous identification of the two processes. The values $\lambda_2 = 0.75 L_2 = 0.16$ and $\lambda_1 = 0.75(L_1 + L_2) = 5.32$ are selected. By applying the tuning rules (9.41)–(9.44) we obtain $\alpha = 0.43$, $\beta = 7.29$, $K_{p2} = 4.93$,

$T_{i2} = 0.51$, $T_{d2} = 0.07$, $K_{p1} = 0.93$, $T_{i1} = 9.75$, and $T_{d1} = 1.86$. The process output obtained is shown in Figure 9.6. It appears that a high-performance load disturbance response is achieved. With respect to the method presented in the previous section, as expected because of the presence of the set-point filters, the rise time in the set-point step response is (significantly) increased and the overshoot is reduced to zero.

9.2.7 Use of a Smith Predictor in the Outer Loop

A scheme based on a Smith predictor has been proposed in (Kaya, 2001) and it is depicted in Figure 9.7. It appears that a Smith predictor scheme (Palmor, 1996) is employed in order to compensate for the delay term in the primary controller. An automatic tuning procedure, based on the use of two sequential relay-feedback tests is proposed. In particular, a FOPDT transfer function is first estimated for the secondary process by employing, for example, an asymmetrical relay (see Section 7.2.2). The parameters of a PI controller are selected consequently according to the tuning rules proposed in (Zhuang and Atherton, 1993), which allows a minimisation of the ISTE integral criterion. Then, again by adopting an asymmetrical relay, the process seen by the primary controller C_1 is identified. In particular, a SOPDT transfer function parameters are estimated (see Section 7.3.2). In this phase the Smith predictor scheme is not employed. Then, the delay free part of the model is denoted by P_{12} and the dead time term is denoted by L. The parameters of the PID primary controller (in ideal form) are finally selected in order to minimise again the ISTE criterion and the Smith predictor based cascade control scheme is implemented.

In order to verify the effectiveness of the devised scheme, it has been tested on the same process (9.18)–(9.19) as before. The resulting PI/PID parameters are $K_{p2} = 3.17$, $T_{i2} = 1.05$, $K_{p1} = 1.5$, $T_{i1} = 8.22$, and $T_{d1} = 0.91$. The process output obtained is shown in Figure 9.8. Obviously, the control architecture is more effective when a high normalised dead time is present in the secondary process, although it has to be stressed that mismatches in the estimation of the dead time term should be carefully handled in order to avoid a significant degradation of the overall performance.

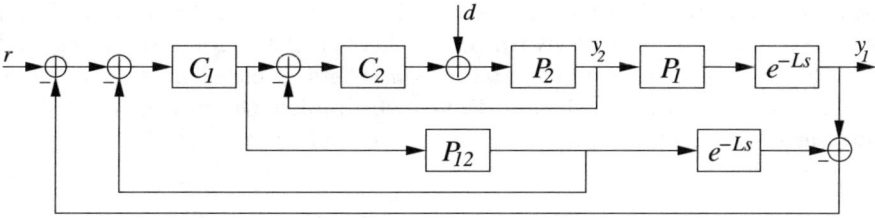

Fig. 9.7. Smith predictor based cascade control structure

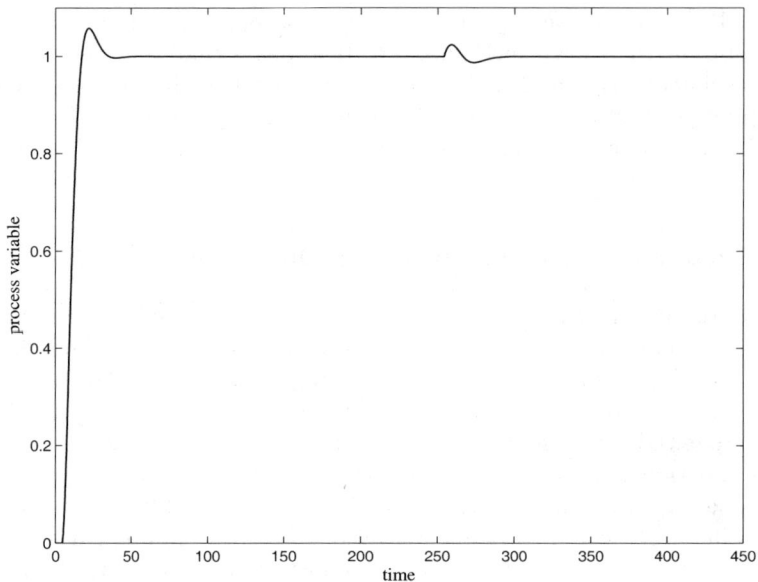

Fig. 9.8. Process variable obtained with the Smith predictor based cascade control scheme

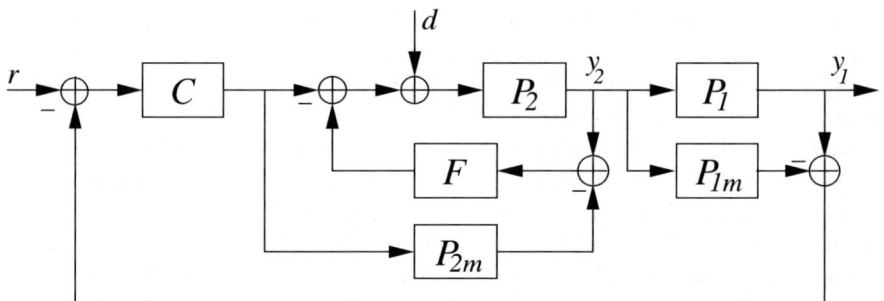

Fig. 9.9. Two-degree-of-freedom cascade control structure

The proposed control scheme has been further developed in (Kaya et al., 2005) for stable processes and in (Kaya and Atherton, 2005) for integrating and unstable processes. The tuning of the controllers is based on the so-called standard forms (Dorf and Bishop, 1995), which allow the direct synthesis of controllers that minimise integral performance indexes.

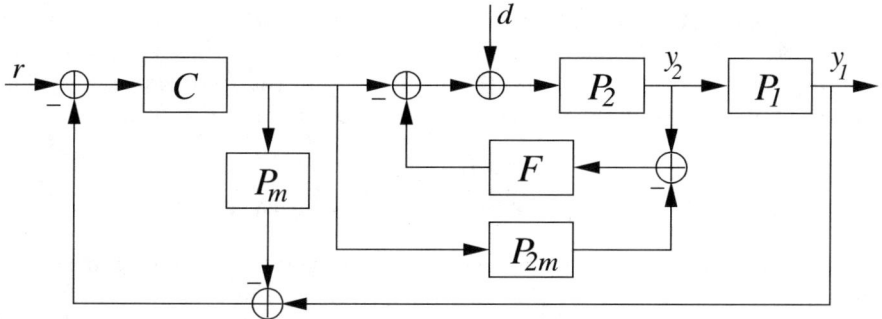

Fig. 9.10. Alternative two-degree-of-freedom cascade control structure

9.2.8 Two Degree-of-freedom Control Structure

A two-degree-of-freedom cascade control structure, aiming at decoupling the set-point tracking and the load disturbance rejection tasks has been proposed in (Liu et al., 2005). Two schemes can be implemented in this context. They are reported in Figures 9.9 and 9.10. Note that P_{1m} and P_{2m} denote models of the primary and secondary processes P_1 and P_2 respectively, while P_m denotes the model of the overall process P. Then F is a load disturbance estimator (indeed, it acts as a secondary controller) and C is the primary controller used for set-point tracking.

From Figure 9.9 it can be deduced that in the nominal case, *i.e.*, when P_{1m} and P_{2m} are perfect models of P_1 and P_2, there is an open-loop control from the set-point r to the primary output y_1 so that the nominal set-point response and the inner loop load disturbance response are decoupled. The scheme of Figure 9.10 has the advantage that an explicit model of the primary process P_1 is not required but, on the other side, when a load disturbance d occurs, both the load disturbance estimator F and the primary controller C concur in compensating it and therefore a performance degradation might appear. In other words, the possible performance degradation in the load disturbance response has to be accepted for a possibly easier and more effective implementation of the control architecture. In any case, the design of the two controllers C and F is the same for both schemes.

The design of C is based on the minimisation of the $\mathcal{H}_2$ performance objective, as in the IMC approach. The transfer functions of the two processes are rewritten as

$$P_1(s) = K_1 \frac{A_{1+}(s) A_{1-}(s)}{B_1(s)} e^{-L_1 s} \qquad (9.49)$$

and

$$P_2(s) = K_2 \frac{A_{2+}(s) A_{2-}(s)}{B_2(s)} e^{-L_2 s} \qquad (9.50)$$

where $A_{1+}(0) = A_{1-}(0) = B_1(0) = 1$, $A_{2+}(0) = A_{2-}(0) = B_2(0) = 1$, and all zeros of $A_{1-}(s)$, $A_{2-}(s)$, $B_1(s)$ and $B_2(s)$ are located in the left half plane, while all zeros of $A_{1+}(s)$ and $A_{2+}(s)$ are located in the right half plane. The primary controller transfer function can therefore be derived as

$$C(s) = K_2 \frac{B_1(s)B_2(s)}{K_1 K_2 A_{1+}^*(s) A_{2+}^*(s) A_{1-}(s) A_{2-}(s)(\lambda_c s + 1)^{n_c}} \quad (9.51)$$

where $A_{1+}^*(s)$ and $A_{2+}^*(s)$ are the complex conjugate of $A_{1+}(s)$ and $A_{2+}(s)$ respectively (i.e., $A_{1+}(s)/A_{1+}^*(s)$ and $A_{2+}(s)/A_{2+}^*(s)$ are all-pass filters), λ_c is a tuning parameter and n_c is the order of the filter to be selected in order to make the controller transfer function proper. Indeed, λ_c allows the handling of the trade-off between aggressiveness and robustness. A good starting point is to select λ_c equal to the overall time delay of the process to be controlled. The design of the load disturbance estimator F is performed by proposing the desired complementary sensitivity function of the inner loop, denoted as $\bar{T}(s)$. In particular, by considering again the $\mathcal{H}_2$ optimal performance objective of the IMC theory, the expression of $\bar{T}(s)$ is selected as

$$\bar{T}(s) = \frac{1}{(\lambda_f s + 1)^{n_f}} \frac{A_{2+}(s)}{A_{2+}^*(s)} e^{-L_2 s} \quad (9.52)$$

where the filter order n_f is chosen appropriately. Thus, the expression of $F(s)$ can be determined as

$$F(s) = \frac{F_1(s)}{1 - F_1(s)P_2(s)} \quad (9.53)$$

where

$$F_1(s) = \frac{B_2(s)}{K_2 A_{2+}^*(s) A_{2-}(s)(\lambda_f s + 1)^{n_f}}. \quad (9.54)$$

The design parameter λ_f allows again the handling of the trade-off between aggressiveness and robustness and it is suggested to select it, as a first guess, equal to the secondary process estimated dead time.

It is worth stressing at this point that the methodology fully exploits an accurate (possibly high-order) modelling of the process, but the two controllers are not of PID type in general and the overall scheme has to be implemented in a different way with respect to the standard scheme of Figure 9.1.

The same process (9.18)–(9.19) has been adopted to illustrate the methodology. Based again on the simultaneous identification method of Section 9.2.4, the model of the primary and secondary processes are chosen respectively as

$$P_{1m}(s) = \frac{1}{7.55s + 1} e^{-6.88s} \quad (9.55)$$

and

$$P_{2m}(s) = \frac{1}{0.99s + 1} e^{-0.21s}. \quad (9.56)$$

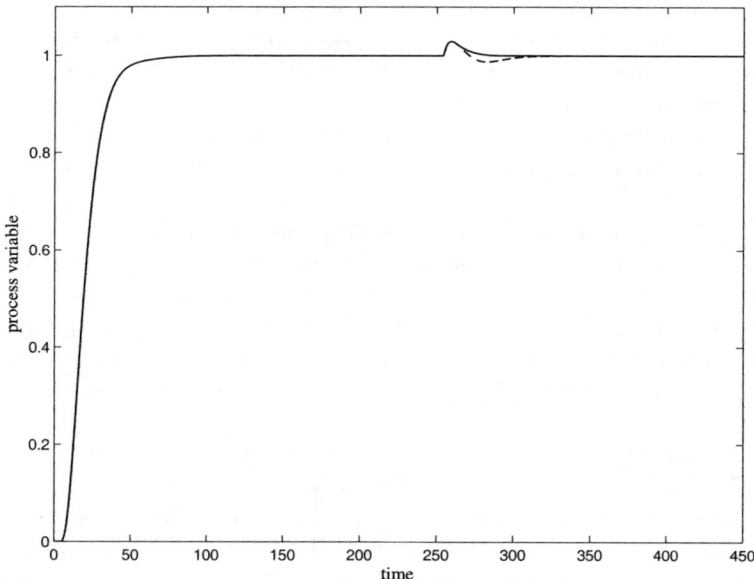

Fig. 9.11. Process variable obtained with the two degree-of-freedom control structure of Figure 9.9 (solid line) and of Figure 9.10 (solid line)

Consequently, the two filters time constants are selected as $\lambda_c = 7.09$ and $\lambda_f = 0.21$. It results:

$$C(s) = \frac{(0.99s + 1)(7.55s + 1)}{(7.09s + 1)^2} \tag{9.57}$$

and

$$F_1(s) = \frac{0.99s + 1}{0.21s + 1}. \tag{9.58}$$

The process variables obtained by considering the two control schemes are plotted in Figure 9.11 (note that a unit step load disturbance is applied again to the process at time $t = 250$). The effectiveness of both schemes is apparent.

9.3 Ratio Control

9.3.1 Generalities

Ratio control, which consists in keeping a constant ratio between two (or more) process variables, irrespective of possible set-point changes and load disturbances that might occur on the plant, is of concern in a variety of industrial applications such as chemical dosing, water treatment, chlorination, mixing vessels and waste incinerators. For example, in combustion systems it is necessary to control accurately the air-to-fuel ratio in order to obtain a high efficiency, and in blending processes a selected ratio of different flows has to be maintained to keep a constant product composition. In this latter case, both flows can be controlled or, alternatively, one of them can be measured only (the so-called *wild flow*) and the other is regulated in order to achieve the desired ratio.

Formally, denote by a the desired ratio to be kept between the values of two process variables y_1 and y_2. For this purpose, the control scheme shown in Figure 9.12 (also termed *series metered control*) can be implemented. Each variable is controlled by two separate controllers C_1 and C_2 (typically of PI type) and the output y_1 of the first process is multiplied by a and adopted as the set-point signal of the closed-loop control system of the second process, *i.e.*, it is $r_2(t) = ay_1(t)$.

The main disadvantage of this scheme is related to its transient response to a change in the set-point r_1, since the output y_2 is necessarily delayed with respect to y_1, due to the closed-loop dynamics of the second loop. In general, the second loop is chosen as the one with the fastest dynamics. However, in order to keep the ratio close to the desired value, it might be necessary to detune the first loop and therefore the performance obtained in the set-point following task and in the rejection of the load disturbance d_1 decreases.

A possible alternative scheme is the one shown in Figure 9.13 (termed *parallel metered control*). In this case, provided that the two closed-loop systems have the same dynamics, a high performance can be achieved in the set-point following task, but, obviously, a disturbance acting on the first process can cause a large error in the ratio value. For this reason, this approach has to be employed in those applications where load disturbances are unlikely to occur. Finally, it is worth remembering that in the particular case of combustion control, where a selected air-to-fuel ratio has to be maintained, it is often essential to prevent the occurrence of a fuel rich environment, since this might lead to a furnace explosion. In this context, the so-called cross-limiting control (also known as lead-lag control) shown in Figure 9.14 can be adopted (Gomes, 1985). The two loops are interlocked by using a low and a high selectors that force the fuel to follow the air flow when the set-point increases and that force the air to follow the fuel when the set-point decreases.

Techniques for the improvement of these standard techniques have been recently proposed. They are presented in the following sections.

9.3.2 The Blend Station

Methodology

As already mentioned, the use of the series metered control scheme of Figure 9.12 has the disadvantage that the output y_2 is actually delayed with respect to y_1 because of the closed-loop dynamics of the second loop. In order to

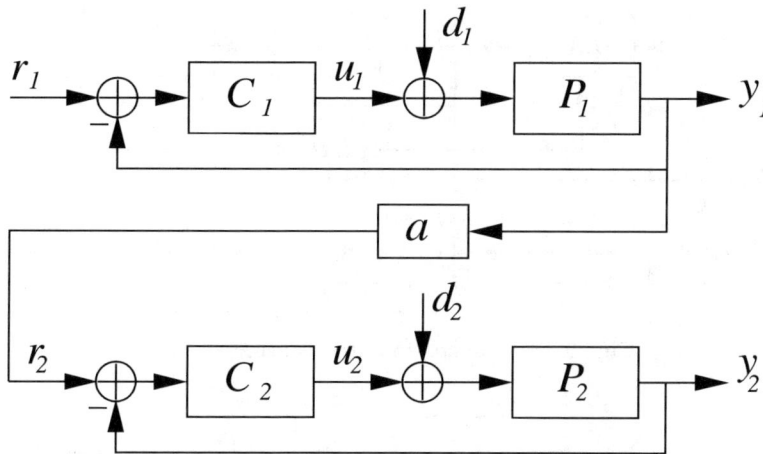

Fig. 9.12. The typical ratio control scheme (series metered control)

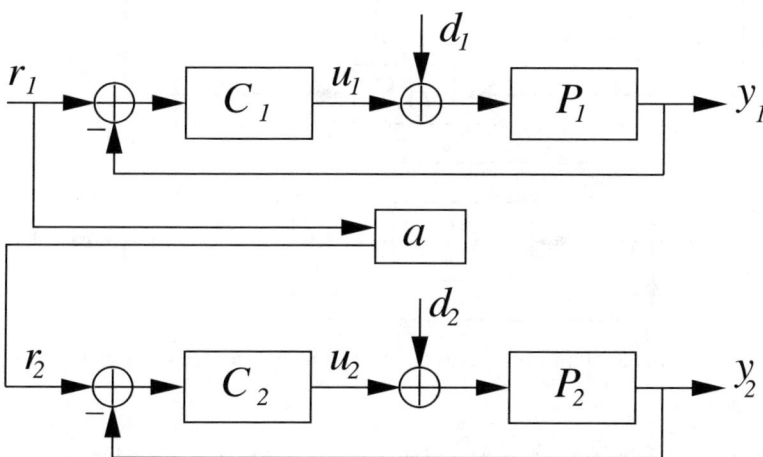

Fig. 9.13. An alternative ratio control scheme (parallel metered control)

address this problem, an alternative architecture, called the Blend Station and shown in Figure 9.15, has been proposed in (Hägglund, 2001). Its use is suggested when no disturbances are likely to occur in the processes and when the two processes exhibit a different dynamics (thus, it has to be considered a valid alternative to the parallel metered control scheme of Figure 9.13).
The main feature of the scheme is that the value of the set-point r_2 depends both on the value of the process output y_1 and on the value of the set-point r_1, according to the expression

$$r_2(t) = a(\gamma r_1(t) + (1-\gamma) y_1(t)). \tag{9.59}$$

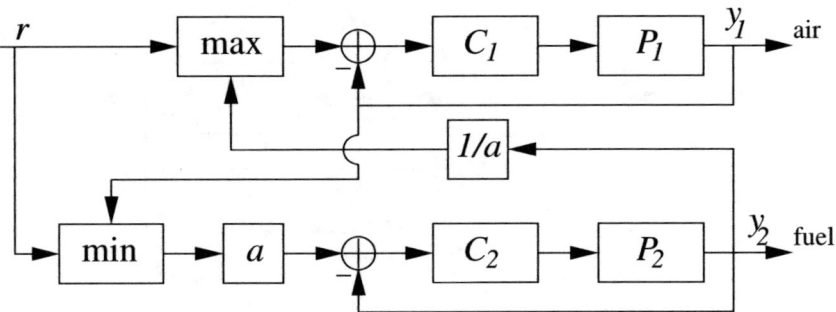

Fig. 9.14. The cross-limited control scheme

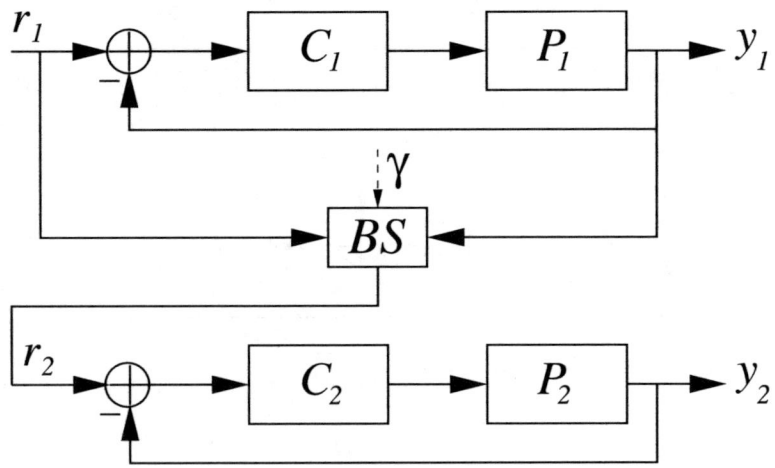

Fig. 9.15. The ratio control scheme using the Blend Station

Note that γ is a constant parameter that weights the relative influence of the set-point r_1 on r_2 with respect to y_1 (for $\gamma = 0$ the classical scheme of Figure 9.12 is obtained). The value of γ can be selected as the ratio of the time constants of the two closed-loop systems (or, if they are not available, as the ratio of the integral time constants of the two controllers T_{i2}/T_{i1}) or, alternatively, by applying a suitable adaptive procedure, *i.e.*, by applying the following formula (Hägglund, 2001):

$$\frac{d\gamma}{dt} = \frac{S}{T_a}(ay_1 - y_2) \tag{9.60}$$

where $S \in \{-1, 0, 1\}$ is a sign parameter that takes into account if the set-point step is positive or negative. In (Hägglund, 2001) it is suggested selecting the value of the adaptation rate T_a as a factor times the longest integral time of the two loops.

A methodology where a time-varying parameter $\gamma(t)$ is adopted has been proposed in (Visioli, 2005a). Assume that a transition from the initial value y_1^i to the final value y_1^f is required to be performed at time $t = t_0$ for the process variable y_1 (i.e., a step set-point signal of amplitude $y_1^f - y_1^i$ is applied to the set-point signal $r_1(t)$ at time $t = t_0$). Without loss of generality, in the following it will be assumed that a positive step signal is applied, i.e., $y_1^f > y_1^i$. First, as usual, the second loop has to be selected as the one with the fastest dynamics, i.e., the dynamics of process P_2 is faster than the one of P_1. Processes P_1 and P_2 are then modelled with FOPDT transfer functions:

$$P_1(s) = \frac{K_1}{T_1 s + 1} e^{-L_1 s}, \qquad (9.61)$$

$$P_2(s) = \frac{K_2}{T_2 s + 1} e^{-L_2 s}. \qquad (9.62)$$

Based on these models, the two single-loop controllers C_1 and C_2 are selected as PI controllers with set-point weighting, i.e., the manipulated variables u_1 and u_2 are expressed as:

$$u_1(t) = K_{p1}\left(\beta_1 r_1(t) - y_1(t) + \frac{1}{T_{i1}}\int_0^t (r_1(\tau) - y_1(\tau))d\tau\right), \qquad (9.63)$$

$$u_2(t) = K_{p2}\left(\beta_2 r_2(t) - y_2(t) + \frac{1}{T_{i2}}\int_0^t (r_2(\tau) - y_2(\tau))d\tau\right). \qquad (9.64)$$

The value of γ is chosen as the output of a PI controller as well, whose input is the current ratio error, added to a constant value γ^*. An additional condition has to be set to account for the case in which $L_1 > L_2$, in order to avoid that at the beginning of the transient response the condition $y_2(t) > ay_1(t)$ holds, i.e., the output y_2 starts its transient before that of y_1. Formally, it is:

$$\gamma(t) = \begin{cases} 0 & \text{if } L_1 > L_2 \text{ and } t < t_0 + L_1 - L_2 \\ \gamma^* + K_p\left(e_r(t) + \frac{1}{T_i}\int_0^t e_r(\tau)d\tau\right) & \text{elsewhere} \end{cases} \qquad (9.65)$$

where

$$e_r(t) = y_2(t) - ay_1(t). \qquad (9.66)$$

In this way, the two process outputs are forced to start their transient response at the same time instant.

It appears that the adoption of a time-varying parameter γ aims actually at "shaping" the reference function $r_2(t)$ in such a way that the response of the second closed-loop system is as equal as possible to that of the first one, despite their possible different dynamics.

A tuning procedure for the overall scheme has been also proposed in (Visioli,

Table 9.1. Tuning rule of the proposed ratio controller

K_{p1}	T_{i1}	β_1	K_{p2}	T_{i2}	β_2	γ^*	K_p	T_i
$\dfrac{0.9T_1}{K_1 L_1}$	$3L_1$	0	$\dfrac{0.9T_2}{K_2 L_2}$	$3L_2$	0	$\dfrac{T_{i2}}{T_{i1}}$	$0.5\dfrac{L_2 T_1}{T_2 L_1}$	$\dfrac{T_1}{L_1}$

2005a). The two PI controllers C_1 and C_2 (see (9.63)–(9.64)) are tuned according to the Ziegler–Nichols formula (Åström and Hägglund, 1995) and the set-point weights β_1 and β_2 are set to zero in order to avoid significant overshoots. Finally, γ^* is chosen as T_{i2}/T_{i1} and the gains of the PI controller that provides the current value of γ (see (9.65)) are selected according to the following formula:

$$K_p = 0.5\frac{L_2 T_1}{T_2 L_1}, \qquad T_i = \frac{T_1}{L_1}. \tag{9.67}$$

The overall tuning rule is summarised in Table 9.1.

It appears that, being based on a simple identification experiment and on the direct application of simple formulae, the tuning procedure can be easily performed automatically. It is worth noting again that, although the Ziegler–Nichols rules are known to provide large overshoot, the use of the set-point weight fixed to zero prevents this fact and extends the range of processes for which they provide satisfactory results. Obviously, this implies also that the rise time increases, but this can be accepted in a ratio control framework, where keeping the desired ratio is of major concern, rather than obtaining a high-performance step response. However, in case the dynamics of a process is not suitable for the Ziegler–Nichols formulae, *i.e.*, the dead time is greater than the dominant time constant, then a more appropriate (though more complex) tuning rule such as the Kappa–Tau (Åström and Hägglund, 1995) should be used for the two PI controllers C_1 and C_2, while Formula (9.67) is maintained.

Note also that the value of γ^* is selected according to the considerations made in (Hägglund, 2001), since γ^* is actually the initial value of γ.

In any case, being based on a standard PI controller (9.65), the user can easily modify the performance of the ratio controller by increasing or decreasing the value of K_p and T_i according to its typical know-how, which is therefore conveniently fully retained.

Simulation Results

In order to verify the effectiveness of the original Blend Station approach and of its modification, some simulation results are presented. In all cases a unit step is applied to the set-point signal $r_1(t)$ at time $t_0 = 0$ s (*i.e.*, $y_1^i = 0$ and $y_1^f = 1$). Further, the value $a = 1$ has been fixed.

As a first example, the following two FOPDT processes have been considered:

$$P_1(s) = \frac{1}{6s+1}e^{-2s} \qquad P_2(s) = \frac{1}{2s+1}e^{-2s}. \qquad (9.68)$$

By applying the tuning procedure, the values of $K_{p1} = 2.7$, $T_{i1} = 6$, $K_{p2} = 0.9$ and $T_{i2} = 6$ result. The original Blend Station has then been designed with $\gamma = T_{i2}/T_{i1} = 1$ (note that the scheme of Figure 9.13 naturally results) and the adaptive version has been implemented by setting $T_a = 10 \cdot \max\{T_{i1}, T_{i2}\} = 60$. In order to provide the best achievable result in this context, a large number of set-point steps has been applied to the ratio control architecture, until the value of γ converges around its optimal value. Then, regarding the control scheme where the value of γ is determined by means of Formula (9.65), the values of $\gamma^* = 1$, $K_p = 1.5$, $T_i = 3$ are selected. The process variables resulting for the different schemes considered are shown in Figure 9.16 (the classic scheme of Figure 9.12 has been also considered). The corresponding control variables are shown in Figure 9.17. Note that no particular aspects emerge for the control signals and therefore they will be not shown for the other examples for the sake of brevity.

The overall performance achieved with the different approaches can be compared by calculating the following performance:

$$J = \int_0^\infty |ay_1(t) - y_2(t)|dt. \qquad (9.69)$$

The value $J = 4.44$ results for the original Blend Station, $J = 0.35$ for the adaptive Blend Station, $J = 2.07$ for the modified Blend Station, $J = 12.65$ for the standard ratio controller.

It appears that the adaptive Blend Station provides the best performance, although this occurs after many set-point changes are applied in order to allow the convergence to the best value of γ.

In order to understand better the different approaches, the γ signals are plotted in Figure 9.18 together with the reference signal obtained for the second loop in the case where the modified Blend Station is adopted. It can be seen that the variation of γ for the case of the adaptive Blend Station is hardly visible (*i.e.*, $\gamma(t)$ is nearly constant during the transient response), and this indicates that the best result that can be obtained with this scheme has been achieved.

As a second example, the following two FOPDT processes have been considered:

$$P_1(s) = \frac{1}{4s+1}e^{-3s} \qquad P_2(s) = \frac{1}{8s+1}e^{-2s}. \qquad (9.70)$$

The tuning procedure results in $K_{p1} = 1.2$, $T_{i1} = 9$, $K_{p2} = 3.6$ and $T_{i2} = 6$. Then, for the original Blend Station approach the value of $\gamma = 0.67$ is selected and by applying the adaptive procedure (with $T_a = 90$) on a sequence of set-point steps, the value of γ converges around 0.49. Finally, for the modified approach, the value of $\gamma^* = 0.67$, $K_p = 0.17$ and $T_i = 1.33$ are determined (see Table 9.1).

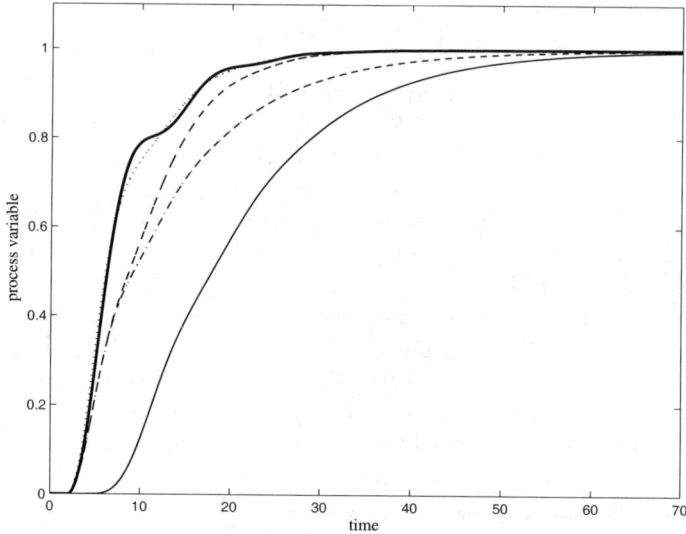

Fig. 9.16. Process variables obtained for the first example of the Blend Station approach. Thick solid line: y_1; dash-dot line: y_2 with the original Blend Station; dotted line: y_2 with the adaptive Blend Station; dashed line: y_2 with the modified Blend Station; solid line: y_2 with the classic approach ($\gamma = 0$).

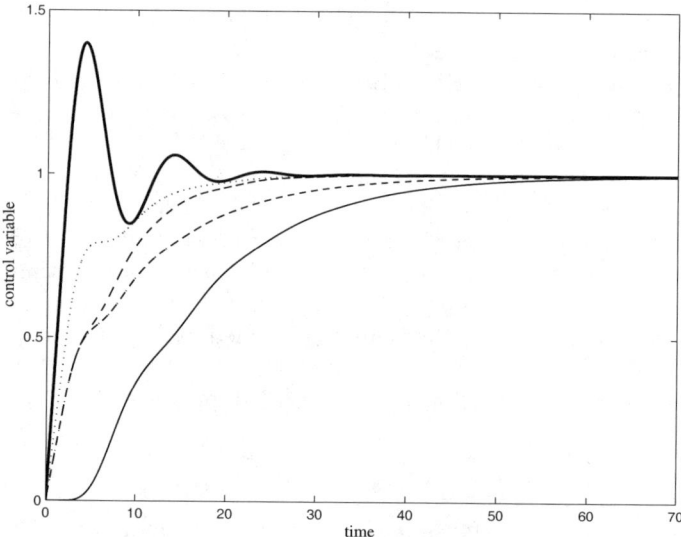

Fig. 9.17. Control variables obtained for the first example of the Blend Station approach. Thick solid line: u_1; dash-dot line: u_2 with the original Blend Station; dotted line: u_2 with the adaptive Blend Station; dashed line: u_2 with the modified Blend Station; solid line: u_2 with the classic approach ($\gamma = 0$).

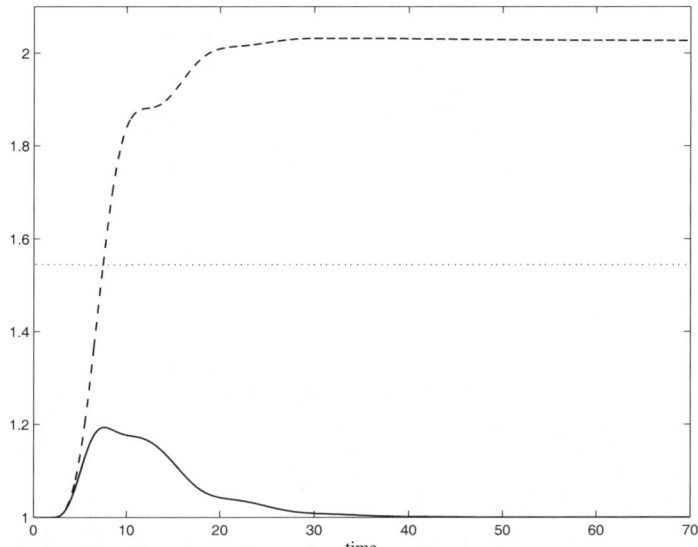

Fig. 9.18. Different signals obtained for the first example of the Blend Station approach. Solid line: $r_2(t)$ for the modified Blend Station; dashed line: $\gamma(t)$ for the modified Blend Station; dotted line: $\gamma(t)$ for the adaptive Blend Station.

The resulting process variables are reported in Figure 9.19, where again the standard approach has been also considered. The value of γ for the case of the Blend Station with the adaptive procedure has been plotted together with $\gamma(t)$ and $r_2(t)$ for the modified method in Figure 9.20. The resulting values of the performance index are $J = 3.33$ for the original Blend Station, $J = 1.044$ for the adaptive Blend Station, $J = 1.37$ for the modified approach and $J = 7.67$ for the standard ratio controller. Indeed, the same considerations made for the first example can be made also for this one.

As a third example the same processes of the previous example are considered, but their position has been swapped in the overall control scheme, namely,

$$P_1(s) = \frac{1}{8s+1}e^{-2s} \qquad P_2(s) = \frac{1}{4s+1}e^{-3s}. \tag{9.71}$$

In this case the values $K_{p1} = 3.6$, $T_{i1} = 6$, $K_{p2} = 1.2$, $T_{i2} = 9$, $\gamma^* = 1.5$, $K_p = 1.5$ and $T_i = 4$ evidently result. Results obtained with the different control architectures are reported in Figure 9.21. Note that in this case the adaptive procedure (again with $T_a = 90$) for the Blend Station, applied when a series of set-point steps occurs, converges around a value of $\gamma = 2.12$. This value has been adopted as initial condition for the process output $y_2(t)$ obtained with the adaptive Blend Station.

As for the previous examples, in Figure 9.22 the value of γ for the case of the Blend Station with the adaptive procedure has been plotted together with

276   9 Control Structures

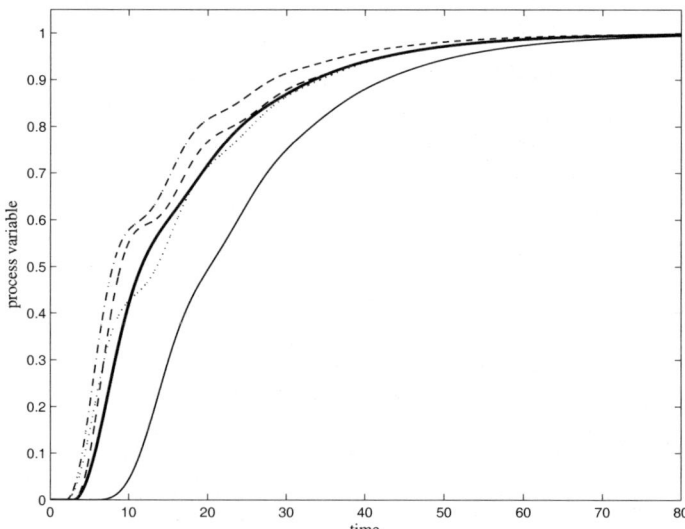

Fig. 9.19. Process variables obtained for the second example of the Blend Station approach. Thick solid line: y_1; dash-dot line: y_2 with the original Blend Station; dotted line: y_2 with the adaptive Blend Station; dashed line: y_2 with the modified Blend Station; solid line: y_2 with the classic approach ($\gamma = 0$).

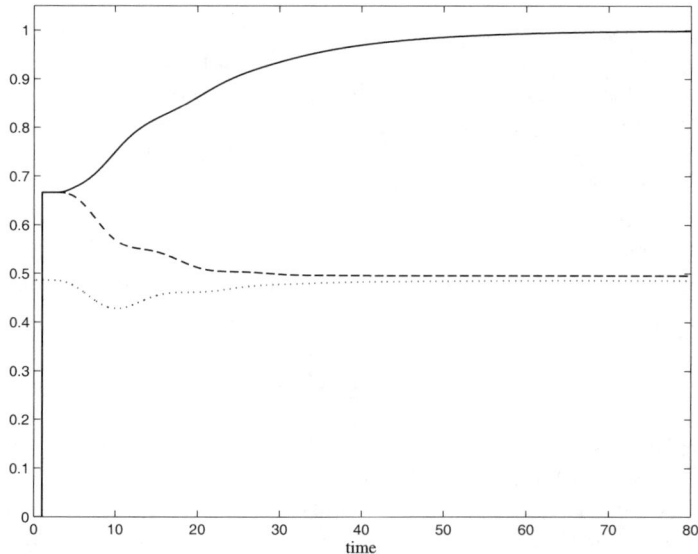

Fig. 9.20. Different signals obtained for the second example of the Blend Station approach. Solid line: $r_2(t)$ for the modified Blend Station; dashed line: $\gamma(t)$ for the modified Blend Station; dotted line: $\gamma(t)$ for the adaptive Blend Station.

$\gamma(t)$ and $r_2(t)$ for the method where the current value of γ is determined by a PI controller. The resulting values of the performance index are $J = 5$ for the original Blend Station, $J = 2.07$ for the adaptive Blend Station, $J = 2.37$ for the modified Blend Station, and $J = 16.5$ for the standard ratio controller.

By comparing the results obtained for this example with those obtained for the previous one, it appears that, as already mentioned, it is more sensible to choose as a first process that with the fastest dynamics but in any case the performance obtained with the proposed ratio controllers are still satisfactory (indeed the same conclusions of the previous examples can be drawn also for this example).

Finally, as a last example, two high-order processes have been considered:

$$P_1(s) = \frac{1}{(s+1)^8} \qquad P_2(s) = \frac{1}{(0.25s+1)^8}. \tag{9.72}$$

The two processes are modelled with FOPDT transfer functions with the following parameters: $K_1 = 1$, $T_1 = 2.99$, $L_1 = 5.55$, $K_2 = 1$, $T_2 = 0.71$, $L_2 = 1.82$.

Being the dead time of the two processes significantly greater than the corresponding dominant time constant, the Kappa–Tau tuning rules (Åström and Hägglund, 1995) have been adopted instead of the Ziegler–Nichols ones. Thus, it results: $K_{p1} = 0.13$, $T_{i1} = 2.62$, $b_1 = 2.91$, $K_{p2} = 0.11$, $T_{i2} = 0.71$, $b_2 = 3.67$ $\gamma^* = 0.27$, $K_p = 0.69$, $T_i = 0.54$.

Resulting process variables are shown in Figure 9.23, where the process output $y_2(t)$ for the adaptive Blend Station has been obtained by starting with a value of γ equal to 0.32, which results after the application of a sequence of set-point steps with $T_a = 26.2$. In Figure 9.24 the value of γ for the case of the Blend Station with the adaptive procedure has been plotted together with $\gamma(t)$ and $r_2(t)$ for the modified Blend Station method.

The resulting values of the performance index are $J = 1.81$ for the original Blend Station, $J = 1.70$ for the adaptive Blend Station, $J = 0.52$ for the modified Blend Station and $J = 4.56$ for the standard ratio controller.

It appears that in this case the scheme with an additional PI controller provides a performance even better than the adaptive Blend Station. This is possibly explained by the fact that $\gamma(t) = 0$ when $t < 3.73$ (see (9.65)), thus allowing the two process outputs to start their transient almost at the same time so that a very satisfactory result is achieved.

278   9 Control Structures

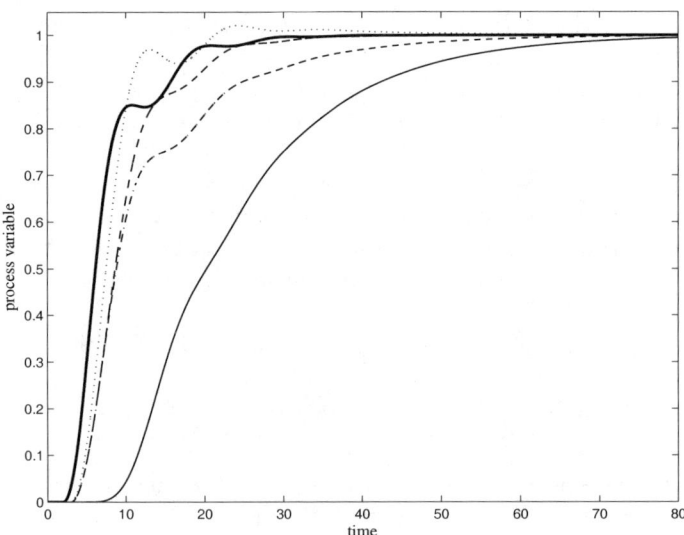

Fig. 9.21. Process variables obtained for the third example of the Blend Station approach. Thick solid line: y_1; dash-dot line: y_2 with the original Blend Station; dotted line: y_2 with the adaptive Blend Station; dashed line: y_2 with the modified Blend Station; solid line: y_2 with the classic approach ($\gamma = 0$).

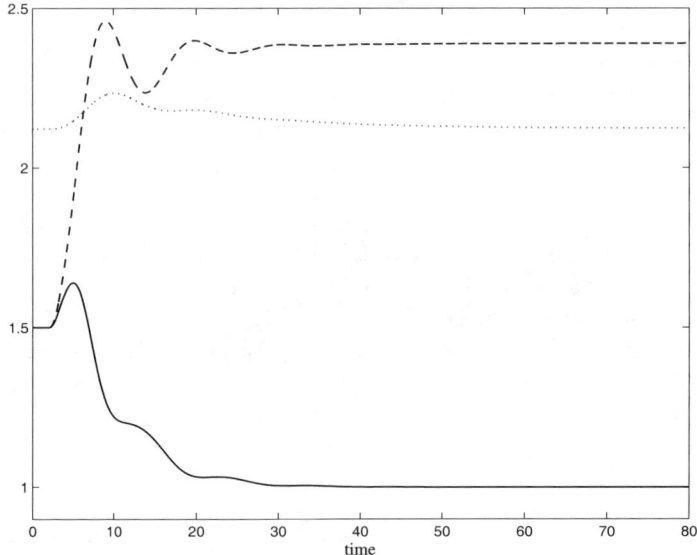

Fig. 9.22. Different signals obtained for the third example of the Blend Station approach. Solid line: $r_2(t)$ for the modified Blend Station; dashed line: $\gamma(t)$ for the modified Blend Station; dotted line: $\gamma(t)$ for the adaptive Blend Station.

9.3 Ratio Control   279

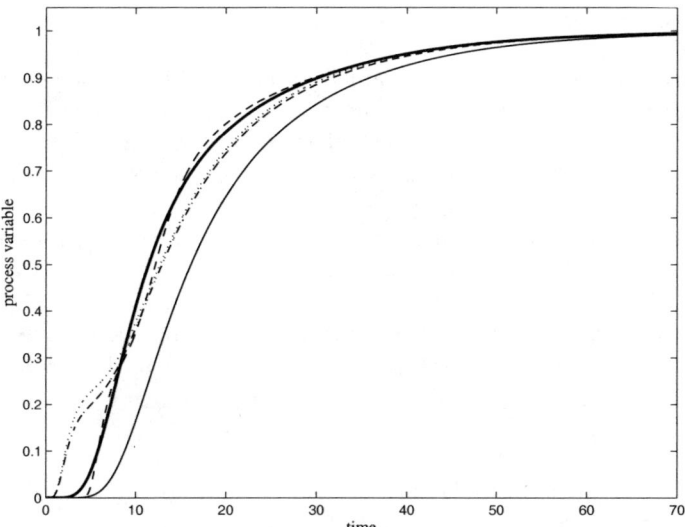

Fig. 9.23. Process variables obtained for the fourth example of the Blend Station approach. Thick solid line: y_1; dash-dot line: y_2 with the original Blend Station; dotted line: y_2 with the adaptive Blend Station; dashed line: y_2 with the modified Blend Station; solid line: y_2 with the classic approach ($\gamma = 0$).

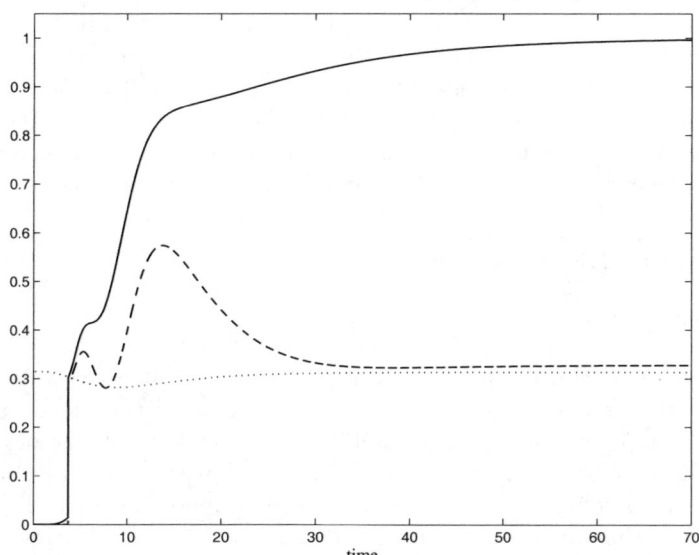

Fig. 9.24. Different signals obtained for the fourth example of the Blend Station approach. Solid line: $r_2(t)$ for the modified Blend Station; dashed line: $\gamma(t)$ for the modified Blend Station; dotted line: $\gamma(t)$ for the adaptive Blend Station.

Experimental Results

Experimental results have been obtained by considering the level control apparatus described in Section A.1, where both tanks have been adopted. In particular, the task to be accomplished is to perform an output transition from 2 V to 3 V for the first tank as well as for the second one, maintaining a desired ratio value $a = 1$ during the whole transient response.

In order to diversify the dynamics of the two level control loops, as a first experiment, a software time delay of 10 s and 5 s has been added to the measure of the level of the first and of the second tank respectively. A FOPDT model of each process has been estimated separately by applying the area method (see Section 7.2.1) to the open-loop step response. The transfer functions obtained are

$$P_1(s) = \frac{1.98}{25s+1}e^{-11s} \qquad P_2(s) = \frac{2.27}{25s+1}e^{-6s}. \qquad (9.73)$$

The application of the formulae of Table 9.1 results in $K_{p1} = 1.03$, $T_{i1} = 33$, $K_{p2} = 1.65$, $T_{i2} = 18$, $\gamma^* = 0.55$, $K_p = 0.27$, $T_i = 2.27$.

Results are shown in Figure 9.25. The resulting performance index (9.69) (calculated over the time interval from $t = 0$ s to $t = 195$ s) is $J = 7.41$ for the Blend Station, $J = 5.32$ for the modified Blend Station and $J = 23.59$ for the standard ratio controller.

A second experiment has been performed by modifying the added dead time of the second process, decreasing it to 4 s. In this case the control parameters are $K_{p1} = 1.03$, $T_{i1} = 33$, $K_{p2} = 1.98$, $T_{i2} = 15$, $\gamma^* = 0.45$, $K_p = 0.23$, $T_i = 2.27$. Results are shown in Figure 9.26 and the calculated performance indexes are $J = 22.60$ for the Blend Station, $J = 12.18$ for the modified Blend Station and $J = 20.35$ for the standard ratio controller. By comparing the results obtained, it appears that making the second loop dynamics faster implies a better performance obtained by the standard controller, as expected.

9.3.3 Dynamic Blend Station

Methodology

In the previous sections it has been stressed that the Blend Station and its modified version have not to be employed when load disturbances are likely to occur in the plant, since they are not able to provide a satisfactory performance in this case. In order to address also the load disturbance rejection task, a different ratio control architecture has been proposed in (Visioli, 2005d), which extends the idea of the original Blend Station by substituting the constant parameter γ in Expression (9.59) by a dynamic system F. The control scheme is shown in Figure 9.27. The transfer function $F(s)$ is determined in such a way that the transfer function from r_1 to y_1 is the same as the transfer function from r_1 to y_2, scaled by a. By analysing the control scheme shown in Figure 9.27 it can be deduced that:

9.3 Ratio Control   281

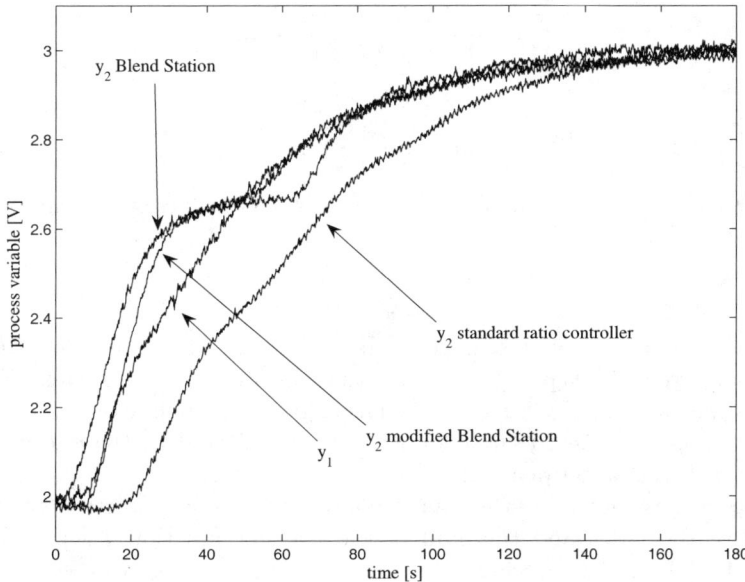

Fig. 9.25. Process variables obtained with the double tank apparatus

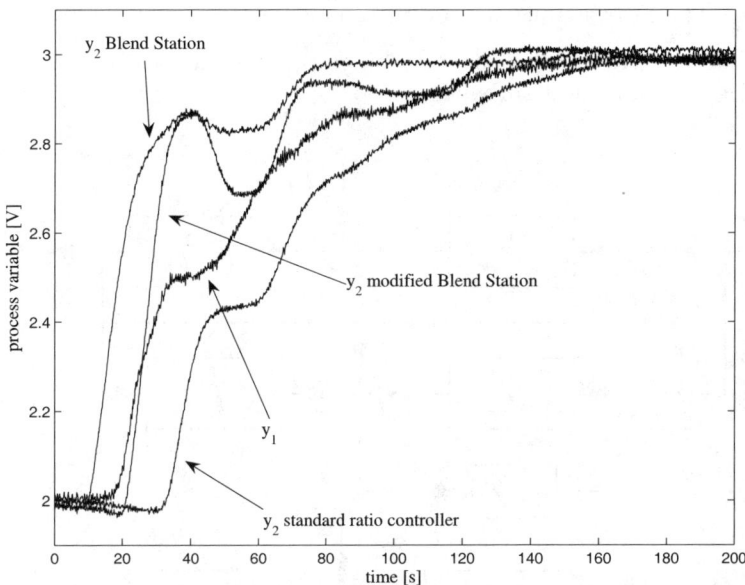

Fig. 9.26. Process variables obtained with the double tank apparatus and modified dead time

$$y_2 = \frac{C_2 P_2}{1 + C_2 P_2} \frac{F + C_1 P_1}{1 + C_1 P_1} a r_1 \qquad (9.74)$$

and

$$y_1 = \frac{C_1 P_1}{1 + C_1 P_1} r_1 \qquad (9.75)$$

After trivial calculations, it results that the two loops have identical dynamic response if:

$$F(s) = \frac{C_1(s) P_1(s)}{C_2(s) P_2(s)}. \qquad (9.76)$$

If the two processes are assumed to have FOPDT dynamics and the two controllers are of PI type, it turns out that, in order for the system $F(s)$ to be causal, it must be $L_1 \geq L_2$. This relation gives a guideline on how to select the first loop, that is, the first loop has to be selected as the one with the process with the larger dead time.

For the purpose of tuning the overall control system, it is worth determining the zeros, the poles and the gain of the transfer function $F(s)$. They are reported in Table 9.2. Because the control architecture guarantees that the desired ratio is obtained along the whole transient response when a set-point change is required (provided that the two processes have actually a FOPDT dynamics), the selection of the parameters of the two controllers has to be done according to the following intuitive guidelines:

- the parameters of $C_2(s)$ have to be chosen in order to provide the best rejection of a load disturbance d_2;

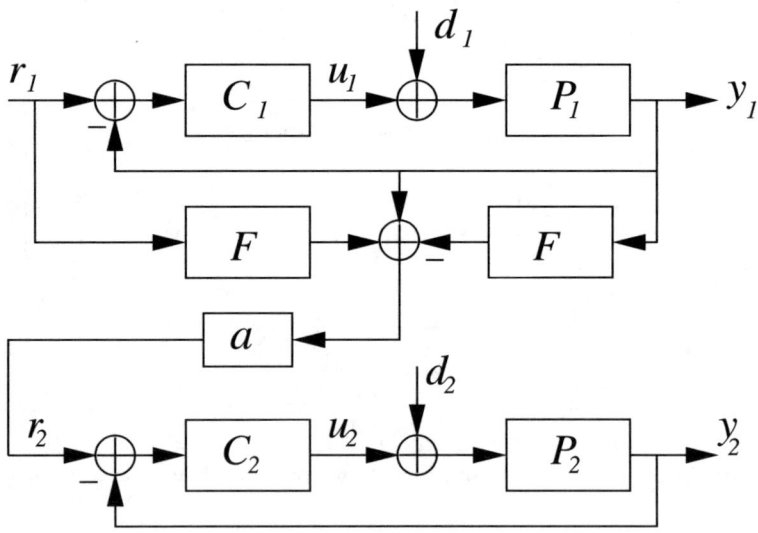

Fig. 9.27. The dynamic Blend Station control scheme

Table 9.2. Zeros, poles, and gain of transfer function $F(s)$

zeros	$-\dfrac{1}{T_{i1}}$,	$-\dfrac{1}{T_2}$
poles	$-\dfrac{1}{T_{i2}}$,	$-\dfrac{1}{T_1}$
gain	$\dfrac{K_1 K_{p1} T_{i2}}{K_2 K_{p2} T_{i1}}$	

- the parameters of $C_1(s)$ have to be chosen in order to provide the best rejection of a load disturbance d_1;
- the parameters of $C_1(s)$ and $C_2(s)$ have to be chosen in order to have a frequency response of $F(s)$ as low as possible. In this way, the reference r_2 of the second loop is determined mainly by output y_1 of the first loop instead of the reference r_1 of the first loop (note that the transfer function from y_1 and r_2 is $a(1 - F(s))$). Thus, a high performance on the desired ratio is obtained when a load disturbance is acting on the first process.

In order to achieve the mentioned goals, the following procedure is suggested. First, the PI controller C_1 is tuned according to the analytical method proposed in (Chen and Seborg, 2002), whose purpose is to obtain a desired specification on the load disturbance rejection task. In this context, it is convenient to choose the value of the time constant of the desired transfer function between the load disturbance d_1 and the process output y_1 equal to the value of the time constant T_1 of the process. This results in the following tuning rule:

$$K_{p1} = \frac{1}{K_1} \frac{T_1}{T_1 + L_1} \qquad T_{i1} = T_1. \qquad (9.77)$$

In this way, in addition to a good degree of robustness, a low value of the ratio K_{p1}/T_{i1} is achieved, which is important in order to ensure a low frequency response of $F(s)$ (see Table 9.2).

Subsequently, the PI controller C_2 is tuned by first imposing again a pole-zero cancellation in the second loop, *i.e.*, by setting $T_{i2} = T_2$. In other words, with the previous choices, we have simply

$$F(s) = K e^{-(L_1 - L_2)s} \qquad (9.78)$$

where

$$K = \frac{K_1 K_{p1} T_{i2}}{K_2 K_{p2} T_{i1}}. \qquad (9.79)$$

Then, parameter K_{p2} is selected by following basically the same idea described in (Åström *et al.*, 1998). Thus, in order to have good load disturbance rejection performance, K_{p2} is fixed, after solving an optimization problem, as the maximum value that guarantees that the closed-loop system is stable and that the largest value M_s of the sensitivity function is constrained.

It should be noted that M_s can be considered as a useful tuning parameter because it allows the handling of the trade-off between performance and robustness. In this context, typical values of M_s are chosen in the range 1.2-2.0, in order to ensure a sufficient damping of the closed-loop system. However, since in this case the set-point response is not of concern, because it is equal to the one of the first loop (because of the ratio control architecture), it might be convenient to choose a higher value of M_s, for example $M_s = 2.5$, in order to obtain a higher value of K_{p2} and therefore a lower value of the gain of $F(s)$. Actually, choosing a value of $M_s = 2.5$ is still sensible, as a higher value is generally obtain by applying the well-known Ziegler–Nichols tuning formula (Åström and Hägglund, 1995).

It has to be stressed that, in any case, by exploiting Expression (9.76), the proposed control architecture can be adopted with different tuning strategies for the two controllers C_1 and C_2, achieving in any case the desired ratio in the presence of a set-point change. Thus, the user might apply its know-how in tuning the two controllers, without impairing the effectiveness of the methodology.

For example, detuning the controller of the first loop implies that when a load disturbance occurs on the first process, a slower rejection is obtained, but the desired ratio is kept better during the transient. In addition, if a process is lag-dominant, the use of a pole-zero cancellation in the feedback control design provides a slow load disturbance response. In this case an alternative tuning rule can be conveniently employed, for example the one proposed in (Skogestad, 2003). Besides, an available more accurate model of the processes can be fully exploited by adopting Expression (9.76).

Simulation Results

As a first example to illustrate the use of the Dynamic Blend Station approach, the following two FOPDT processes are considered ($a = 1$):

$$P_1(s) = \frac{1}{4s+1}e^{-3s} \qquad P_2(s) = \frac{1}{6s+1}e^{-2s}. \tag{9.80}$$

By applying the proposed method and the proposed tuning procedure (with $M_s = 2.5$), the following parameters of the controllers are determined: $K_{p1} = 0.57$, $T_{i1} = 4$, $K_{p2} = 2.58$, $T_{i2} = 6$. Consequently, we obtain

$$F(s) = 0.33e^{-s}.$$

A unit step has been applied to the set-point signal at time $t = 0$ and then to the load disturbance signals d_1 and d_2 at time $t = 30$ and $t = 70$ respectively. The two process outputs are shown in Figure 9.28, while the two manipulated variables are plotted in Figure 9.29. It appears that a perfect ratio control is

9.3 Ratio Control   285

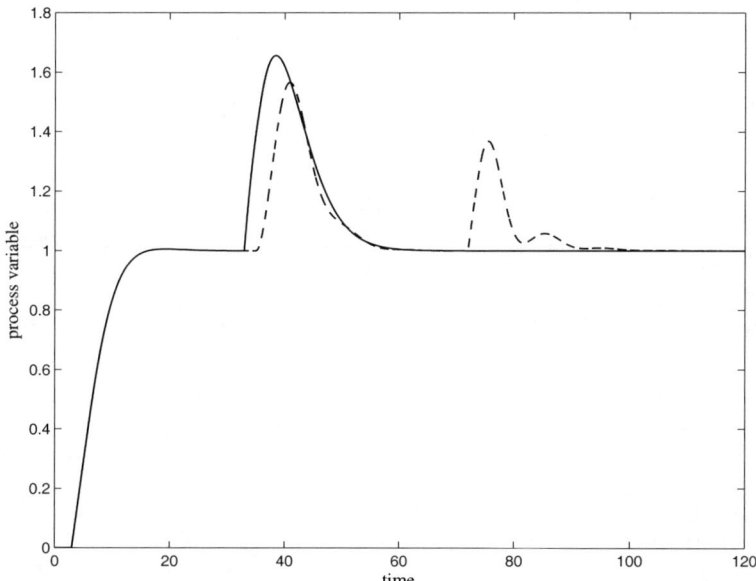

Fig. 9.28. Process variables obtained for the first example of the Dynamic Blend Station approach. Solid line: $y_1(t)$; dashed line: $y_2(t)$.

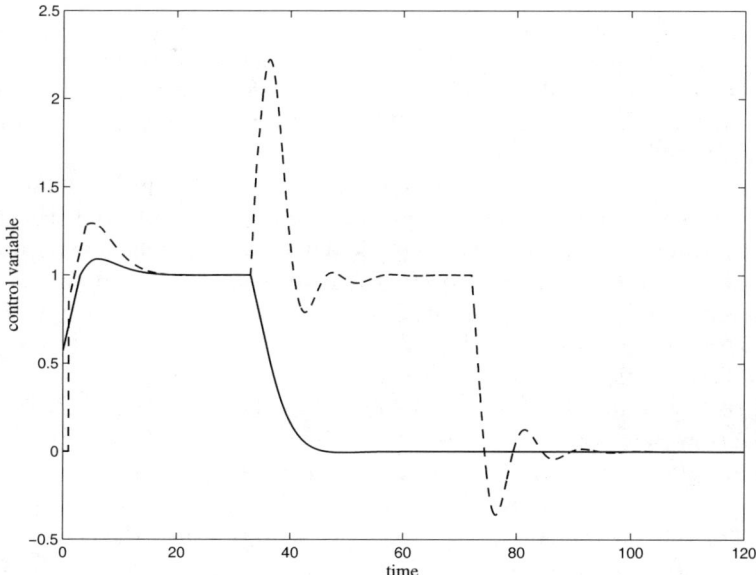

Fig. 9.29. Control variables obtained for the first example of the Dynamic Blend Station approach. Solid line: $u_1(t)$; dashed line: $u_2(t)$.

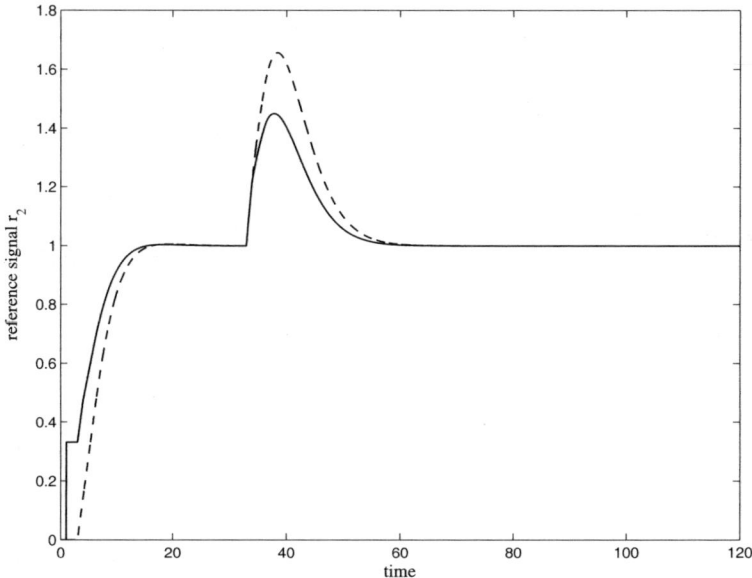

Fig. 9.30. Reference signal $r_2(t)$ obtained for the first example Solid line: dynamic Blend Station; dashed line: standard ratio control.

achieved in the presence of a set-point change, as expected, but the performance is very satisfactory even in the presence of load disturbances. For the sake of comparison, results obtained by the classic ratio control scheme (see Figure 9.13) are shown in Figures 9.31 and 9.32 (note that the PI controllers have been tuned as for the new method).

It appears that the worst performance obtained for the set-point change is not counterbalanced by a better performance in the presence of load disturbances. The value of the performance index J calculated over the whole experiment is 4.80 for the dynamic Blend Station (actually, it is $J = 0$ for just the set-point step response and $J = 2.46$ for the load disturbance d_1 transient response) and 8.55 for the classic one (with $J = 2.44$ for just the set-point step response and $J = 3.77$ for the load disturbance d_1 response).

Indeed, the dynamic Blend Station scheme provides in this case a better load disturbance rejection than the classic one. This is explained by the fact that a somewhat aggressive tuning has been selected for the second loop and, when a load disturbance occurs in the first loop, the dynamic Blend Station scheme provides for the second loop a lower reference signal r_2 than the classic scheme (see Figure 9.30, where obviously r_2 is equal to y_1 for the classic scheme). This is also the reason for the lower control effort that is required by the dynamic Blend Station scheme.

It has to be stressed that it cannot be claimed that the dynamic Blend Station technique provides in general a performance better than the classic one when a load disturbance occurs in the first loop. Indeed, with a proper tuning (*i.e.*, by detuning the second loop at the expense of performance in the set-point following task and in the rejection of a load disturbance affecting process P_2) the typical scheme could provide a smaller value of J.

However, it can be noted that the proposed tuning method provides practically the best performance with the control scheme of Figure 9.27. If the gain K is selected in order to minimise J when a step load disturbance occurs on the process P_1, the values $K = 0.30$ and $J = 2.44$ are obtained, which are very close to the value obtained with Formula (9.79).

As a second example, two systems that are not of first order are considered:

$$P_1(s) = \frac{1}{(s+1)^4} \qquad P_2(s) = \frac{1}{(s+1)^2}e^{-s}. \qquad (9.81)$$

An estimate of a FOPDT transfer function has been obtained by means of the area method (see Section 7.2.1). The values $K_1 = 1$, $T_1 = 1.84$, $L_1 = 1.92$, $K_2 = 1$, $T_2 = 1.39$ and $L_2 = 1.57$ result. It appears that these processes are quite difficult to control as they have a large normalised dead time (*i.e.*, the ratio between the dead time and the time constant of the process).

The tuning procedure described previously (again with $M_s = 2.5$) has been applied by considering the estimated process models, resulting in $K_{p1} = 0.51$, $T_{i1} = 1.84$, $K_{p2} = 0.77$, $T_{i2} = 1.39$, and

$$F(s) = 0.48e^{-0.35s}.$$

A unit step has been applied to the set-point signal at time $t = 0$ s and then to the load disturbance signals d_1 and d_2 at time $t = 30$ s and $t = 70$ s respectively. The two process outputs and the two control variables are shown in Figures 9.33 and 9.34 respectively for the dynamic Blend Station and in Figures 9.35 and 9.36 for the standard approach.

It turns out that, despite the fact that the two processes are not FOPDT, and therefore a perfect ratio control cannot be achieved, the performance achieved with the dynamic Blend Station scheme is still very satisfactory.

Basically, the same considerations made for the previous example applies also in this case. Indeed, with respect to the whole experiment, the performance index (9.69) results to be $J = 5.77$ for the dynamic Blend Station scheme (where the set-point step response contributes for 0.51 and the load disturbance d_1 response for 2.33), while for the classic one it is $J = 9.31$ (where the set-point step response contributes for 2.86 and the load disturbance d_1 response for 3.52).

Note that the value of K that minimises J in the presence of a step load disturbance d_1 only is $K = 0.53$, which results in $J = 2.32$.

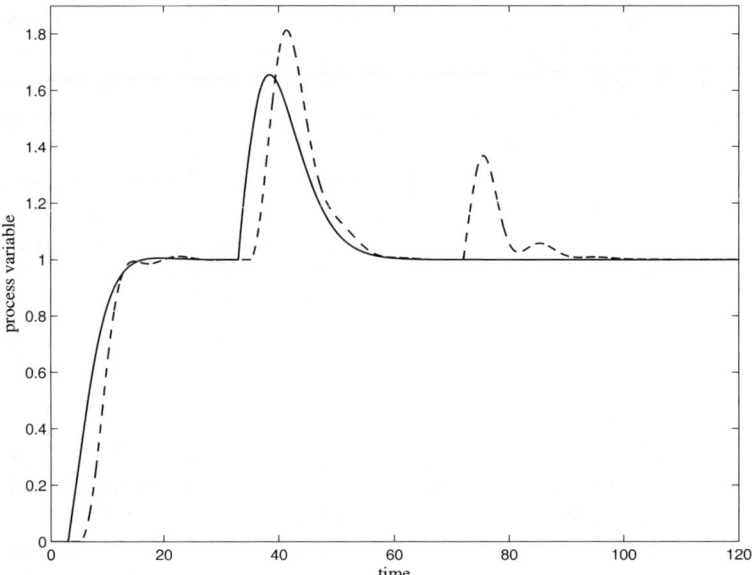

Fig. 9.31. Process variables obtained for the first example with the standard ratio control approach. Solid line: $y_1(t)$; dashed line: $y_2(t)$.

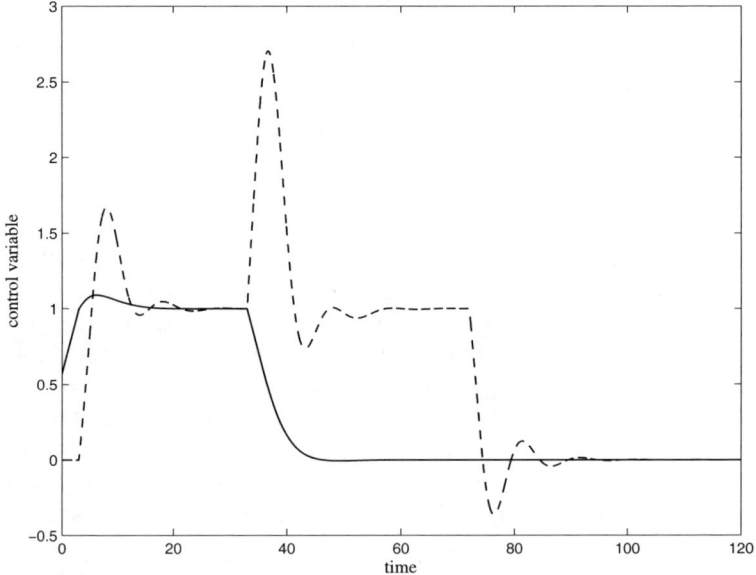

Fig. 9.32. Control variables obtained for the first example with the standard ratio control. Solid line: $u_1(t)$; dashed line: $u_2(t)$.

9.3 Ratio Control 289

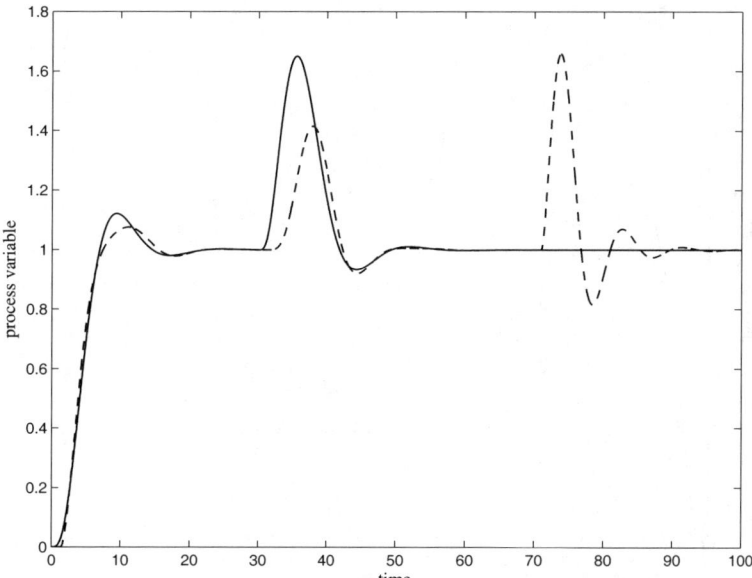

Fig. 9.33. Process variables obtained for the second example of the Dynamic Blend Station approach. Solid line: $y_1(t)$; dashed line: $y_2(t)$.

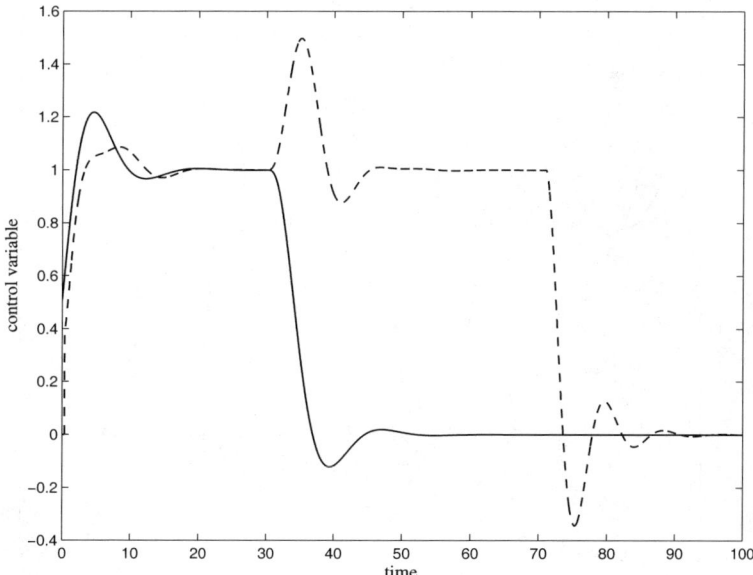

Fig. 9.34. Control variables obtained for the second example of the Dynamic Blend Station approach. Solid line: $u_1(t)$; dashed line: $u_2(t)$.

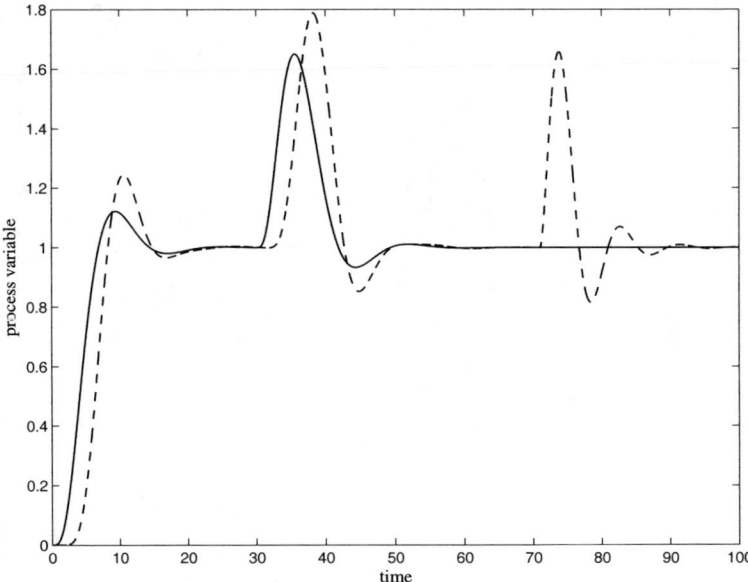

Fig. 9.35. Process variables obtained for the second example with the standard ratio control approach. Solid line: $y_1(t)$; dashed line: $y_2(t)$.

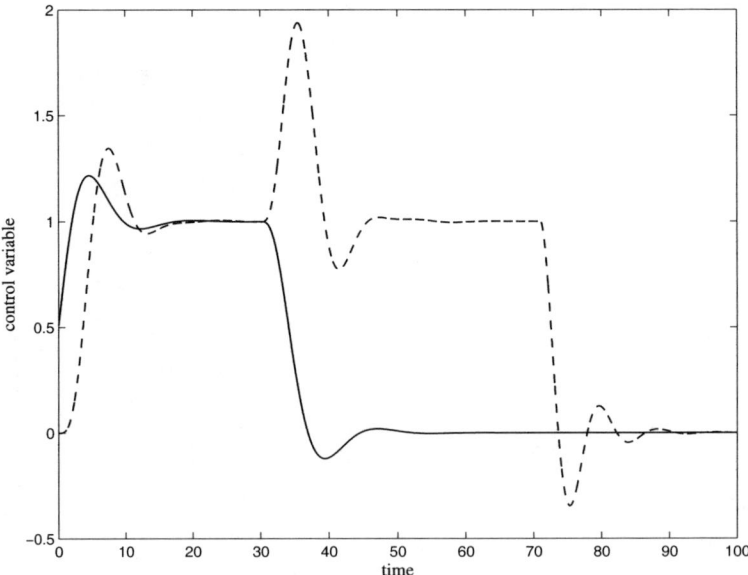

Fig. 9.36. Control variables obtained for the second example with the standard ratio control. Solid line: $u_1(t)$; dashed line: $u_2(t)$.

Experimental Results

As in the previous section, experimental results have been obtained by considering the level control apparatus described in Section A.1. A time delay of 10 s and 5 s has been added via software at the input of the first and second process respectively, but the two tanks have been swapped with respect to the previous case. Hence, the two FOPDT models adopted are

$$P_1(s) = \frac{2.27}{25s+1}e^{-11s} \qquad P_2(s) = \frac{1.98}{25s+1}e^{-6s}. \tag{9.82}$$

The task to be accomplished is to perform an output transition from 2 V to 2.5 V for the first tank as well as for the second one, maintaining a desired ratio value $a = 1$ during the whole transient response. According to the tuning procedure described above, the PI controller of the first process has been tuned with $K_{p1} = 0.31$ and $T_{i1} = 25$, while that of the second process with $K_{p2} = 1.82$ and $T_{i2} = 25$. Results are shown in Figures 9.37–9.39 for the dynamic Blend Station scheme and in Figures 9.40–9.42 for the classic one (see Figure 9.13). The resulting value of the performance index (obviously the result in both cases is biased due to the noise) is $J = 26.7$ for the dynamic Blend Station (note that $J = 7.4$ for the set-point step response and $J = 9.5$ for the load disturbance d_1 response) and $J = 32.7$ for the classic method (in this case $J = 13.2$ the set-point step response and $J = 9.9$ for the load disturbance d_1 response). Thus, experimental results confirm the effectiveness of the dynamic Blend Station approach.

9.4 Conclusions

Cascade control and ratio control schemes based on PID controllers have been addressed in this chapter. It has been shown that the adoption of recent design methodologies and modifications of the classic control architectures can lead to a faster commissioning of the overall control system and to a significant improvement in the control performance, although the general know-how and the simplicity of use of the basic PID control algorithm is retained.

292   9 Control Structures

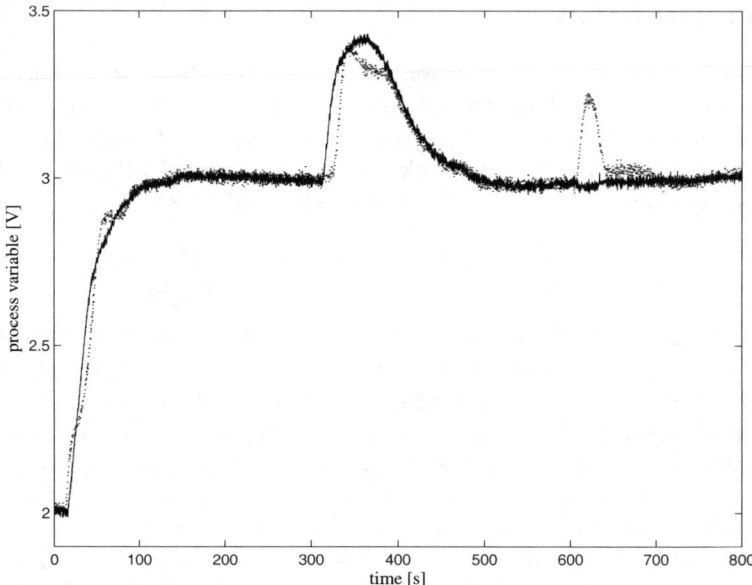

Fig. 9.37. Process variables obtained with the double tank apparatus and the dynamic Blend Station. Solid line: $y_1(t)$; dotted line: $y_2(t)$.

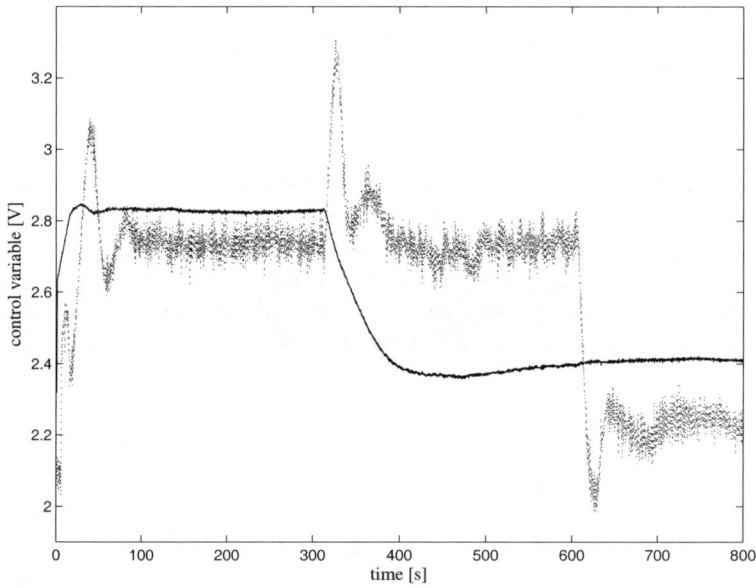

Fig. 9.38. Control variables obtained with the double tank apparatus and the dynamic Blend Station. Solid line: $u_1(t)$; dotted line: $u_2(t)$.

9.4 Conclusions 293

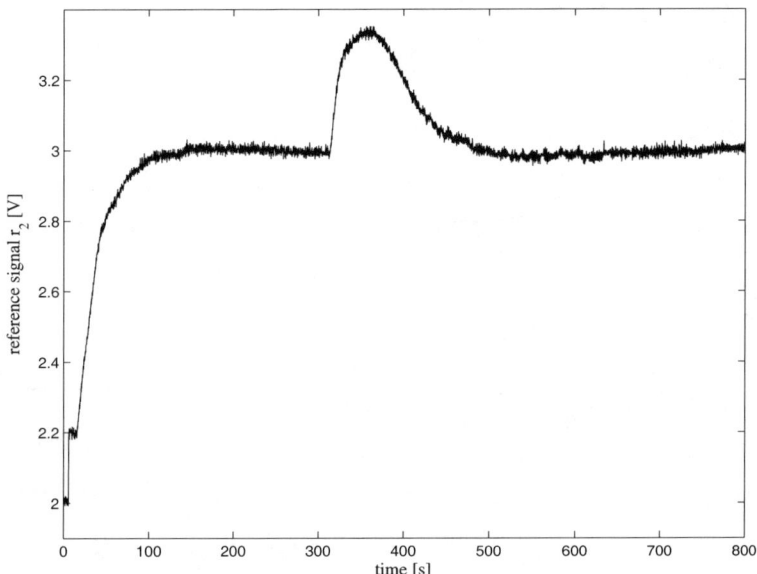

Fig. 9.39. Reference signal $r_2(t)$ obtained with the double tank apparatus and the dynamic Blend Station

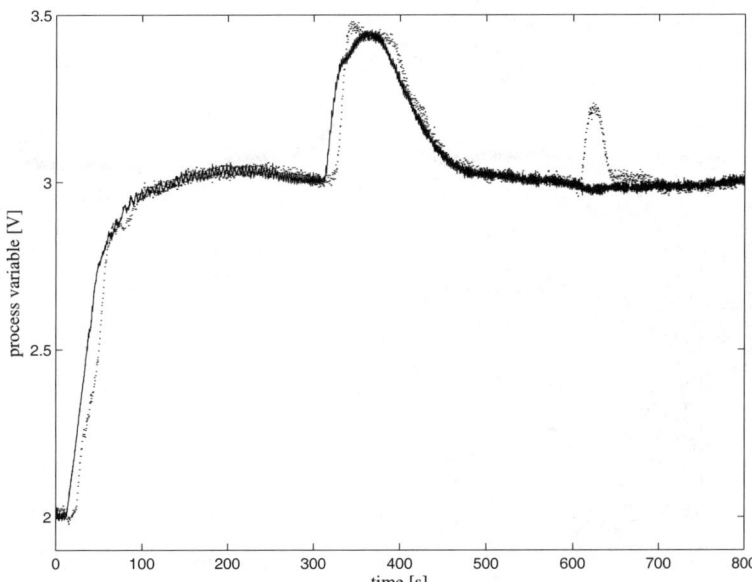

Fig. 9.40. Process variables obtained with the double tank apparatus and the standard ratio controller. Solid line: $y_1(t)$; dotted line: $y_2(t)$.

294   9 Control Structures

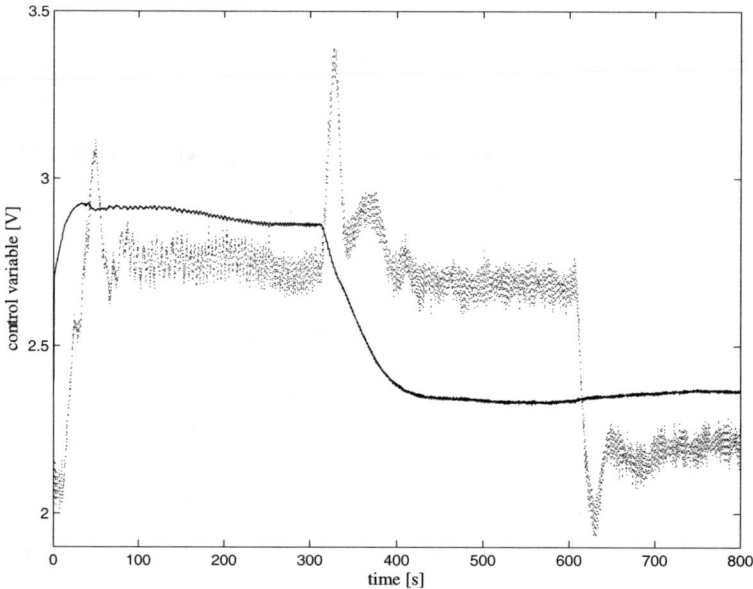

Fig. 9.41. Control variables obtained with the double tank apparatus and the standard ratio controller. Solid line: $u_1(t)$; dotted line: $u_2(t)$.

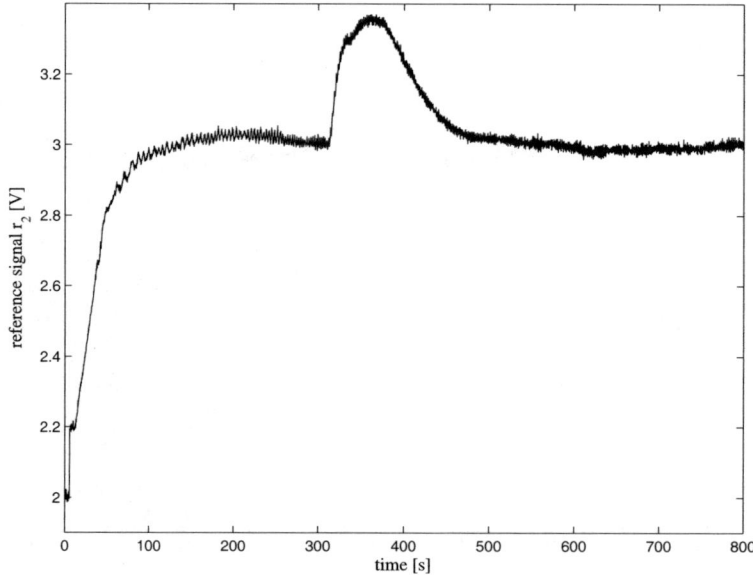

Fig. 9.42. Reference signal $r_2(t)$ obtained with the double tank apparatus and the standard ratio controller

A
Experimental Setups

The laboratory equipment that has been employed to obtain the experimental results shown in the book is described in this appendix. In particular, a double tank apparatus and a oven have been used to implement level control and a temperature control tasks, respectively.

A.1 Level Control Apparatus

The double tank apparatus (made by KentRidge Instruments) adopted for level control experiments is shown in Figure A.1. Although the setup consists of two small perspex tower-type tanks (whose area is $A = 40$ cm^2), only one at a time has been adopted in the experiments. Each tank is filled with water by means of a pump whose speed is set by a DC voltage (the manipulated variable), in the range 0-5 V, through a Pulse Width Modulation (PWM) circuit. The tank is fitted with an outlet at the base in order for the water to return to a reservoir. The measure of the level h of the water is given by a capacitive-type probe that provides an output signal between 0 (empty tank) and 5 V (full tank). For the sake of simplicity, the level variable is expressed in Volts.

The process can be modelled by the following differential equation:

$$A\frac{dh}{dt} = Q_i - Q_o \tag{A.1}$$

where Q_i and Q_o are the input (manipulated variable) and output flow rate respectively. Note that the system is actually nonlinear, since the output flow rate depends on the square root of the level, *i.e.*,

$$Q_o = a\sqrt{2gh}$$

where a is the cross sectional area of the outflow orifice and g is the gravitational constant. The employed control systems are implemented by means of a PC-based controller whose sampling time is 5 ms.

It is worth noting that the two tanks (both of them have been employed in the experiments shown in the book) have a different dynamics. Further, different models arise depending on the adopted identification experiment (see Chapter 7).

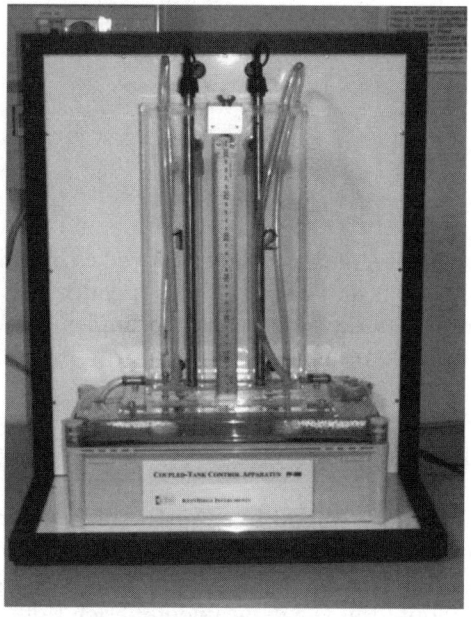

Fig. A.1. The double tank apparatus employed for level control experiments

A.2 Temperature Control Apparatus

A laboratory scale oven has been employed to implement temperature control tasks. It consists of an aluminium plate that is heated by two resistors attached to it. The plate is inserted in an insulating box (whose dimensions are 33 × 21 × 16.5 cm). A fan (which has not been adopted in the experiments) is present in order to provide a fast cooling of the apparatus. The temperature of the plate is measured by means of a thermocouple. The overall process is sketched in Figure A.2. The same PC-based controller (with a sampling time of 5 ms) of the level control experiments has been employed.

The temperature process can be modelled by the following equations:

$$\begin{aligned} C_p \dot{\tau}_p &= P - G_{pb}(\tau_p - \tau_b) \\ C_b \dot{\tau}_b &= G_{pb}(\tau_p - \tau_b) + G_{be}(\tau_e - \tau_b) \end{aligned} \tag{A.2}$$

where τ_p is the temperature of the plate, τ_b is the temperature of the box, τ_e is the temperature of the external environment, C_p and C_c are the heat capacities of the plate and of the box respectively, and G_{pb} and G_{be} are the thermal conductances between the plate and the box and between the box and the external environment, respectively. Finally, P is the thermal power provided by the heating elements.

The control task consists of controlling the temperature of the plate by acting on the thermal power of the heating elements, namely, by manipulating the voltage across the resistors. As for the level control task, for the sake of simplicity the input and output are expressed in Volts (both in the range 0-5 V). It has to be stressed that there is not active cooling and therefore the process is asymmetric (the dynamics is different depending on the fact that the plate has to be heated or cooled). Because of this significant nonlinearity, the apparatus has not been adopted for those methods that relies on a linear dynamics of the process.

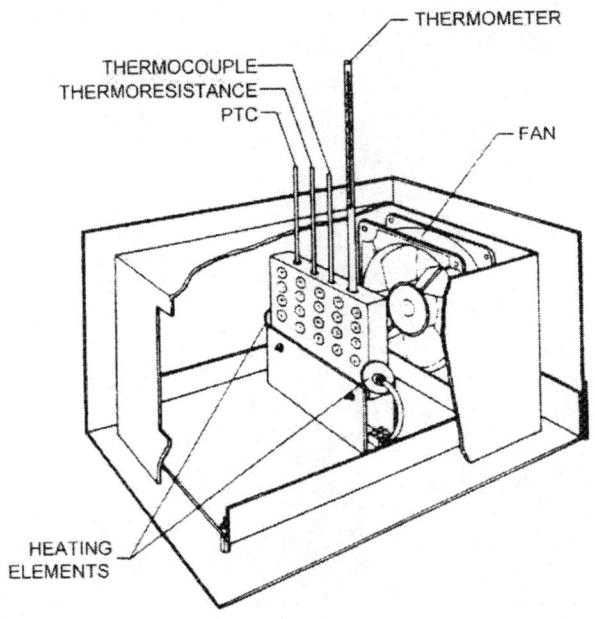

Fig. A.2. The oven employed for level control experiments

References

Ahmed, S., B. Huang and S. L. Shah (2006). Parameter and delay estimation of continuous-time models using a linear filter. *Journal of Process Control* **16**(4), 323–331.

Altmann, W. (2005). *Practical Process Control for Engineers and Technicians*. Newnes. The Netherlands.

Ang, K. H., G. Chong and Y. Li (2005). PID control systems analysis, design, and technology. *IEEE Transactions on Control Systems Technology* **13**, 559–576.

Araki, M. (1988). Two degree-of-freedom PID controller. *Systems, Control, and Information* **42**, 18–25.

Åström, K. and T. Hägglund (2000a). Benchmark systems for PID control. In: *Preprints IFAC Workshop on Digital Control PID'00*. Terrassa, E. pp. 181–182.

Åström, K. J. and B. Wittenmark (1995). *Adaptive Control*. Addison-Wesley.

Åström, K. J. and B. Wittenmark (1997). *Computer-Controlled Systems - Theory and Design*. Prentice Hall. Upper Saddle River, USA.

Åström, K. J. and T. Hägglund (1984). Automatic tuning of simple controllers with specification on phase and amplitude margins. *Automatica* **20**(5), 655–651.

Åström, K. J. and T. Hägglund (1995). *PID Controllers: Theory, Design and Tuning*. ISA Press. Research Triangle Park, USA.

Åström, K. J. and T. Hägglund (2000b). The future of PID control. In: *Preprints IFAC Workshop on Digital Control PID'00*. Terrassa, E. pp. 19–30.

Åström, K. J. and T. Hägglund (2004). Revisiting the Ziegler–Nichols step response method for PID control. *Journal of Process Control* **14**, 635–650.

Åström, K. J. and T. Hägglund (2006). *Advanced PID Control*. ISA Press. Research Triangle Park, USA.

Åström, K. J., H. Panagopoulos and T. Hägglund (1998). Design of PI controllers based on non-convex optimization. *Automatica* **34**(5), 585–601.

Åström, K. J., T. Hägglund, C. C. Hang and W. K. Ho (1993). Automatic tuning and adaptation for PID controllers - a survey. *Control Engineering Practice* **1**, 699–714.

Atherton, D. P. (2000). Relay autotuning: a use of old ideas in a new setting. *Transactions of the Institute of Measurement and Control* **22**(1), 103–122.

Bennett, S. (2000). The past of PID controllers. In: *Preprints IFAC Workshop on Digital Control PID'00*. Terrassa, E. pp. 3–13.

Bequette, B. W. (2003). *Process Control - Modeling, Design, and Simulation.* Prentice Hall. USA.

Björklund, S. (2003). A survey and comparison of time-delay estimation methods in linear systems. Technical Report Licentiate Thesis no. 1061. Department of Electrical Engineering, Linkping University.

Bohn, C. and D. P. Atherton (1995). An analysis package comparing PID antiwindup strategies. *IEEE Control Systems Magazine* pp. 34–40.

Chen, D. and D. E. Seborg (2002). PI/PID controller design based on direct synthesis and disturbance rejection. *Industrial and Engineering Chemistry Research* **41**, 4807–4822.

Choudhury, M. A. A. S., N. F. Thornhill and S. Shah (2005). Modelling valve stiction. *Control Engineering Practice* **13**(5), 641–658.

Choudhury, M. A. A. S., S. L. Shah and N. F. Thornhill (2004). Detection and diagnosis of system nonlinearities using higher order statistics. *Automatica* **40**, 1719–1728.

Corriou, J.-P. (2004). *Process Control - Theory and Applications.* Springer. USA.

Corripio, A. B. (2001). *Tuning of Industrial Control Systems.* ISA Press. USA.

Daubechies, I. (1992). *Ten Lectures on Wavelets.* SIAM Press. USA.

Devasia, S. (2002). Should model-based inverse inputs be used as feedforward under plant uncertainties?. *IEEE Transactions on Automatic Control* **47**(11), 1865–1871.

Devasia, S., D. Chen and B. Paden (1996). Nonlinear inversion-based output tracking. *IEEE Transactions on Automatic Control* **41**, 930–943.

Dorf, R. C. and R. H. Bishop (1995). *Modern Control Systems.* Addison-Wesley.

Driankov, D., H. Hellendoorn and M. Reinfrank (1993). *An Introduction to Fuzzy Control.* Springer-Verlag. USA.

Ellis, G. (2004). *Control System Design Guide.* Elsevier. USA.

Eriksson, P.-G. and A. J. Isaksson (1994). Some aspects of control loop performance monitoring. In: *Proceedings IEEE International Conference on Control Applications.* Glasgow, UK. pp. 1029–1034.

Forsman, K. and A. Stattin (1999). A new criterion for detecting oscillations in control loops. In: *Proceedings 5th European Control Conference.* Karlsruhe, D.

Friman, M. and K. V. Waller (1997). A two-channel relay for autotuning. *Industrial and Engineering Chemistry Research* **36**, 2662–2671.

Gerry, J. and F. G. Shinskey (2005). PID controller specification (white paper). *available on-line at www.expertune.com/PIDspec.htm.*

Gomes, V. G. (1985). Controlling fired heaters. *Chemical Engineering* pp. 63–68.

Hägglund, T. (1995). A control-loop performance monitor. *Control Engineering Practice* **3**, 1543–1551.

Hägglund, T. (1999). Automatic detection of sluggish control loops. *Control Engineering Practice* **7**, 1505–1511.

Hägglund, T. (2001). The blend station - a new ratio control structure. *Control Engineering Practice* **9**, 1215–1220.

Hägglund, T. (2002). A friction compensator for pneumatic control valves. *Journal of Process Control* **12**, 897–904.

Hägglund, T. (2005). Industrial implementation of on-line performance monitoring tool. *Control Engineering Practice* **13**(11), 1383–1390.

Hägglund, T. and K. J. Åström (2000). Supervision of adaptive control algorithms. *Automatica* **36**(2), 1171–1180.

Hang, C. C., A. P. Loh and V. U. Vasnani (1994). Relay feedback auto-tuning of cascade controllers. *IEEE Transactions on Control Systems Technology* **2**, 42–45.

Hang, C.-C. and L. Cao (1996). Improvement of transient response by means of variable set point weighting. *IEEE Transactions on Industrial Electronics* **43**, 477–484.

Hang, C. C., K. J. Åström and Q. G. Wang (2002). Relay feedback auto-tuning of process controllers - a tutorial review. *Journal of Process Control* **12**, 143–162.

Hang, C. C., K. J. Åström and W. K. Ho (1991). Refinements of the Ziegler–Nichols tuning formula. *IEE Proceedings - Control Theory and Applications* **138**(2), 111–118.

Hansson, A., P. Gruber and J. Todtli (1994). Fuzzy anti-reset windup for PID controllers. *Control Engineering Practice* **2**(3), 389–396.

Hanus, R., M. Kinnaert and J.-L. Henrotte (1987). Conditioning technique, a general anti-windup and bumpless transfer method. *Automatica* **23**(6), 729–739.

Harriot, P. (1964). *Process control*. McGraw-Hill. USA.

Harris, T. J. (1989). Assessment of control loop performance. *The Canadian Journal of Chemical Engineering* **67**, 856–861.

Harris, T. J., C. T. Seppala and L. D. Desborough (1999). A review of performance monitoring and assessment techniques for univariate and multivariate control systems. *Journal of Process Control* **9**(1), 1–17.

Ho, W. K. and W. Xu (1998). PID tuning for unstable processes based on gain and phase-margin specifications. *IEE Proceedings - Control Theory and Applications* **145**(5), 392–396.

Hodel, A. Scottedward and C. E. Hall (2001). Variable-structure PID control to prevent integrator windup. *IEEE Transactions on Industrial Electronics* **48**(2), 442–451.

Homaifar, A. and E. McCormick (1995). Simultaneous design of membership functions and rule sets for fuzzy controllers using genetic algorithms. *IEEE Transactions on Fuzzy Systems* **3**(2), 129–139.

Horch, A. (1999). A simple method for detection of stiction in control valves. *Control Engineering Practice* **7**, 1221–1231.

Horch, A. (2000). Condition monitoring of control loops. PhD thesis. Royal Institute of Technology. Stockholm, S.

Horch, A. (2001). Detection of valve stiction in integrating processes. In: *Proceedings 6th European Control Conference*. Porto, P. pp. 1327–1332.

Huang, B. (2003). A pragmatic approach towards assessment of control loop performance. *International Journal of Adaptive Control and Signal Processing* **17**, 589–608.

Huang, B. and S. L. Shah (1999). *Performance Assessment of Control Loop*. Springer. UK.

Huang, C.-T. and C.-J. Chou (1994). Estimation of the underdamped second-order parameters from the system transient. *Industrial and Engineering Chemistry Research* **33**, 174–176.

Huang, C.-T. and M.-F. Huang (1993). Estimation of the second-order parameters from the process transient by simple calculation. *Industrial and Engineering Chemistry Research* **32**, 228–230.

Huang, H.-P. and J.-C. Jeng (2002). Monitoring and assessment of control performance for single loop systems. *Industrial and Engineering Chemistry Research* **41**, 1297–1309.

Huang, H.-P., M.-W. Lee and C.-L. Chen (2001). A system of procedures for identification of simple models using transient step response. *Industrial and Engineering Chemistry Research* **40**, 1903–1915.

Hunt, L. R. and G. Meyer (1997). Stable inversion for nonlinear systems. *Automatica* **33**(8), 1549–1554.

Hunt, L. R., G. Meyer and R. Su (1996). Noncausal inverses for linear systems. *IEEE Transactions on Automatic Control* **41**, 608–611.

Ingimundarson, A. and T. Hägglund (2002). Performance comparison between PID and dead-time compensating controllers. *Journal of Process Control* **12**, 887–895.

Isaksson, A. J. and S. F. Graebe (1999). Analytical PID parameter expressions for higher order systems. *Automatica* **35**, 1121–1130.

Isaksson, A. J. and S. F. Graebe (2002). Derivative filter is an integral part of PID design. *IEE Proceedings - Control Theory and Applications* **149**(1), 41–45.

Jelali, M. (2006). An overview of control performance assessment technology and industrial applications. *Control Engineering Practice* **14**, 441–466.

Johnson, M. A. and M. H. Moradi (eds.) (2005). *PID control - New Identification and Design Methods*. Springer-Verlag. London, Great Britain.

Katebi, M. R. and A. W. Ordys (1996). Minimum variance control. In: *The Control Handbook (W. S. Levine ed.)*. CRC Press. Boca Raton, FL. pp. 1089–1096.

Kaya, I. (2001). Improving performance using cascade control and a smith predictor. *ISA Transactions* **40**, 223–234.

Kaya, I. and D. P. Atherton (2005). Improved cascade control structure for controlling unstable and integrating processes. In: *Proceedings IEEE International Conference on Decision and Control - European Control Conference*. Sevilla, E. pp. 7133–7138.

Kaya, I., N. Tan and D. P. Atherton (2005). Improved cascade control structure and controller design. In: *Proceedings IEEE International Conference on Decision and Control - European Control Conference*. Sevilla, E. pp. 3055–3060.

Ko, B.-S. and T. F. Edgar (2004). PID control performance assessment: the single-loop case. *AIChE Journal* **50**, 1211–1218.

Kothare, M. V., P. J. Campo, M. Morari and C. N. Nett (1994). A unified framework for the study of anti-windup design. *Automatica* **30**(12), 1869–1883.

Kozub, D. J. (2002). Controller performance monitoring and diagnosis: industrial perspective. In: *Preprints of the 15th IFAC World Congress on Automatic Control*. Barcelona, E.

Krishnaswami, P. R., G. P. Rangaiah, R. K. Jha and P. D. Deshpande (1990). When to use cascade control. *Industrial and Engineering Chemistry Research* **29**, 2163–2166.

Kristiansson, B. and B. Lennartson (2001). Robust and optimal tuning of PI and PID controllers. *IEE Proceedings - Control Theory and Applications* **149**(1), 17–25.

Kristiansson, B. and B. Lennartson (2006). Evaluation and simple tuning of PID controllers with high-frequency robustness. *Journal of Process Control* **16**, 91–102.

Kuehl, P. and A. Horch (2005). Detection of sluggish control loops - experiences and improvements. *Control Engineering Practice* **13**, 1019–1025.

Kwak, H. J., S. W. Sung and I.-B. Lee (1997). On-line process identification and autotuning for integrating processes. *Industrial and Engineering Chemistry Research* **36**, 5329–5338.

Lee, Y., S. Oh and S. Park (2002). Enhanced control with a general cascade control structure. *Industrial and Engineering Chemistry Research* **41**, 2679–2688.
Lee, Y., S. Park and M. Lee (1998a). PID controller tuning to obtain desired closed loop responses for cascade control systems. *Industrial and Engineering Chemistry Research* **37**, 1859–1865.
Lee, Y., S. Park, M. Lee and C. Brosilow (1998b). PID controller tuning for desired closed-loop responses for SI/SO systems. *AIChE Journal* **44**(1), 106–115.
Leva, A. (1993). PID autotuning algorithm based on relay feedback. *IEE Proceedings - Control Theory and Applications* **140**(5), 328–338.
Leva, A. (2005). Autotuning process controller with improved load disturbance rejection. *Journal of Process Control* **15**, 223–234.
Leva, A. and A. M. Colombo (1999). Methods for optimising set-point weights in ISA-PID autotuners. *IEE Proceedings - Control Theory and Applications* **146**(2), 137–146.
Leva, A. and A. M. Colombo (2001). IMC-based synthesis of the feedback block of ISA-PID regulators. In: *Proceedings 6th European Control Conference*. Porto, P. pp. 196–201.
Leva, A., C. Cox and A. Ruano (2001). Hands-on PID autotuning: a guide to better utilisation. Technical report. IFAC Technical Brief, available at www.ifac-control.org.
Lewin, D. R. and C. Scali (1988). Feedforward control in the presence of uncertainty. *Industrial and Engineering Chemistry Research* **27**, 2323–2331.
Lewis, F. L. (1996). Optimal control. In: *The Control Handbook (W. S. Levine ed.)*. CRC Press. Boca Raton, FL. pp. 759–778.
Li, W., E. Eskinat and W. L. Luyben (1991). An improved autotune identification method. *Industrial and Engineering Chemistry Research* **30**, 1530–1541.
Liu, T., D. Gu and W. Zhang (2005). Decoupling two-degree-of-freedom control strategy for cascade control systems. *Journal of Process Control* **15**, 159–167.
Ljung, L. (1996). System identification. In: *The Control Handbook (W. S. Levine ed.)*. CRC Press. Boca Raton, FL. pp. 1033–1054.
Luyben, M. L. and W. L. Luyben (1997). *Essentials of Process Control*. McGraw-Hill. USA.
Luyben, W. L. (1987). Derivation of transfer function model for highly nonlinear distillation columns. *Industrial and Engineering Chemistry Research* **26**, 2490–2495.
Luyben, W. L. (2001a). Effect of derivative algorithm and tuning selection on the PID control of dead-time processes. *Industrial and Engineering Chemistry Research* **40**, 3605–3611.
Luyben, W. L. (2001b). Getting more information from relay-feedback tests. *Industrial and Engineering Chemistry Research* **40**, 4391–4402.
Macvicar-Whelan, P. J. (1976). Fuzzy sets for man-machine interaction. *International Journal of Man-Machine Studies* **8**, 687–697.
Majhi, S. and D. P. Atherton (2000). Online tuning of controllers for an unstable FOPDT process. *IEE Proceedings - Control Theory and Applications* **147**(4), 421–427.
Marlin, T. E. (2000). *Process Control: Designing Processes and Control Systems for Dynamic Performance*. McGraw-Hill. USA.
Miao, T. and D. E. Seborg (1999). Automatic detection of excessively oscillatory feedback loops. In: *Proceedings IEEE International Conference on Control Applications*. Kohala Coast, HW. pp. 359–364.

Mitchell, M. (1998). *An Introduction to Genetic Algorithms*. MIT Press. USA.
Morari, M. and E. Zafiriou (1989). *Robust Process Control*. Prentice Hall. Englewood Cliffs, NJ.
O'Dwyer, A. (2006). *Handbook of PI and PID Tuning Rules*. Imperial College Press.
Ogunnaike, B. A. and W. H. Ray (1994). *Process Dynamics, Modeling, and Control*. Oxford University Press. USA.
Palmor, Z. J. (1996). Time-delay compensation - Smith predictor and its modifications. In: *The control handbook (W. S. Levine ed.)*, CRC Press. pp. 224–237.
Panagopoulos, H., K. J. Åström and T. Hägglund (2002). Design of PID controllers based on constrained optimisation. *IEE Proceedings - Control Theory and Applications* **149**(1), 32–40.
Panda, R. C. (2006). Estimation of parameters of under-damped second order plus dead time processes using relay feedback. *Computers and Chemical Engineering* **30**(5), 832–837.
Panda, R. C. and C.-C. Yu (2003). Analytical expressions for relay feed back responses. *Journal of Process Control* **13**, 489–501.
Panda, R. C. and C.-C. Yu (2005). Shape factor of relay response curves and its use in autotuning. *Journal of Process Control* **15**, 893–906.
Panda, R. C., C.-C. Yu and H.-P. Huang (2004). PID tuning rules for SOPDT systems: review and some new results. *ISA Transactions* **43**, 283–295.
Park, H. I., S. W. Sung, I.-B. Lee and J. Lee (1997). On-line process identification using Laguerre series for automatic tuning of the proportional–integral–derivative controller. *Industrial and Engineering Chemistry Research* **36**, 101–111.
Patwardhan, R. S. and S. L. Shah (2002). Issues in performance diagnostics of model-based controllers. *Journal of Process Control* **12**, 413–417.
Paulonis, M. A. and J. W. Cox (2003). A practical approach for large-scale controller performance assessment, diagnosis, and improvement. *Journal of Process Control* **13**, 155–168.
Peng, Y., D. Vrancic and R. Hanus (1996). Anti-windup, bumpless, and conditioned transfer techniques for PID controllers. *IEEE Control Systems Magazine* pp. 48–57.
Perez, H. and S. Devasia (2003). Optimal output transitions for linear systems. *Automatica* **39**, 181–192.
Petersson, M., K.-E. Årzèn and T. Hägglund (2001). Assessing measurements for feedforward control. In: *Proceedings of the 6th European Control Conference*. Porto, P.
Petersson, M., K.-E. Arzen and T. Hägglund (2003). A comparison of two feedforward control structure assessment methods. *International Journal of Adaptive Control and Signal Processing* **17**, 609–624.
Pfeiffer, B.-M. (1999). Towards Plug&Control: selftuning temperature controller for plc. In: *Proceedings 5th European Control Conference*. Karlsruhe, D.
Pfeiffer, B.-M. (2000). Towards 'plug and control': self-tuning temperature controller for PLC. *International Journal of Adaptive Control and Signal Processing* **14**, 519–532.
Piazzi, A. and A. Visioli (2000). Minimum-time system-inversion-based motion planning for residual vibration reduction. *IEEE/ASME Transactions on Mechatronics* **5**(1), 12–22.
Piazzi, A. and A. Visioli (2001a). Optimal inversion-based control for the set-point regulation of nonminimum-phase uncertain scalar systems. *IEEE Transactions on Automatic Control* **46**, 1654–1659.

Piazzi, A. and A. Visioli (2001b). Optimal noncausal set-point regulation of scalar systems. *Automatica* **37**(1), 121–127.
Piazzi, A. and A. Visioli (2001c). Robust set-point constrained regulation via dynamic inversion. *International Journal of Robust and Nonlinear Control* **11**, 1–22.
Piazzi, A. and A. Visioli (2005). Using stable input-output inversion for minimum-time feedforward constrained regulation of scalar systems. *Automatica* **41**(2), 305–313.
Piazzi, A. and A. Visioli (2006). A noncausal approach for PID control. *Journal of Process Control* **16**, 831–843.
Press, W. H., S. A. Teukolsky, W. T. Vetterling and B. P. Flannery (1995). *Numerical Recipes: The Art of Scientific Computing.* Cambridge University Press. Cambridge, UK.
Qin, S. J. (1998). Control performance monitoring - a review and assessment. *Computers and Chemical Engineering* **23**, 173–186.
Ramakrishnan, V. and M. Chidambaram (2003). Estimation of a SOPDT transfer function model using a single asymmetrical relay feedback test. *Computers and Chemical Engineering* **27**, 1779–1784.
Rangaiah, G. P. and P. R. Krishnaswamy (1994). Estimating second-order plus dead time model parameters. *Industrial and Engineering Chemistry Research* **33**, 1867–1871.
Rangaiah, G. P. and P. R. Krishnaswamy (1996). Estimating second-order dead time parameters from underdamped process transient. *Chemical Engineering Science* **51**(7), 1149–1155.
Rivera, D. E., S. Skogestad and M. Morari (1986). Internal model control. 4. PID controller design. *Industrial and Engineering Chemistry Process Design and Development* **25**(1), 252–265.
Rossi, M. and C. Scali (2005). A comparison of techniques for automatic detection of stiction: simulation and application to industrial data. *Journal of Process Control* **15**(5), 505–514.
Salsbury, T. I. (2005). A practical method for assessing the performance of control loops subject to random load changes. *Journal of Process Control* **15**(4), 393–405.
Scali, C. and D. Semino (1991). Performance of optimal and standard controllers for disturbance rejection in industrial processes. In: *Proceedings IEEE International Conference on Industrial Electronics, Control, and Instrumentation.* Kobe, J. pp. 2033–2038.
Scali, C., G. Marchetti and D. Semino (1999). Relay with additional delay for identification and autotuning of completely unknown processes. *Industrial and Engineering Chemistry Research* **38**, 1987–1997.
Seborg, D. E., T. E. Edgar and D. A. Mellichamp (2004). *Process Dynamics and Control - 2nd edition.* Wiley. USA.
Shen, S.-H., J.-S. Wu and C.-C. Yu (1996). Use of a biased relay feedback for system identification. *American Institute of Chemical Engineering Journal* **42**, 1174–1180.
Shinskey, F. G. (1994). *Feedback Controllers for the Process Industries.* McGraw-Hill. New York.
Shinskey, F. G. (1996). *Process Control Systems - Application, Design, and Tuning.* McGraw-Hill. New York, USA.

Shinskey, F. G. (2000). PID-deadtime control of distributed processes. In: *Preprints IFAC Workshop on Digital Control PID'00*. pp. 14–18.

Silva, G. J., A. Datta and S. P. Bhattacharyya (2002). Robust control design using the PID controller. In: *Proceedings IEEE International Conference on Decision and Control*. Las Vegas, USA. pp. 1313–1318.

Singhal, A. and T. I. Salsbury (2005). A simple method for detecting valve stiction in oscillating control loops. *Journal of Process Control* **15**(4), 371–382.

Skogestad, S. (2003). Simple analytic rules for model reduction and PID controller tuning. *Journal of Process Control* **13**, 291–309.

Srinivasan, K. and M. Chidambaram (2003). Modified relay feedback method for improved system identification. *Computers and Chemical Engineering* **27**, 727–732.

Sundaresan, K. R. and P. R. Krishnaswamy (1978). Estimation of time delay, time constant parameters in time, frequency and Laplace domains. *Canadian Journal of Chemical Engineering* **56**, 257–262.

Sundaresan, K. R., C. Chandra Prasad and P. R. Krishnaswamy (1978). Evaluating parameters from process transients. *Industrial and Engineering Chemistry Process Design and Development* **17**, 237–241.

Sung, S. W., I.-B. Lee and B.-K. Lee (1998). On-line process identification and automatic tuning method for PID controllers. *Chemical Engineering Science* **53**, 1847–1859.

Swanda, A. P. and D. E. Seborg (1999). Controller performance assessment based on set-point response data. In: *Proceedings American Control Conference*. San Diego, CA. pp. 3863–3867.

Taha, O., G. A. Dumont and M. S. Davies (1996). Detection and diagnosis of oscillations in control loops. In: *Proceedings IEEE International Conference on Decision and Control*. Kobe, J. pp. 2432–2437.

Tan, K. K., Q.-G. Wang, C. C. Hang and T. Hägglund (1999). *Advances in PID control*. Springer-Verlag. London, Great Britain.

Tan, K. K., T. H. Lee and R. Ferdous (2000). Simultaneous online automatic tuning of cascade control for open loop stable processes. *ISA Transactions* **39**, 233–242.

Thornhill, N. F. and T. Hägglund (1997). Detection and diagnosis of oscillation in control loop. *Control Engineering Practice* **5**(10), 1343–1354.

Thornhill, N. F., B. Huang and S. L. Shah (2003). Controller performance assessment in set point tracking and regulatory control. *International Journal of Adaptive Control and Signal Processing* **17**, 709–727.

Thyagarajan, T. and C.-C. Yu (2003). Improved autotuning using the shape factor from relay feedback. *Industrial and Engineering Chemistry Research* **42**, 4425–4440.

Thyagarajan, T., C.-C. Yu and H.-P. Huang (2003). Assessment of controller performance: a relay feedback approach. *Chemical Engineering Science* **58**, 497–512.

Tzafestas, S. G. (1994). Fuzzy systems and fuzzy expert control: an overview. *The Knowledge Engineering Review* **9**(3), 229–268.

Visioli, A. (1999). Fuzzy logic based set-point weight tuning of PID controllers. *IEEE Transactions on Systems, Man, and Cybernetics - Part A* **29**, 587–592.

Visioli, A. (2000). Adaptive tuning of fuzzy set-point weighting for PID controllers. In: *Preprints IFAC Workshop on Digital Control PID'00*. Terrassa, E. pp. 513–518.

Visioli, A. (2001a). Optimal tuning of PID controllers for integral and unstable processes. *IEE Proceedings - Control Theory and Applications* **148**(2), 180–184.

Visioli, A. (2001b). Tuning of PID controllers with fuzzy logic. *IEE Proceedings - Control Theory and Applications* **148**(1), 1–8.
Visioli, A. (2003a). Modified anti-windup scheme for PID controllers. *IEE Proceedings - Control Theory and Applications* **150**(1), 49–54.
Visioli, A. (2003b). Time-optimal plug&control for integrating and FOPDT processes. *Journal of Process Control* **13**, 195–202.
Visioli, A. (2004). A new design for a PID plus feedforward controller. *Journal of Process Control* **14**, 455–461.
Visioli, A. (2005a). Design and tuning of a ratio controller. *Control Engineering Practice* **13**, 485–497.
Visioli, A. (2005b). Experimental evaluation of a Plug&Control strategy for level control. In: *Preprints 16th IFAC World Congress on Automatic Control*. Prague, CZ.
Visioli, A. (2005c). Model-based PID tuning for high-order processes: when to approximate. In: *Proceedings IEEE International Conference on Decision and Control - European Control Conference*. Sevilla, E. pp. 7127–7132.
Visioli, A. (2005d). A new ratio control architecture. *Industrial and Engineering Chemistry Research* **44**, 4617–4624.
Visioli, A. (2006). Method for proportional-integral controller tuning assessment. *Industrial and Engineering Chemistry Research* **45**, 2741–2747.
Visioli, A. and A. Piazzi (2003). Improving set-point following performance of industrial controllers with a fast dynamic inversion algorithm. *Industrial and Engineering Chemistry Research* **42**, 1357–1362.
Visioli, A. and A. Piazzi (2005). On the use of dynamic inversion for the improvement of PID control. In: *Preprints 16th IFAC World Congress on Automatic Control*. Prague, CZ.
Visioli, A. and A. Piazzi (2006). An automatic tuning method for cascade control systems. In: *Proceedings IEEE International Conference on Control Applications*. Munich, D.
Visioli, A. and M. Veronesi (1999). Nuove funzionalita' per controllori PID (in italian). *Automazione e Strumentazione* (October), 149–155.
Vivek, S. and M. Chidambaram (2005a). Identification using single symmetrical relay feedback test. *Computers and Chemical Engineering* **29**, 1625–1630.
Vivek, S. and M. Chidambaram (2005b). An improved relay autotuning of PID controllers for unstable FOPDT systems. *Computers and Chemical Engineering* **29**, 2060–2068.
Vrancic, D. (1997). Design of anti-windup and bumpless transfer protection. PhD thesis. University of Ljubljana. Ljubljana, SLO.
Walgama, K. S. and J. Sternby (1990). Inherent observer property in a class of anti-windup compensators. *International Journal of Control* **52**(3), 705–724.
Walgama, K. S., S. Ronnback and J. Sternby (1991). Generalisation of conditioning technique for anti-windup compensator. *IEE Proceedings - Control Theory and Applications* **139**(2), 109–118.
Wallen, A. (2000). Tools for autonomous process control. PhD thesis. Lund Institute of Technology. Lund, S.
Wallen, A. and K. J. Åström (2002). Pulse-step control. In: *Preprints of the 15th IFAC World Congress on Automatic Control*. Barcelona, E.
Wang, L. and W. R. Cluett (1994). Optimal choice of time-scaling factor for linear system approximations using laguerre models. *IEEE Transactions on Automatic Control* **39**(7), 1463–1467.

Wang, L. and W. R. Cluett (2000). *From Plant Data to Process Control*. Trevor and Francis. London (UK).

Wang, Q.-G. and Y. Zhang (2001). Robust identification of continuous systems with dead-time from step responses. *Automatica* **37**, 377–390.

Wang, Q.-G., C.-C. Hang and B. Zou (1997). Low-order modeling from relay feedback. *Industrial and Engineering Chemistry Research* **36**, 375–381.

Wang, Q.-G., C.-C. Hang and Q. Bi (1999a). A technique for frequency response identification from relay feedback. *IEEE Transactions on Control Systems Technology* **7**(1), 122–128.

Wang, Q.-G., C.-C. Hang, S.-A. Zhu and Q. Bi (1999b). Implementation and testing of an advanced relay auto-tuner. *Journal of Process Control* **9**, 291–300.

Wang, Q.-G., H.-W. Fung and Y. Zhang (1999c). Robust estimation of process frequency response from relay feedback. *ISA Transactions* **38**, 3–9.

Wang, Q.-G., T. H. Lee and C. Lin (2003). *Relay Feedback: Analysis, Identification and Control*. Springer. UK.

Wang, Q.-G., X. Guo and Y. Zhang (2001). Direct identification of continuous time delay systems from step responses. *Journal of Process Control* **11**, 531–542.

Yamashita, Y. (2006). An automatic method for detection of valve stiction in process control loops. *Control Engineering Practice* **14**, 503–510.

Yu, C. C. (1999). *Autotuning of PID Controllers*. Springer-Verlag. London, Great Britain.

Zheng, L. (1992). A practical guide to tune of proportional and integral (pi) like fuzzy controllers. In: *Proceedings of the IEEE International Conference on Fuzzy Systems*. pp. 633–640.

Zhuang, M. and D. P. Atherton (1993). Automatic tuning of optimum PID controllers. *IEE Proceedings - Control Theory and Applications* **140**(3), 216–224.

Ziegler, J. G. and N. B. Nichols (1942). Optimum setting for automatic controllers. *ASME Transactions* pp. 759–768.

Zou, Q. and S. Devasia (1999). Preview-based stable-inversion for output tracking of linear systems. *ASME Journal of Dynamic Systems, Measurements, and Control* **121**, 625–630.

Index

actuator saturation, 35, 94, 96
anti-windup, **35**
anticipatory control, 6
area index, 227, 240
area method, 99, 103, 107, 116, 124, **166**, 184, 257, 280
autocorrelation function, 214, 227
automatic reset, 5, 10, 42, 44, 50
automatic tuning, **18**, 145
average residence time, 142

back-calculation, **38**, 41, 44, 50
beta-gamma controller, 13
bias term, 5
blend station, 268
bumpless transfer, 39, 152

cascade control, 15, **251**
command signal generator, 109
conditional integration, **38**, 41, 44, 50
conditioning technique, 39
control effort, 20, 97, 99, 116, 131
control variable, 2
controller zeros, 20
cross limited control, 268

dead zone, 3
dead-time compensator, 16
decay ratio, 227
derivative action, **6**, 8, 15, 16, 19
derivative filter, 9, **19**, 148
derivative gain, 6
derivative kick, 10
derivative term, 14

derivative time constant, 7, 16
digital implementation, 13

feedback control, **2**
feedforward control, 12, 61, **93**, 140
  linear causal, **93**
  linear noncausal, **109**, **130**
  nonlinear causal, **96**
feedforward filter, 2, 62, 97
FOPDT systems, 37, 96, 110, 149, **165**, 233, 240
fuzzy control, 72, 90

gain scheduling, 90
genetic algorithms, 74, 173

half rule, 195
Harris index, 213
hysteresis, 3, 176, 177

ideal form, 7, 9, 13, 19, 22, 28
identification, 18, 145, 149, **165**, 257
idle index, 231, 240
incremental algorithm, 14, 37
integral action, **5**, 15, 35
integral gain, 5
integral term, 13, 45
integral time constant, 7, 16
integrator clamping, *see* conditional integration
integrator windup, 5, **35**, 253
interacting form, 7
Internal Model Control, 33, **194**, 234, 257, 266

inversion, 109, 111, 132
IPDT systems, 110, 149
ISA form, 13

Laguerre functions, 171
lead-lag controller, 141
least-squares, 150, 151, 170, 189, 257
level control, 50, 82, 103, 126, 136, 157, 247, 280, 291
load disturbance, 2, 5, 10
load disturbance detection, 223, 226
load disturbance rejection, 11, 16, 20, 25, 28, 33, 37, 61, 98, 140, 146, 207, 240, 251, 265

Maclaurin series expansion, 198, 257
manipulated variable, 2
master controller, 251
measurement noise, 2, 9, 19, 29, 42, 132, 192
minimum variance control, **210**
model reduction, 170, **193**

noise band, 42, 152, 230
non-interacting form, 7

on–off control, **3**
optimisation, 149, 172, 191, 199, 232, 236, 240, 263
oscillation detection, 223, 227
oscillation diagnosis, 225
output index, 242
output transition time, 96, 100, 111, 115, 124, 132

Padè approximation, 110, 112, 195
parallel form, 8, 9
parallel metered control, 268
parameters estimation, *see* identification
performance assessment, **209**
    deterministic, **222**
    stochastic, **210**
plug&control, **145**
pole placement, 22
pole-zero cancellation, 33, 146
positional algorithm, 14
postactuation, 114
pre-act, 6
preactuation, 114
preloading, 38, 42, 44, 50

primary controller, 251
process monitoring, 209
proportional action, **3**, 11, 15, 29
proportional band, 5
proportional gain, 3, 7, 16
proportional kick, 12

rate action, 6
ratio control, **267**
realisable reference, 46
relay feedback, 65, 173
relay-feedback, 191, 235, 253, 263
reset term, 5

secondary controller, 251
self-tuning, 18
series form, 7, 9, 22, 29
series metered control, 268
set-point filter, *see* feedforward filter
set-point following, 11, 16, 25, 36, 93, 207, 219, 234, 265
set-point weight, **11**, **61**
slave controller, 251
Smith predictor, 16, 215, 263
SOPDT systems, **180**, 233
standard form, 264
steady-state error, 4, 5, 15, 38
stiction, 225

tangent method, 166, 184
temperature control, 55, 85, 107, 145, 160
three-state controller, 3, 149
tracking mode, 40
tracking time constant, 39
transition polynomial, 111, 131
tuning, 16, 33, 46, 61, 149, 234, 253, 282
tuning rules
    Kappa–Tau, 28, 277
    refined Ziegler–Nichols, 66
    Ziegler–Nichols, 17, 20, 27, 272
two-degree-of-freedom control, 11, 61, 265

ultimate gain, 65, 173
ultimate period, 65, 170, 173, 191, 254
underdamped systems, 180

velocity algorithm, *see* incremental algorithm